D1029129

TECHNIQUES OF CHEMISTRY

ARNOLD WEISSBERGER , *Editor*

VOLUME IX

CHEMICAL EXPERIMENTATION UNDER EXTREME CONDITIONS

TECHNIQUES OF CHEMISTRY

VOLUME IX

CHEMICAL EXPERIMENTATION UNDER EXTREME CONDITIONS

Edited by

BRYANT W. ROSSITER

Research Laboratories
Eastman Kodak Company
Rochester, New York

A WILEY-INTERSCIENCE PUBLICATION

JOHN WILEY & SONS
New York · Chichester · Brisbane · Toronto

Library of Congress Cataloging in Publication Data

Main entry under title:

Chemical experimentation under extreme conditions.

(Techniques of chemistry; v. 9)
"A Wiley-Interscience publication."
Includes index.
1. Chemistry—Experiments. I. Rossiter, Bryant W., 1931–

QD61.T4   vol. 9   [QD43]   542'.08 s   [542]   79-10962
ISBN 0-471-93269-8

Printed in the United States of America

10 9 8 7 6 5 4 3 2 1

AUTHORS

EDWARD M. EYRING

Department of Chemistry, University of Utah, Salt Lake City, Utah

H. TRACY HALL

Chemistry Department, Brigham Young University, Provo, Utah

JOHN F. HAMILTON

Research Laboratories, Eastman Kodak Company, Rochester, New York

ROBERT H. HAUGE

Department of Chemistry, Rice University, Houston, Texas

JOHN L. MARGRAVE

Department of Chemistry, Rice University, Houston, Texas

NEIL PURDIE

Department of Chemistry, Oklahoma State University, Stillwater, Oklahoma

STEPHEN C. PYKE

Department of Chemistry, Washington State University, Pullman, Washington

LICESIO RODRIQUEZ

Department of Chemistry, University of Utah, Salt Lake City, Utah

MAURICE W. WINDSOR

Department of Chemistry, Washington State University, Pullman, Washington

v

INTRODUCTION TO THE SERIES

Techniques of Chemistry is the successor to the Technique of Organic Chemistry Series and its companion—Technique of Inorganic Chemistry. Because many of the methods are employed in all branches of chemical science, the division into techniques for organic and inorganic chemistry has become increasingly artificial. Accordingly, the new series reflects the wider application of techniques, and the component volumes for the most part provide complete treatments of the methods covered. Volumes in which limited areas of application are discussed can easily be recognized by their titles.

Like its predecessors, the series is devoted to a comprehensive presentation of the respective techniques. The authors give theoretical background for an understanding of the various methods and operations and describe the techniques and tools, their modifications, their merits and limitations, and their handling. It is hoped that the series will contribute to a better understanding and a more rational and effective application of the respective techniques.

Authors and editors hope that readers will find the volumes in this series useful and will communicate to them any criticisms and suggestions for improvements.

ARNOLD WEISSBERGER

Research Laboratories
Eastman Kodak Company
Rochester, New York

PREFACE

The first recorded example of a planned chemical manipulation was taken from an Egyptian hieroglyphic demonstrating the separation of gold from alluvial river mud. From this early beginning until Torricelli, in the mid-seventeenth century, successfully achieved a vacuum by inverting a tube filled with mercury, the chemist was restricted to experimental conditions achievable under the most rudimentary laboratory conditions. Later, Cronstedt devised the blow pipe to increase the temperature of a flame, and this advancement was to be followed by a long succession of developments which allowed the scientist to achieve in the laboratory extremes of temperature and pressure previously believed impossible.

In the drive to further understand the substances of nature and to create new compositions of matter, scientists have put newly found techniques to work and over the decades have opened up entirely new frontiers some of which, even today, have been barely probed. In fact, many scientific observers believe that many of the most significant advancements of the next quarter century will result from chemical experimentation employing extreme conditions, and it is for this reason the editors feel that a volume dealing with this subject is both important and timely.

The purpose of this volume is to provide a working laboratory guide for achieving, controlling, measuring, and using the extreme limits of some intensive and extensive variables such as temperature, pressure, time, and material quantity. It is the aim of the authors to define the value, potential, and limitation of the respective techniques in terms of their application to the study of chemical substances and their transformations. This work is not intended as a comprehensive survey of application of the literature, although extensive literature references are provided. The material is self-contained in the sense that it provides the theory, procedures, and/or references necessary for the successful handling of the technique on a laboratory or experimental basis. Other important subjects such as low-temperature techniques and chemistry under zero gravity conditions will be treated in future volumes.

The definition of an "extreme condition" is arbitrary. For purposes of the present volume, an extreme condition is defined as a condition not readily available or achieved in most modern scientific laboratories. Since circumstances vary widely, authors have been asked to treat, briefly, usual and readily achievable conditions and then move in greater detail to the extreme where the prime interest of this work lies.

Help and encouragement from Dr. Arnold Weissberger was invaluable in the preparation of this initial volume. I also express my gratitude to the authors and to Dr. Ted R. Evans, Dr. Richard L. Reeves, Mrs. Ardelle Kocher, Mrs. Joyce G. Tydings, Mrs. Janet R. Brown, and the staff at Wiley–Interscience for their helpful and significant contributions.

BRYANT W. ROSSITER

Eastman Kodak Research Laboratories
Rochester, New York
August 1979

CONTENTS

TECHNIQUES OF CHEMISTRY

ARNOLD WEISSBERGER, *Editor*

VOLUME IX

CHEMICAL EXPERIMENTATION UNDER EXTREME CONDITIONS

Chapter I

INTRODUCTION

H. Tracy Hall

Chemical transformations are brought about by the transfer of energy from a source to a sink, the latter being the chemical system of interest. Strictly speaking, energy is that which diminishes when work is done and is equivalent to the work done. The dimensional units of energy and work, ml^2/t^2, are identical. Energy is also defined as the capacity for producing effects, said effects being of widely differing character.

Energy may be classified as being stored or in transition. Examples of stored energy are (1) mechanical, as in a flywheel, where the stored energy is given by $E = \frac{1}{2}mv^2$ for each particle of mass m and velocity v; (2) geopotential, due to gravity, where $E = mGh$, G being a gravitational constant and h being the height above some frame of reference; and (3) internal energy stored within matter, such energy being a function of the state of the elementary particles comprising the matter. This includes strain energy as stored in a spring, chemical energy, and energy stored by virtue of temperature.

The following are examples of energy in transition. (1) One is mechanical, a situation in which a force moves its point of application. Mechanical work is defined as $W = \int F \cos \theta \, ds$, where F is the force applied and s represents the course followed. In the field of high pressure (a special form of mechanical energy), the work done is expressed as $W = \int P \, dV$, where P is the pressure and dV is the change in volume. (2) Electrical work or energy due to a potential difference ΔE under which a current flows is another example. The potential E is measured in volts, the current flow in amperes, and the energy (product of current flow and potential difference) is measured in joules. (3) Another is heat, that enigmatic form of energy in transit because of a temperature difference. This energy is not called heat before it starts to flow or after it has ceased to flow. Heat is unidirectional. It only flows from a higher to a lower temperature, and the source and the sink, in an isolated system, will eventually attain the same intermediate temperature.

Energy in transition can often be expressed as the product of two

1

factors: a capacitive or extensive factor and an intensive factor. Some examples are shown below.

Type of Energy	Extensive Factor	×	Intensive Factor	=	Energy, Commonly Measured In
Electrical	Coulombs		Potential		joules
Heat	Heat capacity		Temperature		calories
Radiation	Planck's constant		Frequency		ergs
Kinetic	Mass		$\frac{1}{2}$(Velocity)^2		ergs
Pressure	Volume		Pressure		liter-atmospheres
Chemical	Moles		Chemical potential		calories

The intensive factor associated with energy is usually responsible for inducing phenomenologic changes in matter. Such changes often occur when a certain threshold value of the intensive factor is attained. In connection with electrical energy, we observe ionization potentials, deposition potentials, decomposition potentials, and so on. With heat energy there are melting points, boiling points, triple points, and all kinds of critical points. In the case of pressure energy there are solidification pressures, a large variety of phase transformation pressures, electronic transition pressures, and various types of critical point pressures. Because intensive factors bring about such changes in matter, it is important to vary this parameter over as large a range as possible. By so doing, new phenomena are invariably discovered.

It is worthwhile to reflect on how extreme values of intensive properties may be obtained. As I see it, there are three general methods to employ: (1) disproportionation, (2) gathering or focusing, and (3) in situ energy transformation.

An electrical transformer by having a high turns ratio between primary and secondary windings can take a given amount of low voltage/high current energy at the primary terminals and by disproportionation provide high voltage/low current energy at the secondary terminals, or vice versa. Similarly a lever can disproportionate a given amount of energy as a small force operating through a large distance to yield a tremendous force available over a small distance. "Give me a place to stand and, with a lever, I will move the world," said Archimedes. Hydraulic pumps and rams operate on similar principles to those used by levers and make it possible to exert very large forces on small areas, thus giving rise to high pressures.

Each of the above disproportionators has theoretical capabilities for raising the intensive property to infinitely large values. Practically, however, each is restricted by limitations in the properties of materials of construction. Voltages obtainable with transformers are limited by breakdown of the electrical insulation. Forces obtained with a lever are limited by the tensile strength of the material of which the lever is constructed, and pressures derived by transferring the large thrust of a hydraulic ram to a small area (to generate high pressure) are limited by the compressive strength of the transferring material. Limitations imposed by the strengths of materials can sometimes be circumvented by inventive design. Thus the maximum voltage obtainable from a transformer can be increased considerably by certain spacing and placement of the windings. Similarly geometrical considerations, to be discussed in the chapter on high-pressure techniques, increase the pressures obtainable from high-pressure devices.

The second law of thermodynamics apparently precludes disproportionation as a means for increasing the temperature of a system. Thus in an isolated system of initial temperature T_0 and heat capacity C_0, it would not be possible to increase the temperature to a high value T_h by causing the heat capacity to decrease to a low value C_l such that $T_0 C_0 = T_h C_l$.

An example of a gathering or focusing process to obtain practical use of an already high intensive property is the use of parabolic mirrors or lenses to concentrate radiant energy from the sun. The energy coming from sun to earth arrives in a very diffuse but undegraded state; that is, the radiation spectrum is about the same as that emanating from a blackbody at the same temperature as the sun's surface. In order to make it possible for this energy to interact with a useful quantity of matter, it must be concentrated by some kind of gathering process. Incidentally, it is interesting that the capacitive factor for radiant energy is invariant; namely, it is Planck's constant, the quantum of action.

In situ energy transformation is the most widely used method for obtaining a high value of an intensive property. For example, high temperatures are obtained in an electrical resistance furnace by the in situ transformation of electrical energy into heat. In a particle accelerator, tremendous velocity is imparted to electrically charged particles by extraction of energy from electromagnetic fields. At extremely high velocity, relativistic phenomena come into play, and there is an in situ transformation of kinetic energy into mass. The kinetic energy of a high-speed electron is transformed into radiant energy as the electron impinges on a metallic target, and so on.

In the field of high pressure, the use of disproportionation to obtain high pressures has given the most satisfactory results. However there are some in situ energy transformations that have been used. It is possible to generate a pressure in a fixed volume by increasing the temperature of a material within this confined space. Chemical energy may be utilized to generate pressure by employing a chemical reaction or phase change in which the products of the reaction or the new phase normally occupy a larger volume than that to which they are confined.

So far this discussion has centered around means for obtaining high values of intensive properties. Low values are also important, particularly low values of pressure (high vacuum) and low temperature. One of the most exciting discoveries of all time is the phenomenon of electrical superconductivity at low temperature, discovered by Kamerlingh Onnes in the Netherlands in 1911. The attainment of high vacuum is a necessity for all the achievements of high-energy physics and for a great deal of modern solid-state technology.

To obtain low temperatures, two items are important: (1) thermal insulation to prevent heat leak from ambient conditions into the system, and (2) means for extracting heat from the system (refrigeration). The attainment of high vacuum is also ancipital: means must be provided (1) for preventing leak of gaseous substances into the system and (2) for extracting gases from the system. Materials science and inventiveness, particularly in the mechanical area, have been necessary for the advancement of these fields.

Aside from the general requirements briefly considered herein for attaining extreme conditions and the significance of the intensive factor of energy in transition, there are ancillary items important to chemical experimentation. One is the efficiency with which energy is transferred from the source to the sink. I will illustrate this with a story. During the period 1927 to 1933, I was a skinny, underweight farm boy attending school (grades 1 through 9) at a four-room building in Marriott, Utah. We played baseball (hardball) during lunch hour and two recesses every day, weather permitting. It was fashionable to use the heaviest bat available (an ego-building show of strength). But I could hardly hold such a bat. However, while enduring taunts from my muscular peers, I could handily knock a home run with a small girl's softball bat. I didn't know it then, but it was all a matter of $E = \frac{1}{2}mv^2$ and matching the energy available at the source (myself) to the sink or load (the ball). A heavy bat swung slowly could contain less energy than a light, fast swinging bat because energy increases with the square of the velocity but only linearly with the mass. Only recently have the major baseball leagues recognized this.

In an automobile, the necessity of matching the energy available at the source (the engine) to the sink (the wheels) is manifest by the gear-shift box or automatic transmission. In the field of electricity, the efficient coupling of an energy source to a sink is known as impedance matching. Gears, levers, pulleys, coaxial cables, and parallel twin-lead TV wires all serve the same purpose—the effective transfer of energy from source to sink.

There are parallels for chemistry. All chemical reactions, including phase changes, are connected with the repositioning of atoms relative to other atoms. Thermodynamics gives us a nonmechanistic criterion for ascertaining that rearrangements may or may not occur given P, V, T, n, and C (pressure, volume, temperature, number of moles, and heat capacity) for all substances in the system. But it does not tell us the time it will take. Chemical kinetics, at its present state of development, can predict reaction time in only a few instances. But experimental studies in this field are very valuable for eliciting mechanisms by which reactions take place. Chemical reactions are much more complex than mechanical or electrical systems; but, in the gross, inducing a reaction to proceed in a reasonable length of time (assuming favorable thermodynamics) is an impedance-matching problem. That is, energy from a source must be channeled through an interface matched to the sink for energy in transit to flow freely and rapidly into the system. Often a chemical reaction is "broad band." It will accept energy (heat or otherwise) without benefit of a specific impedance-matching interface. But many reactions, while highly favorable from a thermodynamic standpoint, will not proceed without a catalyst. The catalyst may be regarded as the impedance-matching device—the bat between the boy and the ball. In order to match a source to a sink, something must be known about the characteristics of each. This is easy in electronics and mechanics but difficult in chemistry. Hence choosing a catalyst is often done by trial and error. Exploring under extreme conditions will discover only the broad-band reaction products. Catalysts will be needed to find the diamonds.

Kinetic studies are often directed toward understanding chemical reactions at the atomic and molecular level. At this level, reactions proceed very fast, sometimes in 10^{-15} sec. Since the Greeks, we have wondered what matter was like if we divided and divided it into ever smaller pieces. Fortunately for chemists, it need only be divided into neutrons, protons, and electrons. Physicists, with their greater dividing power, are currently having Technicolor migraines.

Troubled by the increasing complexities of divided matter, some have turned to dividing time, but this presents perplexities also, even for chemists. Time can be divided and measured by means of waves. Useful waves for chemical experimentation are electromagnetic and ultrasound, for which the corresponding particles are photons and phonons. When the energy of waves is high, the uncertainty principle may limit the information obtainable from fast reactions in chemistry. The uncertainty principle states that $\Delta E\, \Delta t \geq h$, where ΔE is the uncertainty in the range of the energy and Δt is the uncertainty in measurement of the time. Planck's constant is represented by h. For a time interval of a femtosecond (10^{-15} sec), the uncertainty of an energy measurement would be at least 10 kcal/mole. Since chemical bond energies lie in the range of 5 to 200 kcal/mole, the uncertainty principle will frustrate accurate measurement of energy parameters in times less than about a femtosecond. Interestingly, time itself may be quantized. The smallest unit of time is the chronon, which is about 4.5×10^{-24} sec. This corresponds to a frequency of 2.2×10^{23} sec^{-1} and a wavelength of 1.4×10^{-23} cm.

Since there is a lower time limit in which chemists can do their kinetics and an upper limit (for the laboratory) based on man's patience and lifetime, it is interesting to ask questions concerning upper and lower limits for pressure and temperature. For temperature, we speak of absolute zero as the theoretical lower limit. Nevertheless, energy content at absolute zero, while a minimum, is not zero and is so expressed by the uncertainty principle. Chemical synthesis near absolute zero would ordinarily be restricted by slow reaction rates, at least in the gross. An insufficient number of molecules would have enough energy to transcend the kinetic barrier. However it seems possible that the order characteristic of systems at low temperature might, through quantum effects or for other reasons, change the nature of the activated state for a chemical reaction. There may be unexplored territory here.

Physical measurements at low temperature have been very enlightening with respect to theory and fundamental problems of chemistry. For example, specific heat measurements have been made to temperatures as low as $0.001°$ K, and it has been possible to sort out contributions due to the lattice, the electrons, and the nuclei. In these measurements investigators have "seen" energy of the nucleus equilibrate with the electrons. This transfer of energy can occur because of the finite possibility that an electron will, in due time, pass through the nucleus.

At extremely low temperatures, radioactive nuclei have been aligned in a magnetic field such that particles or radiation resulting from nuclear disintegration are emitted in preferred directions. Can this order be used to advantage in nuclear fusion? Extremely high temperature (the sledgehammer approach) is the current mode for attempts at sustained nuclear fusion. Are we overlooking an ultralow-temperature finesse?

The upper theoretical temperature limit for chemical experimentation is set only by the fact that chemistry, as ordinarily perceived, ceases to exist. At a sufficiently high temperature, all molecules will become monatomic ions and form a plasma.

Very low pressures (high vacuum) offer opportunity for chemical experimentation with atomic, molecular, and ion beams and for the study of surfaces and other things. There is no intrinsic property of high vacuum, per se, that should limit chemical experimentation in the sense that low temperature does. The practical limit for vacuum may be set, considering perfect pumps and no leaks, by contamination from the ever-present high-energy rays and particles from outer space.

The theoretical limit for high-pressure chemical experimentation is similar to the case of high temperature. For any system there will be a pressure where ordinary chemistry will not be known. Sufficient pressure will make metals of all nonmetallic elements. Regular chemistry will prevail at pressures as high as 10^7 atm. The ultimate galactic pressure is thought to exist in the recently postulated black holes.

Chapter **II**

HIGH-PRESSURE TECHNIQUES

H. Tracy Hall

1   INTRODUCTION

This chapter is written, primarily for the nonspecialist, out of 25 years of experience in designing and constructing all types of high-pressure equipment. During this period there have been only brief intervals in which I have not had a piece of apparatus under construction. My notebooks record about 250 ideas on apparatus design and at least 32 different devices valued at around $2 million have been built under my direction. I have been personally involved in all aspects of the work, including conception, design, drafting, and machining. In the latter I have tried to emulate that pioneer in the field of high pressure of whom it has been said, "There was only one person at Harvard who was a better machinist than Charlie Chase and that was Percy Bridgman." Mr. Charles Chase was Physics Professor Percy W. Bridgman's unusually skilled machinist and craftsman. In this work I too believe in getting my hands dirty. But it has not been work. For me it has been love.

The modern-day interest in high-pressure technology was most certainly sparked by the announcement on February 15, 1955, by the General Electric Company, that diamonds had, at long last, been made by man. Indeed, this had been one of the most intriguing, unsolved problems of science, attempts to synthesize diamond by the conversion of its polymorph graphite having been instigated in 1792. That was the year Lavoisier discovered that diamond consisted only of the element carbon.

Considerable excitement surrounded the General Electric announcement as journalists from around the world assembled to hear the news. There was also disappointment. For while there was a discussion of the unsuccessful attempts of others to synthesize diamond, and the tiny diamonds that had been made were displayed, the details of how these diamonds were made and the details of the equipment that was used were not revealed. Several months later, additional secrecy beyond company interest was imposed by the United States Government. Abatement of the secrecy did not come until five years later.

You can well imagine my anguish, anxiety, and dismay when, having invented (January 2, 1953) the belt apparatus in which the synthesis was achieved and having accomplished the world's first reproducible synthesis of diamond (December 16, 1954), I was not able to publish or claim recognition for an accomplishment that had been sought by eminent scientists, including Nobel laureates, for much more than 100 years; and further, of watching my accomplishments being diluted through various means as the years went by, including leaks of information from company and government sources that allowed others, not hampered by proprietary or governmental interest, to come ever closer to the process and to the design that were original to me. A sampling of some of these problems is found elsewhere [1]. These difficulties, however, led to the development of other high-pressure equipment, specifically the multianvil presses, of which the most important members are the tetrahedral and cubic presses. These are discussed later.

Practical diamond synthesis requires pressures of the order of 50,000 atmospheres (atm) and higher. Having mentioned a practical unit of pressure measurement, I might diverge for a moment. Through the years many units have been used to measure pressure. Harvard Professor P. W. Bridgman, who received the Nobel Prize in 1948 for his work in the field of high pressure, measured pressure in kilograms per square centimeter. Bridgman's first paper on high pressure was published in 1908, and his last, in 1958. Chemists commonly use the atmosphere, or the kiloatmosphere, while geologists and others prefer the bar or the kilobar. Recently the National Bureau of Standards began encouraging the use of the newton per square meter (now called the pascal, abbreviated Pa). Fortunately, these units are closely related: 1 atm = 1.0133 bar = 1.0332 kg/cm^2 = 14.696 lb/in.^{2} = 760 torr (mm Hg) = 1.0133 × 10^5 N/m^2 or Pa.

The rapid growth and apparent decline of high-pressure research since 1950 is graphically represented in Fig. 2.1. This figure shows the number of papers published in the field through 1974. Only papers reporting work above 5000 atm are included in the count.

The United States Bureau of Standards maintains a high-pressure data center at Brigham Young University. This center has the function of evaluating certain aspects of data found in the high-pressure literature. It compiles a "Bibliography on High Pressure Research," a current-awareness bulletin, published in six issues annually, on a subscription

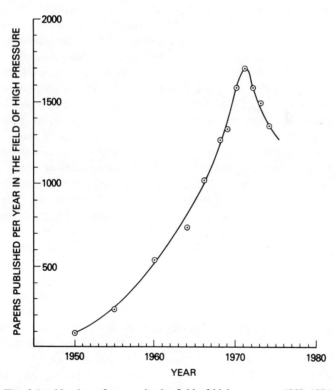

Fig. 2.1. Number of papers in the field of high pressure, 1950–1974.

basis. The center cooperates with foreign countries, notably Japan and Russia, which compile like information for their countries. The center attempts to obtain and file one copy of every article on high pressure published anywhere in the world. Another publication of the center is "The International Directory of Workers in the Field of High Pressure Research," compiled by John F. Cannon and Leo Merrill, April 1971 (a directory was also issued in April 1968). There is also a high-pressure bibliography covering the period from 1900 to 1968. Volume I includes a bibliography and author index and Volume II, a subject index, published in April 1970 under the direction of Leo Merrill, Associate Director of the center. Another publication of note from the center is "Behavior of the Elements at High Pressures," by John Francis Cannon, which is reprint number 55 from *J. Phys. Chem. Ref. Data*, 1974. Reprint number 12, from *J. Phys. Chem. Ref. Data*, by D. L. Decker, W. A. Bassett, L. Merrill, H. T. Hall, and J. D. Barnett, is entitled "High Pressure Calibration, a Critical Review." This is a good source of information concerning the calibration of high-pressure equipment. *The International Directory of Workers in the Field of High Pressure* includes the names of researchers, their affiliation, and the pressure range covered by their research. It also indicates the temperature range and the type of apparatus used, such as the belt, Bridgman anvils, cubic press, diamond anvils, gas.apparatus, liquid systems, shock techniques, piston–cylinder devices, and so on. Also, if the workers have interests in calibration and equipment design, this is indicated. There is also a key to the type of interest that the particular laboratory or investigator has. For example, if the person is interested in crystal chemistry, that is so indicated. Other subject headings are crystal structure, equilibrium studies, inorganic synthesis, lattice defects, organic chemistry, polymorphism, electronic and quantum properties, geological and geophysical properties, magnetic properties, amorphous solids, diamond research, mechanical properties, metallography and metallurgy, method of analysis, infrared, Mossbauer, neutron diffraction, optical properties, thermodynamics, and so on. This publication, then, can be the new interested researcher's entrée into the field of high pressure.

Most high-pressure apparatus used for research is constructed in university shops and laboratories to suit some particular interest of the investigator. Consequently, when considered in specific detail, there are nearly as many high-pressure devices as there are high-pressure researchers. However, when considered in broad perspective, there are a limited number of basic devices. In order of their development, they are (1) piston–cylinder, (2) Bridgman anvil, (3) the belt, and (4) multiple anvils, the principal multianvil types being the tetrahedral press and the cubic press. These are the landmark devices, and all other devices, by whatever name they may be called, are derivatives of these. The origins of the piston–cylinder apparatus are lost in antiquity, but some important adaptations of it for high-pressure, high-temperature use are due to Parsons and Coes, whose work will soon be discussed. Bridgman anvil devices are the invention of Percy W. Bridgman; the belt and multiple-anvil types are the inventions of this author.

The chemical changes that can be induced in a system by perfusing a substance with energy are qualitatively related to the intensity factor (pressure, temperature, volts, etc.) of the energy. As far as chemistry is concerned, the most important energy introduced into the chemical substances of concern is heat. Heat of sufficient potency to bring about changes in chemical bonding has been available for a long time, from fires, flames, torches, and Bunsen burners. Readily obtainable temperatures of a few hundred degrees Celsius cause many chemical changes to occur. In terms of "ideal" energy equivalence, one cubic centimeter kiloatmosphere is equivalent to a $12.19°K$ change in temperature.

Of course the atomic and molecular effects of temperature and pressure on substances are not the same. However, pressure and temperature are diametric opposites in some ways. For example, as temperature is increased, solids transform to liquids, which in turn become gases. Systems then proceed to products of molecular dissociation, and at sufficiently high temperature the atomic nuclei will be separated from the electrons, and so on. With increased pressure, on the other hand, substances originally gaseous become liquids, which in turn transform to solids; and with ever increasing pressure, the collapse of electronic shells will occur. At pressures of billions of atmospheres nuclear fusion occurs. Simultaneous high temperature with pressure is important in the field of chemistry because of the general effect of high pressure on reaction rates. High pressure tends to reduce the rate of a chemical reaction. Therefore, in order to have a chemical reaction take place under high-pressure conditions in a reasonable length of time, it is almost always desirable to use elevated temperature. Some of the thermodynamic and kinetic problems have been considered elsewhere [2].

Tens of thousands of atmospheres are required to produce changes in solids and liquid substances qualitatively equivalent to the changes that can be produced in these same substances by means of heat energy. Temperature can be controlled easily and cheaply. Pressure has been controlled only with recent technology and at great cost. Also, not until recent years has high-pressure equipment been available for purchase.

Researchers have had to devise their own. Consequently heat has been much more universally applied to studying chemical change than has pressure.

Pressure is ordinarily defined as a force being exerted upon a unit of area. Within the universe, the range of natural pressures that exists is enormous. The vast reaches of space between the galaxies contain such a small amount of matter that the vacuum existing there is more nearly perfect than any that can be obtained in the laboratory. The distance between atoms in this space is so great that it does not make sense to talk of gaseous pressure as existing at all.

All pressures of any magnitude occurring in nature on a macroscopic scale are due to gravitational forces. The earth's gravitational field attracts the air molecules that form our atmosphere. The average pressure of this blanket of air at sea level is 14.70 lb/in.^{2} (the normal atmosphere). The pressure existing at the greatest ocean depth is $\sim$1000 atm. Interestingly some of the early pressure experiments were performed by lowering liquid-filled, corked bottles into the sea. Halfway between the earth's surface and its center, at the core boundary, the pressure is about 1.4×10^6 atm, and at the earth's center the pressure is estimated to be 3×10^6 atm. The pressure at the center of the sun is estimated at about 10^{11} atm, and the pressures found at the center of white dwarf stars may be 10^{16} atm. Matter at 10^{16} atm will completely degenerate. It will have an effective temperature of absolute zero and a density of one million grams per cubic centimeter.

Man and his machines are indeed puny with regard to any discussion in galactic dimensions. Other comparisons, however, can place man's accomplishments in flattering light. For example, the laboratory-attained pressure of 200,000 atm is equivalent to that present at the base of a column of granite 400 miles high—4000 Washington Monuments stacked on top of each other.

A point has been reached in the field of scientific endeavor where results can often be achieved if vast sums of money can be spent. There are two aspects to this development. The first aspect concerns the increasing complexity of science. At the turn of the century a particle accelerator in the form of a glow-discharge tube might have cost $100. In 1934 the first cyclotron cost approximately $50,000. Today, a multibillion-volt particle accelerator may cost hundreds of millions of dollars. The second aspect involves expendability. Institutional accounting procedures have for years categorized expenditures into areas of supplies, equipment, travel, and so on, and the category of equipment has usually been defined in terms of cost. That is, anything valued at $100 to

$500 or more has arbitrarily been defined as a major item of equipment. Such property must be given an inventory number and thereafter be properly accounted for, and there have been some difficulties between the scientists and the accountants over this matter. Before the accountants have placed an inventory tag on a certain "equipment item," set up a maintenance procedure, and so on, the scientist may have already destroyed the item in his experiments. The ultimate example of today's willingness to expend is exemplified by the billion-dollar rockets being shot into space. High-pressure research, when probing at the highest pressures, requires a modest willingness to spend and expend. The expendable equipment lost to an active investigator working in the field of high pressure is of the order of $2000 to $20,000 per year. Most of this loss occurs from the breakage of cemented tungsten carbide pistons, anvils, and so on.

The highest pressures attainable by man are attained by explosive means. Pressure and temperature are simultaneously generated by explosions, but the period for which the pressure and temperature exist is of the order of micro- or at the most milliseconds. Much of the apparatus used in these experiments is literally "blown up," and the cost of the expendable materials can be much greater than the tungsten carbide broken in static-pressure apparatus (apparatus that can hold pressure for at least several minutes, but more usually for hours or days). While the generation of pressure by explosive means is not discussed in detail in this treatise, there are some interesting aspects of it that will be mentioned. Pressures as high as 10 million atmospheres have been claimed by some investigators. However most of the literature deals with pressure of the order of a few hundred thousand atmospheres or even less. In most chemical (synthesis) experiments, the sequence of events is as follows. Pressure is applied at room temperature to the substance in which chemical change is wanted. This is followed by increasing the temperature to the point where reaction will occur. Next, the temperature is reduced to room temperature to prevent reversion, and this in turn is followed by reducing the pressure to atmospheric. These events are easily controlled in static-pressure apparatus but are not so readily controlled in an explosion.

The use of explosions to generate pressure was considered at the General Electric Company during the same time that static high-pressure apparatus was being used in attempts to synthesize diamond. It was concluded, however, that the pressure and temperature sequence could not be properly controlled to convert graphite to diamond. In 1961, approximately six years after I had synthesized diamond in static-pressure apparatus, John C. Jamieson, Professor at the University of

Chicago, and P. S. DeCarli of the Stanford Research Institute collaborated in an attempt to make diamond by explosive means and succeeded [3]. Their experiment was eminently elegant. They used a 55-gallon drum half-filled with water, open at the top, with a board across it on which was placed a block of graphite. This in turn had on top of it an explosive device. The explosive was fired from a safe distance, and after a settling time, the powder in the bottom of the drum was recovered, dried, and enriched by density separation. They obtained a definite X-ray diffraction pattern, although the lines were very broad, for diamond. The diamond was up to 10 μm in size, and the yield in the process was something of the order of 2%. This work was financed by Allied Chemical Corporation. Improvements were made in the process, and for a time it was possible to purchase a dark-gray diamond powder from that company. This diamond was not pure and apparently contained around 12% oxygen, which may be the reason that the material is no longer available. The function of the water in Jamieson and DeCarli's experiment was to cool the newly formed diamond so fast that it did not convert back to graphite.

Subsequent to DeCarli and Jamieson's development, personnel at the du Pont Company developed a process in which iron was the coolant [4]. They compressed, by explosive means, a material known as ductile or malleable iron. This iron contains a high proportion of carbon dispersed as graphitic spheroids. If a shaped charge of explosive is applied to ductile iron, a shock wave passes through it (explosive techniques are also called shock wave techniques). A shock wave passing through a solid is like a sound wave. There is a pressure front, followed by a trough in which the pressure falls. The rate at which the pressure front travels through a solid depends on the nature of the material, and there are other things that occur, such as reflections and absorptions, which complicate matters. Temperature also rises and falls in a pattern that depends on several variables. The advance of a front through nonhomogeneous material, such as ductile iron, will be complicated, but the net result of a proper experiment is that part of the graphite in the spheroids is transformed to diamond. The iron, which immediately surrounds the spheroids, serves to take away the heat and quench the diamond rapidly enough that it does not all revert to graphite.

There is a wonderful story, perhaps apocryphal, concerning the du Pont development. There was an abandoned mine tunnel somewhere in one of the north central states, I think Minnesota, in which a large quantity of ductile iron was placed. This was surrounded with appropriate charges of dynamite that were later detonated, resulting in the formation of several tons of impure diamond powder. After the explosion subsided, a front-end loader removed some material and deposited it in a digester where dilute sulfuric acid dissolved the iron, leaving a pasty residue. This residue was dried and subsequently dissolved in one or more oxides of lead, which selectively oxidized the graphite, leaving most of the diamond untouched. Further minor cleaning operations yielded a pure diamond powder of submicron size that under the microscope had a spongy appearance. When most of this batch had been sold, the front-end loader again went into the man-made diamond mine, retrieved more material, and so on. Supposedly there is enough diamond in this mine to last the world for many years!

But there are now improvements in the process, and the mine has been abandoned. A rather complicated cylindrical device is now used onto which appropriately shaped charges are fastened. Rather than ductile iron being used as a charge, graphite containing copper particles is used. The purpose of the copper is to appropriately absorb the heat to prevent reversion of the newly formed diamond. Recovery of the diamond in an explosion could of course be a problem, but there is a momentum-absorbing plug in the end of the cylindrical device which gives way at the appropriate time, allowing the charge to be less violently blown into a retrieval container. Apparently the submicron diamond powder after retrieval and purification is again placed into an explosive device and reshocked, whereupon an agglomerization of the diamond takes place, producing particles of up to 40 or 50 μm. These larger sizes are more in demand than the original submicron sizes.

2   METHODS FOR GENERATING STATIC PRESSURES

In the introduction to this volume, I gave three general methods for obtaining high values of intensive properties such as pressure, temperature, voltage, and the like. They were disproportionation, gathering or focusing, and in situ energy transformation. At the present time the use of disproportionation to obtain high pressure has given the most satisfactory result. However there are energy-transformation processes that have also been used, some of which are attractive for further development. It is possible to generate pressure by increasing the temperature of a material which fills a confined space, for example, the heating of a liquid in a metal tube. Chemical energy may also be utilized to generate pressure by employing a chemical reaction or a phase change in which the products of the reaction or the new phases formed will tend to occupy a larger volume than that to which they are confined. The freezing of water in a confined space is a common example of this principle, as many know who have frozen radiators or pipes in the wintertime. The freezing of water in a confined space can generate a theoretical maximum pressure of about 2000 atm. The solidification of liquid bismuth in a confined volume can generate a theoretical maximum pressure of about 17,000 atm. Likewise, the freezing of molten germanium can generate a theoretical pressure of the order of 100,000 atm. These theoretical pressures can be approached providing the container volume increases only slightly as the freezing liquid presses against it. It is also necessary that the volume of the germanium, bismuth, or water is large compared to the volume of the material being subjected to compression by the expansion process. At the present time, germanium has been demonstrated to be capable of generating higher pressures by a freezing process than any other known substance. However its use in a pressure-generating device has not yet been fully exploited.

As mentioned, disproportionation provides the most satisfactory means of obtaining high pressures at the present time. The most obvious form taken by a high-pressure device is that of a piston and cylinder. In such a device, a sample confined by the cylinder is compressed by advance of the piston. A larger piston and cylinder drives the smaller piston into its cylinder, this being the disproportionation principle (see Fig. 2.2). In the field of hydraulics this type of device is called an intensifier.

3 PISTON–CYLINDER DEVICES

Today's piston–cylinder devices are operable to pressures around 50,000 atm with solid media [5]. In these devices both the piston and the cylinder are constructed of cemented tungsten carbide. The cemented tungsten carbide used is made of pure, virgin material, formula WC, cemented together with approximately 3 to 12% cobalt. For general-purpose use, a tungsten carbide containing 6 to 8% cobalt is the most desirable. Tungsten carbides have the highest compressive strength of any engineering material readily available at the present time. Although cemented carbides possess tremendous compressive strengths, they have rather low tensile strengths and are brittle. Their tensile strength increases with increasing cobalt content. However, the compressive strength decreases with increasing cobalt content. Compressive strengths for 3 weight-percent cement tungsten carbide cylinders of 60,000 atm have been reported by some manufacturers. However, the usual commercial product has a compressive strength of only 35,000 atm. These figures are for solid "rounds" in which the length equals the diameter, the round being compressed between carbide blocks. Longer rounds will have lower com-

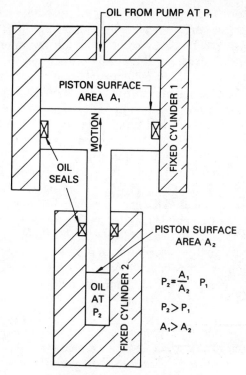

Fig. 2.2. Generation of pressure by disproportionation.

pressive strengths (the so-called column effect). Because of the low tensile strength, cylinders constructed of this material must be supported in such a manner that the carbide is not subjected to a damaging tensile load. This can be accomplished by surrounding the carbide cylinder with massive alloy-steel binding rings.

The upper pressure limit of a piston–cylinder device so constructed is set by failure of the pistons. When materials with compressive strengths greater than those of the cemented carbides become available, the upper pressure limits obtainable in a piston and cylinder device, as well as in other types of devices, will be increased. Sintered diamond, a newly available material, will eventually make it possible to obtain higher pressures [6].

Tremendous pressure could be theoretically obtained by a cascade process wherein one piston-and-cylinder device would be placed inside another, and so on. The failure of high-pressure components is not caused by the absolute pressure to which they are subjected, but rather the differential pressure. Thus a cylinder in which the internal pressure is 100,000 atm and the surrounding external pressure is 50,000 atm is in no more danger of failure than a cylinder wherein the internal pressure is 50,000 atm and the external pressure is zero. There is some indication that tungsten carbide increases in strength under high pressure and can therefore under these conditions withstand a greater differential pressure. At the present time, the experimental problems connected with cascading have made it impossible to utilize this principle beyond two stages. As a matter of fact, the only use of a two-stage device that has been reported in the literature is due to Bridgman [7]. The working volume of Bridgman's apparatus was very small. The final piston had a diameter of approximately 1.5 mm, and the sample length was also about 1.5 mm. This device is complicated and difficult to use. Electrical leads could not be taken into the sample area. A double-ram hydraulic press was required for its operation. The double-ram press has two coaxial thrust generators. The initial program at the General Electric Company for equipment to synthesize diamonds called for the construction of a Bridgman two-stage piston cylinder device of greatly enlarged size. It was hoped that the larger size would make it possible to introduce electrical leads into the inner chamber to provide current for electrical resistance heating of the sample and also to make temperature and other kinds of measurements. However the invention of the belt apparatus, which is infinitely simpler and easier to use, resulted in abandonment of this approach.

The High Pressure Institute in Moscow, Russia, has been constructing a monumental high-pressure apparatus for a number of years. It has been reported that the total investment in this unfinished device is around $20 million. It is expected to be a piston-and-cylinder apparatus of five stages. Drawings of the contemplated apparatus are not available, but the problems of providing thrust from the outside through all five stages and of providing electrical communication to the innermost stage for measurement and to supply heating current are formidable, if not impossible.

Let us now return to providing further details of single-stage piston–cylinder apparatus. Emphasis will be on modern devices that use solids or a combination of solid plus liquid to transmit pressure, since there is a plethora of literature available on fluid pressure apparatus. Pressure transmission by solids allows much higher pressures to be achieved and also allows high temperature simultaneously with high pressure. This is important in chemistry.

Sir Charles Parsons was the first to attack the problem of generating

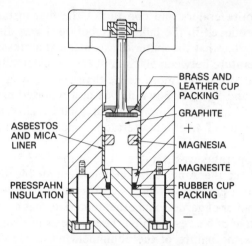

Fig. 2.3. One of Parson's high-pressure, high-temperature devices.

high pressure simultaneously with high temperature [8]. He began a diamond synthesis program about 1880 and ended it about 1928. From his studies he concluded that neither he nor anyone else had, up to that point, succeeded in making diamond. Parsons had excellent laboratory and machine shop facilities and conducted experiments on a grand scale for that period. His pressure apparatus consisted of piston–cylinder devices that used internal electrical resistance heating (see Fig. 2.3). He used a solid pressure-transmitting material which also served as thermal and electrical insulation. He seems to have been the first person to utilize such a combination in a high-pressure, high-temperature device. His cylindrical chambers ranged in diameter from 1 to 15 cm. The maximum pressure at the temperature he reported was of the order of 15,000 atm at 3000°C.

Loring L. Coes, Jr., of the Norton Co., was the first person to develop a piston–cylinder device with capabilities substantially beyond those of the Parsons device. In July 1953, Coes reported the synthesis of a new dense silica which has since been called coesite [9]. Coesite has a density of 3.01, whereas the density of quartz is only 2.65. The refractive index of quartz is 1.50; the refractive index of coesite is 1.60. This material was discovered during a synthesis study of the minerals with which diamonds are associated in the diamond-bearing pipes of South Africa. Although man-made "minerals" are not usually named in honor of their discoverer, this new SiO_2 substance was unique and so interesting that researchers immediately began to call it coesite. Coesite was later found to be present in some meteorite craters and constitutes the first instance in which a mineral was produced in the laboratory before it was discovered in nature. Coes reported that the material was made at a pressure near 35,000 atm at a temperature between 500° and 800°C. Incidentally, another SiO_2 material, called stishovite, more dense than coesite, was synthesized by S. M. Stishov and S. V. Popova in 1961 at a reported pressure of about 115 kilobars and a temperature of about 1300°C. This material has the remarkable density of 4.35 and a refractive index of 1.81 [10]. Stishovite has also been found in meteorite craters.

Coes' first apparatus was not described in the 1953 *Science* article. On April 21, 1954, at the American Ceramic Society Meeting in Chicago, Coes presented a paper entitled "High Pressure Minerals." This paper described a number of garnets and other materials that he had synthesized during the course of his research on the synthesis of diamond. Again, he did not discuss the nature of the equipment used. However at the 7th Symposium on Crystal Chemistry as applied to ceramics at Rutgers University, New Brunswick, New Jersey, on June 4, 1954, Coes described his apparatus. He did not personally publish a description of this equipment, however, until 1962 [11]. A diagram of the apparatus and of the sample chamber is given in Fig. 2.4. The key feature of the device is the use of a hot, molded alumina liner or cylinder. The apparatus is double ended, pressure being generated by pushing a tungsten carbide piston into each end of the alumina cylinder. Because the alumina cylinder is electrically insulating, heating is accomplished, very simply, by passing an electric current from one piston through a sample heating tube and out through the opposite piston. The apparatus was used at pressures as high as 45,000 atm simultaneously with a temperature of 800°C. A temperature of 1000°C could be obtained at a lower pressure of 30,000 atm. Temperature was measured by means of a thermocouple located in a well, as shown in the figure. Although the temperature at this point is lower than the temperature in the sample, comparisons of temperatures measured here were made with a thermocouple inserted in the sample at 1000 atm. From these measurements it was possible to correct the temperature reading of the thermocouple in the well to the approximate temperature of the sample in the chamber.

At 45,000 atm and 800°C, only one run is obtained in this device, the pistons and the alumina cylinder both being expendable. Even at 30,000 atm the alumina cylinder is only useful for a few runs, as is also the case for the tungsten carbide pistons. The expense of using such a device is great, and while it was used extensively at the Norton Co., I am unaware of any use of the device elsewhere except for some limited use of the equipment at Brigham Young University. Coes was kind enough to give a

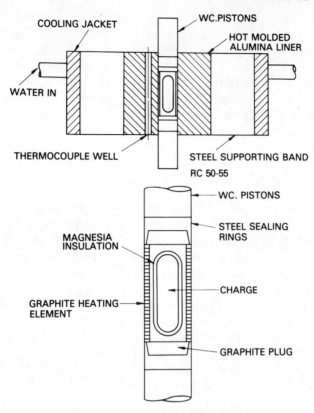

Fig. 2.4. Coes apparatus.

number of alumina cylinders to me sometime around 1957, at which time I
experimented briefly with their use. Coes once related to me that they had
barrels of broken alumina cylinders around the lab. In fact, they had so
many they wondered if a use could be found for them. Consequently, the
broken cylinders were crushed to provide mesh-size abrasive particles
which were subsequently made into grinding wheels. These wheels performed better than the wheels they were then manufacturing. Consequently, this serendipitous discovery led to an important change in the
method of manufacture of aluminum oxide grain for use in grinding
wheels.

Nowadays both the piston and the cylinder are constructed of cemented
tungsten carbide, and electrical insulation is provided in a different manner than in the device of Coes. For the highest pressure the cobalt content
should be only 3%. Such a tungsten carbide is very brittle. Therefore the

alignment of the piston and the cylinder and the hydraulic ram providing the thrust must be nearly perfect to prevent off-axis loading. The tungsten carbide should be fine grained. The industry code number in the United States for this carbide is C-4.

The outside diameter of the cylinder should be about six times its inside diameter. When solid materials are used for transmitting the pressure, the sample cell length should be less than twice the piston diameter. With the cylinder in place in its binding rings, the fit between the piston and the cylinder should be as close as possible. A clearance of 0.0025 mm is appropriate. The surface of the cylinder hole should be highly polished, as should also the outside diameter of the piston; the ends of the piston must be ground absolutely square with the cylindrical axis. A coaxial ram press is used to operate the piston and cylinder device. The inner ram provides thrust to the piston, and the outer ram provides clamping force to the cylinder (Fig. 2.5). The clamping force is necessary to prevent the tungsten carbide cylinder from splitting in two in a plane perpendicular to the cylindrical axis. Such splitting will usually occur in a plane that coincides with the surface of the tip of the piston. The surface roughness values (standard machining practice) of the inside diameter of the cylinder and the outside diameter of the piston should be about 1 μm. The remaining surfaces of the piston and the cylinder may have a higher roughness

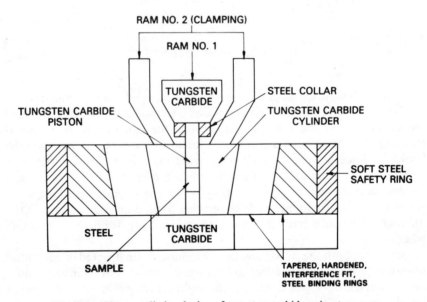

Fig. 2.5. Piston–cylinder device of tungsten carbide using two rams.

value of about 4μm. The smooth surfaces of these components reduce breakage. Fracture under high stress begins on irregularities in the materials, often at surface scratches. In addition, the friction between the piston and the bore of the cylinder as the piston moves within the bore is reduced when the surfaces are smooth. In experiments where it is not necessary to have electrical leads into the sample region, it is best to make the piston-and-cylinder device double-ended, as Coes did. A double-ended piston-and-cylinder device requires extraspecial care in the alignment and in the parallelism of the various components with the moving elements of the hydraulic press.

The piston should be as short as possible and should protrude as little as possible from the bore of the cylinder. The protruding part of the piston is the weak part, and fracture will usually occur there. A tight-fitting collar, as shown in Fig. 2.5, around a portion of the protruding part of the piston is very helpful in preventing breakage on the exposed piston ends. This collar is of low alloy steel such as SAE Number 4340 hardened to Rockwell C55. The interference fit between the collar and the piston should be about 0.003 cm/cm of piston diameter. That is, the inside diameter (I.D.) of the collar is smaller than the outside diameter (O.D.) of the piston by 0.003 cm for each cm of piston diameter. In addition, the I.D. of the collar should be tapered about 0.050 cm/cm length. A matching taper is placed at one end of the piston so that when the collar is pushed over the piston with a forcing press, the interference of about 0.003 cm/cm is developed when the small end of the collar is flush with the small end of the tapered portion of the piston. Molybdenum disulfide paste (such as Molykote G) is used as a lubricant between the mating surfaces as they are forced together.

A fully annealed (dead soft) steel safety ring surrounds the hardened steel collar. This must be in place at all times during and after the forcing operation. Highly hardened steels can break capriciously when under heavy tension, and precautions must be taken. The O.D. of the collar should be about three times the piston diameter and its thickness should about equal the piston diameter. The safety ring I.D. should be the same as the O.D. of the collar. There is no taper here. The O.D. of the safety ring should be about five times the piston diameter. All these parts are built to precision tolerances, the hardened steels and carbides being ground to final dimensions.

At maximum compression of the sample the portion of the piston "exposed to air" should be as small as possible so that all portions of the piston are receiving some kind of "support"; that is, the cylinder supports that portion of the piston within the bore (in compression the piston expands), the reaction of the sample against the tip of the piston within the bore supports the tip, the driving force of the press on the top side of the piston supports the top, and the tight-fitting collar around the piston supports the portion it surrounds. Thus, the only part of the piston unsupported is the small exposed portion that has not entered the cylinder bore.

When operating at the highest pressures, even with the aforementioned support precautions, the piston's lifetime is usually only one or two runs, and those using simple-piston and cylinder devices at the highest pressures must have a large supply of pistons.

After each run the cylinder bore and the outside diameter of the piston are carefully cleaned and coated lightly with molybdenum disulfide powder, a very good dry lubricant. This powder is best applied by dipping the finger into it and then rubbing the coated finger onto the parts.

The steel platens of a commercial press will usually be a rather soft steel with Rockwell C hardness around 20. Consequently the tungsten carbide piston cannot push directly against these platens. Hardened steel, preferably of Rockwell C55, must be placed on top of the platens, and then tungsten carbide must be placed on top of the hardened steel. This tungsten carbide in turn bears on the end of the piston. All of these components must be square and parallel, including the platens of the press, for the slightest cocking of the brittle pistons will cause them to break. The tungsten carbide backing block, which bears on the end of the piston, should also be of grade C4. Its area should be about three to five times that of the area of the end of the piston, and this backing block must be supported by a steel binding ring to keep it under radial compression. In this application, a single binding ring is usually satisfactory. A low-alloy steel forging, such as SAE Number 4340 heat treated to Rc40-43 with a 0.0015 cm/cm diameter interference fit is commonly used. The sides need not be tapered but may be straight at this relatively low interference. The edges of the carbide backing block are rounded and polished, and the steel ring is pushed over the carbide with a forcing press. Molybdenum disulfide paste is again used as a lubricant between the mating parts. The outside diameter of the binding ring is usually about twice the diameter of the carbide backing block. The thickness of the backing block should be about two thirds of its diameter.

In electrical resistance heating of samples under high pressure, the resistance heater is usually of very low electrical resistance. Consequently a high current at low voltage is required. High currents require large electrical conductors. It is difficult to use a double-ended piston cylinder device, where pistons and cylinders are electrical conductors, if electrical resistance heating is required. In this instance it is best to have a closure on one end of the cylinder bore to serve as one electrical contact and the moving piston at the other end of the cylinder serving as the

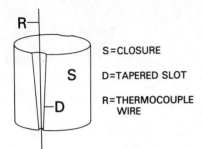

S = CLOSURE

D = TAPERED SLOT

R = THERMOCOUPLE
       WIRE

Fig. 2.6. Cylinder closure.

second electrical contact. The closure must be insulated from the cylinder bore, or there will be a short circuit. It is also desirable to place thermocouples and other electrical leads inside the higher-pressure chamber. The closure can also serve this purpose. A suitable closure is shown in Fig. 2.6. This closure is made about 0.13 mm smaller in diameter than the bore and is wrapped with a strip of electrically insulating paper about 0.05 mm thick. Thermocouple or other electrical leads can gain entrance to the interior of the cylinder, as is shown in Fig. 2.6, where there are two tapered slots (more could be used). The taper is not critical. A taper of about 0.1 cm/cm on a diameter is satisfactory. Electrical leads of about 0.1 to 0.2 mm in diameter are satisfactory for most purposes. The narrow part of the tapered slot should be about 1 mm in diameter. A piece of pyrophyllite is machined to fit the slot, and a tiny hole is drilled along its length to accommodate a wire. This hole can be considerably larger than the wire without causing problems. For example, if a 0.2-mm wire is used, a 0.4-mm hole is satisfactory.

Under high-pressure operation the pyrophyllite is compressed and driven into the taper, and forms a strong frictional grip around the wire. The frictional properties of the pyrophyllite can be increased by painting the surface with a water suspension of red iron oxide and allowing this to dry. Instead of using slots on the exterior surface of the closure, it is possible to put tapered holes within the closure. For example, a tapered axial hole along the center line of the closure would appropriately do the same job. However, such a tapered hole is usually more expensive to make than the ground tapered slots on the surface. The thermocouple wires or electrical leads so introduced into the sample should be covered with an electrical insulating enamel. Provision should be made to prevent these electrical leads from being pinched or cut at places where they emerge from the closure. This requires holes or slots to be formed or ground in the carbide backing block to accommodate their passage. The bottom side of the cylinder and binding ring must be insulated with a sheet

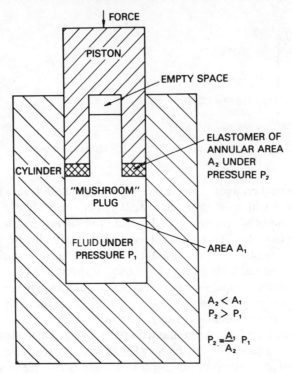

FORCE

PISTON

EMPTY SPACE

CYLINDER

ELASTOMER OF
ANNULAR AREA
A_2 UNDER
PRESSURE P_2

"MUSHROOM"
PLUG

FLUID UNDER
PRESSURE P_1

AREA A_1

$A_2 < A_1$
$P_2 > P_1$

$P_2 = \dfrac{A_1}{A_2} P_1$

Fig. 2.7. Bridgman seal.

of electrical insulation (usually fiber glass epoxy) to prevent contact with the backing block. However the closure must make electrical contact with the backing block. The backing block and steel assemblies must be electrically insulated from the platens of the press in order to prevent short circuiting through the hydraulic press.

An assembly for internal resistance heating of the sample to high temperatures is shown in Fig. 2.8. The entire assembly of Fig. 2.8 is compressed by the piston. Heating current passes from the piston through the metal washer A which it contacts to the electrically conducting heater sample tube E, then out through the metal washer G, the contacting cylinder closure, and the carbide backing block. The electrically conducting tube may be a high-temperature metal such as tantalum or molybdenum or it may be made of graphite. The metal washers should be made of a refractory metal. The solid cylinder C surrounding the heater sample tube must transmit pressure quasi-hydrostatically to the tube and must be electrically and thermally insulating. It must also withstand high temper-

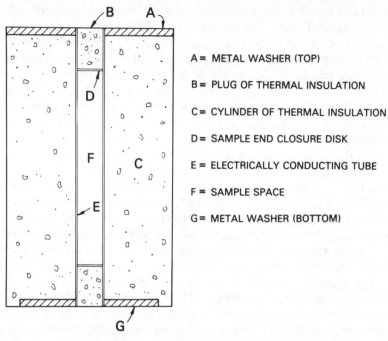

Fig. 2.8. Cell assembly for internal resistance heating.

A = METAL WASHER (TOP)

B = PLUG OF THERMAL INSULATION

C = CYLINDER OF THERMAL INSULATION

D = SAMPLE END CLOSURE DISK

E = ELECTRICALLY CONDUCTING TUBE

F = SAMPLE SPACE

G = METAL WASHER (BOTTOM)

ature. A common material used for this purpose is the naturally occurring mineral pyrophyllite, sometimes called Grade A Lava, or Wonderstone. This material is a hydrous aluminum silicate which will lose water at high temperature and which under high-temperature, high-pressure conditions will transform into other materials of higher density. Pyrophyllite is related to talc. Talc is a hydrous magnesium silicate and is softer than pyrophyllite. Often a liner made of hexagonal boron nitride is used to surround the heater sample tube. This material will of course not give off water. Water from pyrophyllite sometimes penetrates through the heating tube into the sample, or else hydrogen and oxygen will be given off at the high-temperature, high-pressure conditions. As a result, undesired reactions may occur. The BN, at sufficiently high pressures and temperatures, will convert to cubic boron nitride, a high-density polymorph analogous to diamond, and may not transmit the pressure as desired.

There is considerable leeway in the design parameters for the insertion of electrical leads into high-pressure apparatus, and workers in the field tend to develop their own particular preferences. In general, an electrical lead can be inserted by having the wire pass through an unfired ceramic material in a tapered slot or hole and, in some instances, even in a slot or hole with parallel sides. The frictional properties of the ceramic material cause bind-up in the hole or slot and prevent the extrusion of the thermocouple wire. In some instances thermocouple wires are pinched off in passing through the slot if the slot is too short but changing the slope of the taper or the diameter of the hole or the particular frictional material used can, with trial and error, eliminate the difficulty. Pyrophyllite is the most commonly used material. However commercially available unfired or partially sintered alumina, magnesia, and other substances will also serve for this purpose. Also it is possible to merely compact a powder into the hole through which the thermocouple passes. Red iron oxide has higher frictional properties against tungsten carbide surfaces than any material that has been studied and is often used in this application as well as to coat pyrophyllite to increase surface friction where pyrophyllite contacts tungsten carbide. Sheathed pairs of thermocouple wires (or other wires) are readily available today and can be used advantageously in high-pressure high-temperature research for insertion into the sample cell. The sheath is usually stainless steel but may be made of more refractory metals such as tantalum. The two wires (it is also possible to obtain material with more wires) are imbedded in a ceramic powder, usually MgO or Al_2O_3, within the sheath. The total diameter of the assembly is often only 0.25 mm.

Piston-and-cylinder devices were first used in the compression of gases, and the piston usually had a leather cup packing, exactly like the hand-operated air pump used for inflating a bicycle or automobile tire. Pressures attainable in such devices, of course, are small. More sophisticated seals are needed for higher pressures. O-rings serve admirably in many applications, but if pressures cause extrusion of the O-ring, the "Bridgman seal" can be used [12]. The Bridgman seal is shown in Fig. 2.7. A plastic or elastomeric substance, in the form of a washer, goes between the hardened steel mushroom head and the piston tip which has a central hole to accept the stem of the mushroom plug. Gas or liquid pressure is exerted against the mushroom head of the seal. The area of the mushroom plug head exceeds the area of the flexible washer. Consequently, the pressure against the washer will always exceed the pressure of the fluid under pressure. This being the case, there is no way that this seal can leak. The washer will just push harder and harder against the walls of the cylinder as oil pressure increases. Professor Bridgman has told the amusing story of applying for a patent on this seal. The idea was original and unique to him, but in applying for the patent he found that the same principle had been used in a patented sausage grinder. Consequently he was not able to obtain a patent. Yet workers in the field of high pressure invariably refer to the seal as the Bridgman seal, the sausage grinder inventor having been lost to history.

The direct heating of gases and liquids in apparatus where solid pressure-transmitting media are not employed can only be achieved by external heating of the entire chamber. Temperatures of about 300°C are obtainable in externally heated piston–cylinder devices. Higher temperatures are precluded by loss of strength of steel and tungsten carbide above 300°C.

In any experiment at pressure it would be desirable to have the specimen that is the object of the research subjected to what is termed hydrostatic pressure; that is, pressure impinges on the macroscopic object of study from every direction on a microscopic scale. This occurs when a true fluid surrounds the specimen under test. In the high-pressure devices invented to date, the pressure-transmitting medium must be a solid to obtain the highest static pressure. Thus the pressure on the specimen is not truly hydrostatic, but with proper design the pressures can be reasonably hydrostatic. This is often referred to as a quasi-hydrostatic pressure. The highest hydrostatic pressures are obtained using a solid pressure-transmitting substance that presses on a thin-walled metal tube in which liquid is contained. This liquid, in turn, may be exercising its pressure on a sample contained within itself [13]. There is a problem as to the choice of a suitable liquid because most liquid reagents found on the laboratory shelf become solid (freeze) at room temperature at pressures less than 10,000 atm. Liquids that will remain liquid at high pressure are usually mixtures of molecules that do not fit together well and consequently have difficulty in crystallizing. A mixture of 4:1 by volume methanol/ethanol remains hydrostatic to almost 100,000 atm at room temperature [14] and 1:1 by volume pentane/isopentane to about 70,000 atm. The viscosity of liquids increases with pressure, and no doubt the above mixtures have the consistency of heavy molasses as their hydrostatic limit is approached.

There are some solid substances that transmit pressure rather well. Notable among them is indium metal. Among nonconductors, silver chloride transmits pressure effectively. Whether a substance will be a good pressure-transmitting medium (will approach hydrostaticity) or not can be inferred from shear friction measurements made under pressure in Bridgman anvils [15]. Indium and silver chloride have low shear friction at high pressure.

A little philosophy concerning hydrostaticity might be in order at this point. If the object of study at high pressure is a solid material, one might indeed have a true fluid pressing in on this solid specimen from all directions, and one would be inclined to say that the specimen is being subjected to hydrostatic pressure. Well, of course it is on its exterior surface. However the question could be asked, "What about the pressure at some point deep within this solid substance?" The pressure at this point would have to be transmitted by the solid specimen that surrounds that point, and this in turn would depend on the characteristics of the solid and also on its shape. If the specimen is a metal, it would also depend on the particular physical treatment or heat treatment the metal might have had. One would expect that the pressure at the center of a sphere of steel would be higher than at the center of a tungsten carbide sphere of the same size (the modulus of elasticity of tungsten carbide is three times that of steel). Geometrical effects have been observed in such materials as bismuth since this material is often used as a fixed point for the calibration of high-pressure apparatus. With a true hydrostatic liquid surrounding bismuth metal, the transition point will be obtained at a different press loading when the bismuth is used in the form of a thin ribbon than when the bismuth is used as a round wire. Can one really perform "hydrostatic" experiments on solid substances?

When the cell assembly (Fig. 2.8) for a piston–cylinder device is compressed, there is a certain amount of "frictional holdup"; that is, a certain percentage of the pressure at the piston tip is not transmitted to the sample. This frictional holdup is due to two things. There is frictional holdup within the pyrophyllite because it is a solid. There is also frictional holdup at the interface between the cylinder of pyrophyllite and the bore of the tungsten carbide cylinder. The friction at this interface can be reduced by painting the outside of the pyrophyllite with a suspension of molybdenum disulfide or by sheathing the pyrophyllite with In, BN, AgCl, or NaCl. Silver chloride and sodium chloride are corrosive to most metal parts and should not be allowed to remain in contact with them for long periods of time. Substances other than pyrophyllite may be used for the solid portion of the cell in piston–cylinder devices. They have their advantages and disadvantages. For example, boron nitride would have a lower internal friction, but it is much more expensive and has a much higher thermal conductivity, thus requiring higher electrical power to maintain the high temperature inside. Sodium chloride is a rather good substance to use, since its thermal conductivity under pressure seems to be nearly the same as that of pyrophyllite and its pressure-transmitting characteristics are better. However the fabrication of sodium chloride cylinders is difficult.

When polycrystalline refractory substances are used as pressure-transmitting media, a hysteresis exists in which, with increasing pressure external to the solid, the pressure within the solid is less than that outside. Upon reduction of external pressure, however, the pressure inside is higher than that on the outside. A typical hysteresis loop is shown in Fig.

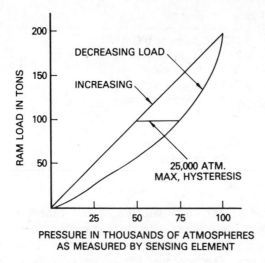

Fig. 2.9. Hysteresis loop for pyrophyllite.

2.9 for a cell assembly such as that of Fig. 2.8, in which pyrophyllite is the pressure-transmitting medium. It is possible to utilize this phenomenon to reduce the load on the pistons or other elements of a high-pressure device and therefore increase the useful life of expensive carbide components. It works in this manner (refer to Fig. 2.9): The press load, that is, the load on the piston, would be increased to a certain value whereupon there would be a given value of pressure at the surface of the sample and a lesser pressure at the center of the sample. The pressure on the piston tip could then be reduced by reducing the pressure on the hydraulic ram that activates the piston. This would result in a reduction of the pressure on the exterior of the pyrophyllite, while the higher pressure would be maintained at its center. Hysteretic phenomena have been observed in which a differential pressure of 25,000 atm has been maintained; that is, the piston would be retracted to the point where the pressure at its tip was 25,000 atm lower than the pressure being maintained by frictional hysteresis in the center of a pyrophyllite cylinder.

Because of the hysteresis in solid pressure-transmitting substances (apparatus friction also enters in), calibration of pressure apparatus is usually made on a basis of ascending (increasing) pressure only. Calibration is usually made in the following fashion: a cylinder of silver chloride is substituted for components B, D, E, and F of Fig. 2.8 and a wire of fixed point calibration material such as bismuth, thallium, or barium is placed through an axial hole in the silver chloride. The pressure-sensing wire

Table 2.1   Fixed-Point Transitions

Hg	7470 ± 2 atm at 0°C
Bi I–II	25.16 ± 0.06 katm at 25°C
Tl II–III	36.2 ± 0.3 katm
Cs II–III	41.9 ± 1 katm (up only)
Cs III–IV	42.4 ± 1 katm (up only)
Ba I–II	54.3 ± 2 katm
Bi III–V	76.0 ± 3 katm
Sn I–II	99 ± 6 atm
Ba	~138 (up only)

makes contact with metal disks (substituted for washers A and G), which in turn make electrical contact with the piston, the closure, the backing blocks, and so on, and thence to an instrument that accurately measures electrical resistance. The accepted fixed-point transition pressures for a number of calibrants, including Bi, Tl, and Ba, are given in Table 2.1 [16]. The shapes of some of the electrical resistance curves are shown in Fig. 2.10.

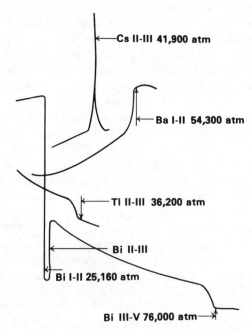

Fig. 2.10.   Electrical resistance curves of some fixed-point pressure calibration metals.

The determination of the pressure at which these electrical discontinuities occur is difficult. Briefly, there is a device called the free piston gauge that can be used to determine fixed points up to about 25,000 atm. Beyond this pressure, techniques become more complicated, but in the main the higher pressure fixed-point transitions are determined against in situ X-ray diffraction measurements in which the electrical resistance of a substance, such as barium, is measured simultaneously with the measurement of the change in the crystal lattice spacings of a substance such as sodium chloride. In some experiments, the detailed nature of the phase change of the fixed-point material (Sn, Ba, etc.) has been determined at the identical place where the corresponding electrical discontinuity occurs and simultaneously with a determination of the lattice spacing in NaCl which surrounds the fixed-point material [17]. The continuous change in lattice parameter of sodium chloride as a function of pressure (semiempirically determined by D. L. Decker) is the currently accepted standard used to determine the value of the fixed points [18].

There is some hysteresis in the metal wire itself as it undergoes transformation on increasing pressure, as compared to undergoing the reverse transformation on decreasing pressure. These hysteretic effects, however, are much smaller than is observed in such substances as pyrophyllite. The silver chloride cylinder in which the fixed-point wire is enclosed is made by compressing powdered silver chloride in a hardened steel die to the required shape. The central hole is made by a tiny spade drill. Care must be exercised to avoid breaking the drill, inasmuch as the silver chloride has a gummy, tough consistency. The bismuth, barium, and tin wires are usually prepared by extrusion through a die. The best die has a conical entrance angle, the exact dimensions of this cone not being too important, and an abrupt 90° exit angle. The usual diameter of the wire is around 0.1 to 0.4 mm. A slug of the metal to be extruded is placed in the die, and a piston pushes on it to cause extrusion. Bismuth and tin are readily extruded into the air. Since heat is generated in extrusion through the die, care must be used in the case of barium. Barium wire is difficult to make because it tends to hold up in the die and then, at a certain pressure, extrudes violently emitting fire and poisonous smoke. It should be extruded with care into paraffin oil. Thallium can be extruded well but, like barium, is poisonous! Barium and thallium oxidize in air and need to be kept under oil.

Another way of obtaining a "wire" of these substances is to simply place a slug of the material between two tungsten carbide blocks and press the material into a flat sheet. A flat wire can then be obtained by cutting it from the sheet with a razor blade. Connecting fixed-point calibrant wires to metal components of the cell can be a problem. Thallium and barium

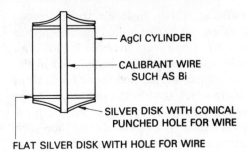

AgCl CYLINDER

CALIBRANT WIRE
SUCH AS Bi

SILVER DISK WITH CONICAL
PUNCHED HOLE FOR WIRE

FLAT SILVER DISK WITH HOLE FOR WIRE

Fig. 2.11. Method for obtaining electrical contact with calibrant metals.

wires must be scraped free of oxide just before use. Barium is quite brittle. Bismuth is very stable in air, but is also brittle. Soldered connections have been used, and it is possible to use bismuth itself as a solder to bismuth wire, but I prefer to make mechanical connections to all calibrant wires. They are made along the lines indicated in Fig. 2.11. A cleanly drilled hole is made in one disk of metal, and a hole is punched with a sharp awl in another disk. The disk is usually made of copper or silver sheet; then a piece of bismuth wire or other wire is placed in this assembly as shown in the figure. When the assembly comes under pressure, the conical hole punched in the copper sheet with the awl is forced to collapse on the wire and make a good connection. A great deal of art is involved in this procedure, and each researcher tends to develop his own methods to overcome the problems. Cesium is a very reactive liquid metal at room temperature, but a "wire" of it can be made by filling a small polyethylene tube with it from a hypodermic needle under paraffin oil. Solid wires of copper or other metal just slightly larger than the inside diameter of the tubing are inserted in each end of the polyethylene tube to make electrical connection to the cesium and to protect it from the atmosphere.

Equipment that utilizes solid pressure-transmitting media is calibrated by noting the hydraulic ram oil pressure required to cause each of the fixed-point transitions to occur. The transitions are made only with increasing pressure, and pressure is increased very slowly over as much as an entire day. A person has to become familiar with his own particular apparatus and cell design by making several calibration runs in order to obtain a useful calibration curve. The hydraulic oil pressure to the hydraulic system, read on a sensitive gauge, is plotted as a function of the accepted fixed-point pressures of the calibrants, and a line is drawn to best fit the points. This is all done at room temperature. Even so, this is often used as the pressure calibration when the press is used at high temperature.

There will certainly be some uncertainty in knowing what the pressure is when the device is used at high temperature. Pressure studies at high temperature (say at 800°C and above) have been few because the problem is so difficult. At high temperature, changes take place in the pyrophyllite or other pressure media that generally increase the density of the pressure-transmitting material and tend to lower the pressure. On the other hand, the elevated temperature will cause expansion and tend to increase pressure. About the best that can be done at present is to assume that these effects approximately cancel each other. Another problem, although minor to the aforementioned problems, is that the reading of a thermocouple is affected by pressure. These difficulties make it imperative that a high-pressure apparatus be described in great detail, listing pertinent dimensions, materials of construction, and procedures in order that others will be able to make sound judgments on the research. Pressure values without this information are not meaningful.

The most common thermocouple used at temperatures above 1000°C in high-pressure work is Pt, Pt–10% Rh. Such a thermocouple will behave well to about 1700°C. With a given cell design and physical arrangement in a given apparatus, a temperature calibration curve can be obtained by plotting thermocouple temperature versus the electrical power in watts. This plot can subsequently be used to indicate the approximate temperature of a sample simply by measuring the current and the voltage imposed upon the heater sample tube. Additional confirmation and extension of the plot may be obtained by observing the wattage required to melt fine wires of tungsten or other high-melting metals. These wires are inserted in the sample in the same manner as the thermocouple leads, melting being detected by the occurrence of an open circuit. When more accurate temperature information is needed than that afforded by the watts-versus-temperature plot, thermocouple leads should be inserted and individual measurements made for the experiment concerned.

Pyrophyllite melts incongruently to a glassy substance at temperatures near 1500°C at pressures of a few thousand atmospheres. Its melting point is increased considerably at higher pressure, but the pyrophyllite does form a hard, white mixture of substances, as has been previously mentioned. This white, hard substance is observable only on removing the cell after an experiment. The exact nature of this material is not known at the high pressure and temperature involved in its formation, and whether or not it seriously hinders transmission of pressure to the sample is also not known.

When the sample located between the moving piston and the closure is subjected to high pressure, the tungsten carbide cylinder in this region is caused to expand (its diameter increases). The maximum increase in diameter occurs at a plane midway between the ends of the sample and tapers off toward the ends of the sample. However the clearance, which initially may only be 0.004 mm on a diameter between the piston and the cylinder bore in the piston tip region, increases sufficiently to allow cell material to extrude between the piston and cylinder. This causes severe binding of the piston, making the piston difficult to remove, and much piston breakage ensues. This is particularly true if it is pyrophyllite that extrudes. This problem can be reduced by placing an outer sleeve around the pyrophyllite that is made of polycrystalline hexagonal boron nitride. Boron nitride tends to be slippery and partially overcomes the problem. Whatever the material, when load on the piston is removed, the trapped material between the piston and the cylinder becomes more tightly wedged because the pressure on the cylinder is now relieved and it contracts. This is a major difficulty with piston-and-cylinder devices at high pressure. Sometimes a hardened steel ring with an outside diameter that fits snugly against the cylinder wall and that has a tapering inside diameter (Fig. 2.12) is placed against the tip of the piston to hinder extrusion [19]. Breaking the piston every single run is to be expected at 50,000 atm.

The belt device, to be described later, does not cost any more than a piston and cylinder device; in fact, the costs are comparable. Furthermore the belt does not require a clamping cylinder. In spite of this, the piston-and-cylinder device is used by many workers in the high-pressure field. Ordinarily, though, the device is used at pressures below 40,000 atm. The difficulties mentioned above are not as important at these pressures, but at 50,000 atm and above, the problems are vexatious.

Many of the items discussed for piston–cylinder apparatus are applicable to other types of apparatus to be considered later.

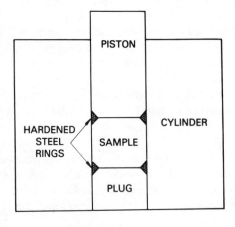

Fig. 2.12. Hardened steel ring to prevent extrusion of material between piston and cylinder wall.

4 BINDING RINGS

Piston–cylinder and belt devices require binding rings to give lateral support to a tungsten carbide cylinder or die. The support required is substantial and is usually provided by a set of compound binding rings, that is, two or more binding rings nested together. There are theories for the construction of these sets [20], but most sets are the result of trial and error. The idea of the compound binding ring set is to squeeze the carbide as strongly as possible but at the same time keep the size of the rings relatively small. In research, the binding ring set is usually lifted in and out of the hydraulic press by hand—often at arm's reach. Not only that, but (as I well remember from my earliest research with the belt) at arm's reach on tiptoe. Thus it is important to keep the weight of the set manageable.

A practical set is shown in Fig. 2.13. This set employs two tension rings, a split shim, and a safety ring. Note the tapers and the interference fits. The smaller diameter given at an interface is the I.D. of the outer ring, and the larger diameter is the O.D. of the inner ring. Assembly begins by placing the outer tension ring within the dead soft, low-carbon safety ring. Next, the inner tension ring is forced into the outer tension ring. Molybdenum disulfide paste lubricant is used on all sliding surfaces during the assembly. At this stage the I.D. of the inner tension ring is taper ground to the dimensions shown. The shim which consists of four equal segments is equispaced around the carbide cylinder or die (there are gaps between the shim segments), and this subassembly is then forced into the inner tension ring. It may require 200 tons of thrust for this final assembly operation.

The design is such that the assembled tension rings are stressed near their ultimate practical capabilities, and occasionally one or both rings will break. I have seen this happen on several occasions, once while holding a newly assembled set in my hands. However the safety ring absorbed the stored energy and prevented mishap in this and in the other instances. If there is going to be breakage, it usually occurs during the first 24 hr after assembly. Ring sets surviving a week will be good for many years.

The primary purpose of the shim is to save the inner tension ring from being scored or damaged from breakage of the tungsten carbide die. Tungsten carbide, being harder than steel, will readily scratch, scuff, and gall at a carbide–steel interface upon breaking but more particularly and severely when a fractured die is forced out of a press-fit assembly. The shim sustains only modest damage on carbide die fracture, and the die and shim are pushed out together. The shim, being in sections, can then be readily removed from the broken carbide. A pushing-out operation is not

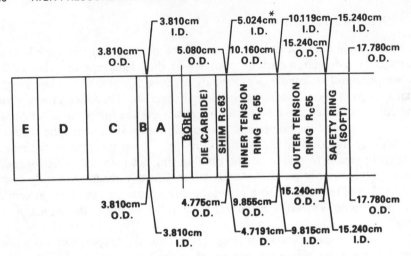

Fig. 2.13. Compound binding ring set. *NOTE—This dimension is ground with ring C in D in E. Other binding-ring and shim diameters are ground with rings disassembled. Shim, after hardening and grinding, is slit longitudinally into four equal sections.

needed. A little touch-up work is usually required on the inner surfaces of the four-piece shim whereupon it is ready to accept a new die and be forced again into the compound binding ring assembly.

For long runs at high temperature, a water cooling jacket is built to surround the safety ring. Hardened, low-alloy steels will lose their strength if heated too hot, and temperature expansion of the steel will decrease the inward thrust on the die. It is best to keep the steel at a temperature below 100°C. Tungsten carbide will also lose strength with increasing temperature.

The tension rings are constructed of AISI-4340 aircraft-quality ring forgings. After being machined to size plus a grind allowance, they are heat treated to Rockwell C-55. Final dimensions are then obtained by grinding. The C-47 to C-53 hardness range in this steel is notch sensitive and should be avoided because parts of such hardness are more likely to fracture. The shim is constructed of an air hardening, nondeforming tool steel. It is ground to size and cut into four equal segments parallel to the cylindrical axis. It is then heat treated to Rockwell C-63 and peened by a steel shot blast primarily to remove scale. Finally, the outer surface is polished to reduce friction during assembly. The taper of the fits is small enough that friction will keep the rings from sliding apart after assembly. The units are rather permanent and stable. I have had some sets for nearly 20 years that are still in good condition. The dimensions of Fig. 2.13 may be used as a base for scaling upward or downward in size.

5 HYDRAULIC PRESSES

Somewhere in a scientific laboratory there is usually a hydraulic press. If so, it can often be inexpensively modified to provide driving thrust for piston–cylinder, belt, and Bridgman anvil apparatus. Such apparatus are consequently popular since a good 300-ton press (a nominal size for research) can cost $25,000 or more.

Presses not specifically designed for high-pressure research usually are not well enough aligned and have too much play in their sliding mechanisms. However this can be overcome by use of a device called a die set. Die sets are used in the precision punch and die stamping of sheet metal and are designed for the very purpose of solving these press problems. The die set is interposed between the platens of the press, and the high-pressure apparatus is placed within the die set.

Ordinary hydraulic presses have only one ram. There is not a second ram for use in clamping the cylinder of a piston–cylinder device. In this event a permanent clamp can be placed on the cylinder. This increases the weight of the cylinder with its compound binding ring assembly and necessitates some changes in design but is satisfactory. The clamp consists of two disks of hardened steel relieved so as to clamp only the carbide cylinder from top to bottom. There is no clamping of the binding rings. A circle of matching holes in these disks just beyond the O.D. of the safety ring accepts a dozen or so high-strength bolts which are heavily but evenly torqued to provide at least 25 tons of clamping load. The central region of the clamp tapers away from its I.D. at a 45° angle to provide room for the piston, its collar, and so on.

If a standard hydraulic press is to be purchased, a *hobbing press* is the least expensive and the best. It has high tonnage in a small size and fairly good alignment. Presses specifically designed for high-pressure use are available from only a few commercial suppliers. The best commercial high-pressure equipment is sold as a completely integrated, turn-key package but is very expensive.

The least expensive hydraulic presses have spring or gravity return cylinders. In the latter case the cylinder pushes upward from the bottom of the press, and gravity pulls it down. The return action of these presses is slow. It is much better to have a press with a double-acting ram (hydraulic cylinder). Oil, under pressure, advances and also returns such a ram.

Pressurized oil for actuating hydraulic presses is usually supplied by an electrically driven two-stage hydraulic pump. This pump automatically supplies a large volume of oil at low pressure—say 35 atm—to move the pressing elements to the point of contact. Then the high-pressure stage of the pump automatically takes over as resistance to motion is sensed. The second stage supplies high-pressure oil at low volume to move the pistons or other pressing elements the short distance required to generate a high pressure in an apparatus which utilizes a solid pressure-transmitting medium. Electric pumps capable of operation at 650 atm are readily available and are of modest cost. Electrically driven pumps for higher pressures are very expensive. If higher oil pressures are desired, it costs less to buy air-driven reciprocating pumps. Such pumps, capable of pressurizing fluids (including hydraulic oil) to 4000 atm, are readily available.

If convenience or automatic controls for the hydraulic press are desired, it is best to use oil pressures below 650 atm. Electrically and pneumatically controlled systems are readily available up to this pressure. For experiments lasting more than a few minutes, automatic oil pressure control is highly desirable. Oil pressure will change with temperature and will fall because of seal leakage in the press.

Oil pressure is measured with a precision bourdon tube gauge. For electronic recording, special pressure transducers are commercially obtainable.

6   ELECTRICAL RESISTANCE HEATING EQUIPMENT

The graphite and metal heater sample tubes have an electrical resistance of the order of 0.001 ohm in a cell such as that of Fig. 2.8. It will require about 500 W (a heating current in the neighborhood of 250 A at a potential around 2 V) to reach a temperature of 1500°C in the sample. There are experiments in which a current of 1000 A may be required, but currents of this magnitude should only be used for a few minutes. Longer periods of time will damage the tungsten carbide. Such high currents cause strong local heating where the current leaves the tungsten carbide to flow into the cell. The combination of pressure plus temperature causes much more carbide breakage than does pressure alone.

Alternating current is usually used for heating. The least expensive low-voltage, high-current transformers that I have found are *stacked core* welding transformers (the so-called *wound core* transformers will not do). These transformers have a limited-duty cycle for their intended purpose at their indicated power rating but can be used in continuous-duty resistance heating at a lower rating. A 35-kVA, 440-V, 50- or 60-Hz transformer is satisfactory for general research but it should be used only at 220 V maximum. At 220 V, the voltage of such a transformer is around 4 V with external series arrangement of the output. The transformers are usually water cooled and a 35-kVA unit weighs about 100 kg. They are available from a number of welding transformer manufacturers.

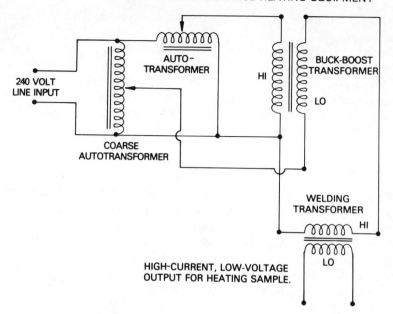

Fig. 2.14. Variable autotransformer control of welding transformer.

It will be necessary to vary the output of the welding transformer. This may be done by using a variable autotransformer arrangement with coarse and fine controls as shown in Fig. 2.14. The coarse-control autotransformer should be rated at about 100 A. The output of the fine-control autotransformer should be rated at about 10 A. The bias transformer is a "buck-boost" transformer of 1 : 10 or 1 : 20 turns ratio capable of carrying 100 A in its low-voltage output. Buck-boost transformers are available from industrial electrical distributors.

The circuit should contain a voltmeter and an ammeter to determine the voltage across the sample heater tube and the current flowing through it. A current transformer rated at 1000 A will be required in the current-measuring circuit. A wattmeter is also desirable. The product of V × A will not always agree with the wattmeter reading because of power factor considerations.

A less expensive control for the output of the welding transformer is a silicon-controlled rectifier (SCR). I have been unable to locate a standard commercial control of 100-A capacity but have had such controls custom-made by an electrical engineer. SCR controls are light in weight and small in size. However, they have the distinct disadvantage of generating a lot of electrical noise which may affect the operation of sensitive electronic measuring instruments being used in a high-pressure experiment.

Direct-current heating of the sample would be desirable for experiments in which measurements would be adversely affected by the presence of the 50- or 60-cycle alternating current. Standard, variable voltage, electronic direct-current supplies are commercially available up to 100 A at about 8 V, and these units may be used in parallel. They are however expensive.

7 OPPOSED-ANVIL DEVICES

The highest static pressures obtained in the laboratory are obtained in opposed-anvil devices. Bridgman relates that he conceived this idea while contemplating the pressure that might exist at the contact point of two knife blades pressed together at right angles to each other. One principle operative in this apparatus is shown in Fig. 2.15. Two truncated tungsten carbide cones pressed together along an axis as shown at (A) will withstand a greater load than two right circular cylinders with the same face area and length as shown at (B). Bridgman called this the *massive support* principle. In Bridgman's first anvil research, a pinch of powder or a solid disk was placed between the faces of the anvils (Fig. 2.16a), and they were then pressed together. Pressure was computed simply as force per unit area of the faces.

Later, an assembly (cell) as shown in Fig. 2.16b was devised to enable the electrical resistance of metals to be measured as a function of pressure. The outer ring E of the cell is pipestone. Pipestone is an impure pyrophyllite obtained near Pipestone, Minnesota, This substance was used by American Indians in making peace pipes. Within the pipestone ring is a silver chloride disk F in which a metallic ribbon G is embedded, as shown. This ribbon makes contact on the top side with the top anvil face and on the bottom side with the bottom anvil face. If the anvils are electrically isolated from each other, an electrical indicating device can be connected to them to measure electrical resistance as a function of pressure. If the anvil circular faces are approximately 10 mm in diameter, the pipestone ring is made about 0.6 mm thick. The outside diameter of the pipestone ring is 10 mm (same as anvil diameter), and the inside diameter is about 7.5 mm. This in essence is a two-dimensional pressure apparatus, and the components within the cell are tiny indeed. Bridgman measured the electrical resistance of many substances at pressures up to what were thought to be 100,000 atm [21]. Later work has shown however that the highest pressure obtained in these experiments was only 65,000 atm.

It is a general principle that the smaller the apparatus, the higher the

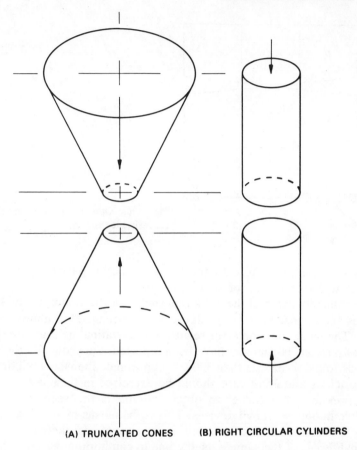

(A) TRUNCATED CONES   **(B) RIGHT CIRCULAR CYLINDERS**

Fig. 2.15. The massive support principle.

pressure that can be generated. This is particularly true of opposed-anvil apparatus. These devices have been miniaturized to the point where diamonds are used for the anvils and cells are assembled under microscopes. The cell (sample) thickness may be as small as 0.025 mm, and the diameter may only be 0.50 mm. Great care must be used in such miniscule devices to avoid error. Slight dimensional errors in fabricated parts (pipestone rings or silver chloride disks, for example) can cause great changes in the pressure obtained with a given anvil load. Also, samples being X-rayed, a common procedure with diamond anvils, may shift in position and an X-ray diffraction pattern of the wrong material may be taken. A serious extant error of this nature occurred when a researcher published the exciting discovery of a new form of carbon of greater

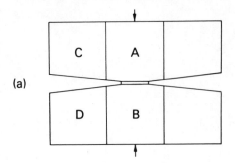

(a)

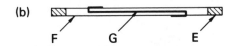

(b)

Fig. 2.16. Opposed anvils (*a*) and electrical resistance cell (*b*).

density than diamond. Actually the X-ray diffraction pattern of the "new carbon" was merely that of silver chloride!

Unfortunately much of the work reported in the literature in which tiny samples and opposed-anvil-type devices have been used contains serious errors. The most common error relates to overstating the pressure. Pressures as high as 700,000 atm have been reported. Subsequently the pressure was found to be less than 300,000 atm. Small opposed-anvil devices are capricious and great care should be exercised in their use.

Compression of a gasket to obtain motion is the second principle operative in opposed-anvil devices. This is the means by which a thicker sample is contained and compressed. In the case of ungasketed anvils, the frictional forces of the sample itself come to equilibrium with the applied load at some thickness characteristic of the substance. This thickness would be of the order of 10% of that of a gasketed sample. The pressure obtained at a given anvil load varies with the substance.

It is amazing that a thin ring of pyrophyllite gasket can contain the sample as the pressure builds up. The gasket not only decreases in thickness as the anvils advance but some of it extrudes until an equilibrium situation is reached for a given pressure. On release of pressure, blowouts are a common phenomenon. Pyrophyllite elasticity is limited and, having partially extruded during pressure buildup, it cannot return to its initial thickness. Consequently, it loses its frictional hold on the anvil faces, and the compressed sample blows out. Accurate alignment of the anvils, particularly parallelism of the faces, is an important consideration. Slight misalignment can cause premature anvil breakage and blowouts, and skew the pressure distribution within the sample.

At pressures of 100,000 atm and beyond, the faces of tungsten carbide anvils are permanently deformed and become concave. Sometimes the deformed anvils are "refigured" (ground flat again with a diamond grinding wheel). After refiguring, anvils are stronger (since they have been cold worked by the deformation) and can be subjected to even higher loads than when new. At the higher pressures, opposed anvils are often used but once, since they usually break on release of pressure.

The sample while under pressure is lens-shaped, being thicker at the center and thinner at the edges. This is due not only to permanent deformation of the anvils but also to the elasticity of the anvils. There is also a pressure gradient in solid samples, the pressure being highest at the center along the anvil axis, trailing off to much smaller pressures at the interface with the gasket. As a matter of fact, this is one of the major disadvantages of the device. Consequently, when measurements of various kinds are to be made, it is necessary to make the measurement in a tiny space as close as possible to the center of the sample. This can be done in X-ray diffraction or optical experiments where a very fine pencil of rays is impinged upon the sample along the centerline axis.

By using a metal, rather than a pipestone or pyrophyllite, gasket, it has been possible to work with liquids to very high pressures, and X-ray diffraction as well as optical observation and measurement of single crystals has been accomplished [22].

Diamond anvils are particularly appealing because they can be used under a microscope, it being most interesting and satisfying to observe directly a phase transformation taking place at a pressure of 100,000 atm. Their use was first described in 1959 [22]. The least expensive way to obtain diamonds for use as anvils is to purchase brilliant cut gem diamonds from a major supplier in New York or Amsterdam. These diamonds are purchased with one slight modification: the culets (bottom tips of the diamonds) are ground off to form small flats parallel to the tables (broad top surfaces). These flats are ground to different surface areas to minimize axial alignment problems when they are pressed together. Alternatively a ground-off culet flat is sometimes pressed against the table of another diamond. The diamonds are commonly of about one-tenth carat (0.020 g). The anvil faces are often about 0.5 mm across. The ground-off culet will not be perfectly round but octagonal. If optical properties are of no great concern, yellowish diamonds, which cost less than water-white diamonds, can be used. If transmission in the infrared is needed, type II diamonds should be purchased. Type I diamonds have an intense absorption band at wavelengths from about 7 to 9 μm. Regardless of color or type, though, the diamonds must be physically sound. Major diamond vendors are familiar with these factors and are prepared to supply the needs of diamond anvil users. Although diamond is the strongest substance (in compression) known to man, it is brittle, and an off-axis load would readily chip it. To assure that the planes of the mating diamond anvils are parallel to each other as they come together, they are mounted in spherical sockets. Alternatively one diamond may be mounted to swivel about the x-axis and the other about the y-axis Cartesian coordinates to accomplish the same purpose. The diamond is commonly set in a metallic mounting (usually stainless steel) by pressing it into the metal to form a seat. It is then held in place by epoxy cement. When the anvils are used in X-ray diffraction work, the metal used is often beryllium because of its relative X-ray transparency.

The area of contact of diamond anvils is so small that it is not necessary to use hydraulic rams to drive them together. Most researchers use either springs, whose tension can be increased or decreased by thumbscrews, or merely the inherent elasticity of steel itself, and clamp the anvils together with socket head capscrews. In spite of the cost of gem-quality diamonds, a diamond anvil apparatus is the least expensive apparatus that a person can acquire. Consequently they are very popular. Some other materials that have been used for opposed anvils are polycrystalline alumina and single-crystal sapphire.

Diamond anvils are so tiny that it would be very difficult to provide internal heating to operate at high temperature simultaneously with high pressure. However Ming and Bassett [23] have used a laser beam to heat samples between the faces of the diamonds and have even converted graphite to diamond by this means. Barnett, Block, and Piermarini [24] have constructed a device for externally heating diamond anvils.

In addition to the pressure gradient problem in opposed anvils, there is the very real problem of accurately determining the pressure. Simultaneous X-ray diffraction of NaCl and a substance embedded therein has been used in diamond anvils [25]. The lattice spacing of the NaCl is used to indicate the pressure on the embedded substance. A drawback of the system is the exposure time required—up to 300 hr.

In some respects the ultimate diamond anvil apparatus is that of Barnett, Block, and Piermarini [24]. They have described an optical system for rapid routine pressure measurement which utilizes a pressure shift in the *sharp R-line* fluorescence of ruby. A metal gasket (simply a relatively large square of Inconel about 0.13 mm thick in which a hole of about 0.20 mm in diameter has been drilled) is centered between the diamond anvils, and a 4:1 mix of methanol and ethanol is placed in the hole. A speck of ruby is added as well as a speck of whatever else is desired, and the anvils are closed on the gasket. Whatever is inside is trapped, and further advance of the anvils subjects the contents to hydrostatic pressure all the

Fig. 2.17. Diamond anvil apparatus for single-crystal X-ray diffraction studies. From Merrill and Bassett [26].

way to 100,000 atm. And the pressure is known by the fluorescence of the speck of ruby!

A recent miniature diamond anvil pressure apparatus for single-crystal X-ray diffraction studies [26] is shown in Fig. 2.17. This device also uses the method of encasing liquids within a small-diameter hole in the metal foil, a technique first used by Van Valkenburg [27]. Beryllium foil is used because of its X-ray transparency. This apparatus is mounted on a standard goniometer head which may be attached to a standard X-ray precession camera or to a single-crystal orienter. The single crystal to be studied is immersed in a true liquid. This device has been used in the study of two high-pressure phases of calcium carbonate, $CaCO_3$ (II) and $CaCO_3$ (III).

8 BELT APPARATUS

Figure 2.18 shows an exploded view of the belt high-temperature, high-pressure apparatus [28]. I invented this device in January of 1953 while employed at the General Electric Research Laboratory in Schenectady, New York. At first the device was not taken too seriously by my colleagues or the management. Consequently it was not protected with company secrecy as one might have expected. For example, on September 16, 1953, a prominent high-pressure researcher, Professor A. Michels of the Van der Waals Laboratory, Amsterdam, visited the G.E. research laboratory as the guest of a G.E. official. Professor Michels' wife accompanied him on this visit. I was told to show them the details of the belt, and while I explained and demonstrated, Mrs. Michels took notes, took measurements, and made detailed sketches. Many people at G.E. who were not connected with the diamond synthesis project knew about the belt, and one young G.E. researcher conveyed the details of the device to his former professor, who was working in the field of high pressure at a Midwest university.

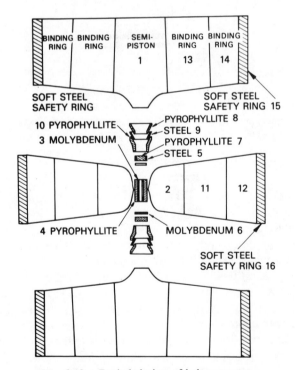

Fig. 2.18. Exploded view of belt apparatus.

The first written report within G.E. on the belt was the usual patent-disclosure letter to the Director of Research. There were subsequent follow-up letters on additional details and miscellaneous memorandum reports that were circulated to about a dozen persons within the company. The most formal company report that I wrote was No. RL-1064 of March 1954 entitled "The Belt Ultra High Pressure, High Temperature Apparatus." This document was labeled "class 4," under which this statement appeared: "This document contains information of importance to G.E. Its distribution is rigidly limited." This report was circulated within the laboratory and also to G.E. personnel elsewhere. It not only contained the general details of the belt and its operation, but included scale drawings and dimensions. This 1954 report, stripped of its detailed drawings and dimensions, became the *Review of Scientific Instruments* article published in February of 1960 [28].

It was not until I had made diamonds in December of 1954 that company secrecy on the belt began to firm up. You can imagine my eagerness as a young scientist to make my mark, upon accomplishing the twin task of developing a high-pressure, high-temperature apparatus capable of synthesizing diamonds and then making them. You can further imagine my anxiety on being unable to have my work published for six years. This anxiety was heightened by the fact that information concerning the belt was at large in the world and that others might publish on this or related devices before I would have the opportunity. The knowledge that the belt existed and the knowledge that diamonds had been synthesized gave others a tremendous advantage in trying to accomplish the same things. There were some other leaks of vital information. After several hundred copies of one of the earliest press releases was distributed in 1955, it was decided that the details and photographs of the belt in this brochure were too revealing. Distribution of this particular brochure was then stopped, and one that was less explicit was substituted. Also, persons in government laboratories professing "need to know" obtained copies of the patent applications.

Although I have not published the full details of the human side of the development of the belt apparatus and the first synthesis of diamonds, I have given some personal glimpses into the events that transpired [29].

The belt was the first apparatus capable of operating simultaneously at high pressure and high temperature at pressures of 100,000 atm. Maximum pressures of the order of 150,000 atm simultaneously with temperatures in excess of 2000°C have been maintained for long periods of time in this device. Referring to Fig. 2.18, the functions of the various parts are as follows: two tapered tungsten carbide pistons (1) push into each side of a tapered tungsten carbide chamber (2). Pressure is transmitted to the sample contained in a molybdenum tube (3). This tube not only serves as an electrical resistance heating element but also contains the sample and is often referred to as the heater sample tube. Graphite and tantalum tubes are also used. Pressure is transmitted to the sample contained in the tube (3) by pyrophyllite (4). The pyrophyllite also serves as thermal and electrical insulation. Electricity passing through the tube (3) serves to heat the sample. Current enters this tube through a molybdenum disk (6), which in turn touches a steel ring (5), which in turn touches the tip of the piston. The short pyrophyllite cylinders (7) provide thermal insulation at the ends of the sample tube. A sandwich gasket, composed of pyrophyllite (8 and 10) and steel (9), maintains the pressure in the chamber.

The hardened steel compound binding rings (11) and (12) provide lateral thrust to the chamber. Binding rings (13) and (14) do the same for the conical pistons. Rings (15) and (16) are made of dead soft low-carbon steel and are intended to absorb the energy released should the inner binding rings fail. These binding ring sets are similar to those already discussed for piston–cylinder apparatus. The chamber (2) and binding rings (11) and (12) form a "torroidal belt" around the sample, and from this I took the code name "belt" for the apparatus. For sustained operation at high temperature, the conical pistons and the corresponding chamber must be cooled. Figure 2.19 shows the belt assembly in closed position. Figure 2.20 shows a photograph of the first belt apparatus.

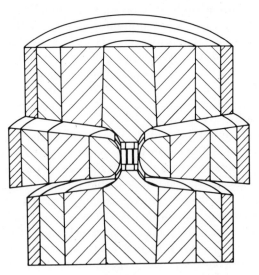

Fig. 2.19. The belt in closed position.

Fig. 2.20. The first belt apparatus.

The belt uses a compressible gasket in a manner akin to that used in Bridgman anvils, but there are some differences. In Bridgman anvils the gasket is a flat ring in a plane 90° from the axis of compression. The motion of the anvils is determined by the amount that the gasket can be compressed. For more compression, one would use a thicker gasket. However there is a limit to the possible thickness of the gasket. If it is too thick, chunks will break from it allowing the sample to escape, and the gasket will not compress symmetrically or evenly. The maximum permissible thickness of the gasket is determined by trial and error. For opposed anvils with 1 cm-diameter faces, the maximum thickness for a pyrophyllite gasket is about 0.25 mm. It would be more for anvils with larger-diameter faces if the gasket width were correspondingly increased.

In the belt the gasket is arranged at an angle less than 90° to the direction of compression, as is shown in the single-ended device (half-belt) of Fig. 2.21. In the belt design the thickness t between the two arrows A–A is substantially the same as would be possible to use in a Bridgman anvil device with faces of this diameter. However in the belt, the relative movement of the tapered piston into the tapered cylinder is along the axis given by B–B. This thickness s is given by $s = t/\cos\theta$. The stroke (maximum motion) in the belt is correspondingly increased. The

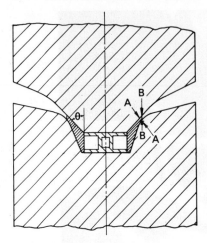

Fig. 2.21. Detail of motion obtained via compressible gasket in the belt.

force acting perpendicular to the contact area between the gasket and the tungsten carbide components (A–A) is reduced by cos θ over the case where the gasket is perpendicular to the direction of compression. Therefore the friction between the gasket and the tungsten carbide will be reduced by cos θ. The smaller the angle θ is made, the larger s becomes. The limit on the smallness of θ is set in part by the frictional forces along the gasket–tungsten carbide interface. When they become too small, the gasket will be blown out by the pressure.

A conical piston has much greater strength than does the cylindrical piston of a piston–cylinder design. Consequently, if sufficient stroke can be obtained to make the apparatus useful, a device with conical pistons should be able to obtain a much higher pressure than is obtained in piston-and-cylinder devices, and this is indeed the case. The sandwich gasket further increases the stroke, hence the sample size and the ultimate pressure of any system utilizing stone gaskets. For example, it will allow an increase in sample thickness in Bridgman anvils. In the belt the thickness of the gasket can be more than doubled by the insertion of the thin metal cone, which is the "meat between the bread" in this array. The metal reduces gasket crumbling, allowing smooth compression and extrusion. The available stroke of the belt apparatus of Fig. 2.18, which utilizes the combined effects of double-ending, tapered gasket, and the sandwich gasket, is approximately 6 mm, compared to a maximum of about 0.7 mm in a Bridgman anvil device of comparable face diameter. This greatly increased stroke makes it possible to have a tremendously large sample size (compared to that available in opposed anvils) and, of course, offers a size practical for the commercial production of diamond.

There are some other advantages of the belt. Very accurate alignment is required for piston-and-cylinder devices. Good alignment, comparatively speaking, is required in Bridgman anvils. The belt is the least sensitive to misalignment, and slight axial misalignment or cocking of the moving members with respect to each other does not interfere with its operation. For certain purposes a plain pyrophyllite gasket (nonsandwich gasket) is perfectly suitable for use in the belt. In other instances two steel cones can be used in the sandwich gasket to increase the stroke of the belt even more. The distance between the two conical piston tips will determine the pressure that can be reached. To obtain the highest pressure, the conical piston tips are placed only about 1 mm apart. Such devices have been called "high compression belts." In addition to placing the piston tips close together, one would miniaturize the apparatus to take advantage of the effect already mentioned that higher pressure can be obtained in smaller devices than in larger devices. There seem to be two effects that make this so. There is the surface-to-volume ratio, which has often been observed to effect changes in engineering problems. Also, in sintered tungsten carbide parts, a part can be made more free from defects in small sizes than in large sizes. In the usual manufacture of cemented tungsten carbide, cobalt powder and tungsten carbide powder are intimately mixed together and then cold pressed to the approximate shape desired. The preformed part is then sintered at a temperature just below the melting point of cobalt. In large parts, cobalt tends to concentrate in the bottom of the part owing to the forces of gravity. This is not so likely to occur in smaller parts. There is also the statistical situation that a large part has a greater probability of containing a flaw than does a small part. In brittle substances, such as cemented tungsten carbide, breakage usually starts at a flaw within or on the surface of the material.

There is a cooperative phenomenon at work in the belt to allow it to operate at higher pressures. Through the gasket, the tapered piston is supported by the tapered conical chamber, and vice versa. An included angle between 60° and 90° is suitable for the conical piston in the belt apparatus. However in some special belts (unpublished work of the author), the angle has been decreased to as low as 20°. Figure 2.22 shows the central region of the belt and the gasket after a high-pressure run.

Thermocouple wires and other electrical leads are inserted into the sample region of the cell by drilling fine holes through the pyrophyllite in a manner such that the wires do not make contact with the tungsten carbide components. Frictional forces will hold the wires in place as pressure is developed within the system. In the belt, the center of the gasket assembly sustains more load than the outer regions, and there is a gradual drop of pressure along the gasket to atmospheric pressure at its outer edge.

Fig. 2.22.   Central region of the belt and the gasket after a run.

This is a kind of multistaging effect and is also a factor in making it possible to sustain such very high pressures in a relatively large volume.

9   MULTIANVIL APPARATUS

Opposed-anvil, piston-and-cylinder, and belt apparatus may all be classified as uniaxial devices. That is, the means for mechanically reducing the volume of a substance is accomplished by moving certain apparatus components toward each other along a line. There are some disadvantages to uniaxial devices. When quasi-hydrostatic pressure-transmitting materials are used, pressure gradients will invariably be set up within the cell in a pattern depending on the movement of the compressing members. The Poisson ratio of the material between the pistons or anvils will also influence these pressure gradients. So will the geometry of the cell components.

In the belt apparatus, the heater sample tube is originally a right circular cylinder; but when the high-pressure, high-temperature run is over, it will have been considerably foreshortened and will have a barrel shape. If one were producing a sintered part, for example, a sintered diamond cylinder, it will be distorted in the sintering process. Furthermore the sintered product will tend to delaminate (break into several disks perpendicular to the axis of compression) after pressure is released. Sometimes the strains resulting from the uniaxial compression are so great the material almost explodes as the layers separate from each other. This problem can be overcome to some extent by surrounding the heater sample tube with a material that is more hydrostatic than pyrophyllite, for example, sodium chloride. Indium metal will also serve this purpose well because of its very low internal friction. However it is metallic, has a low melting point, and also has a relatively high thermal conductivity, and these characteristics restrict its use.

Uniaxial hydraulic presses are rather standard equipment in research laboratories and in industry. Consequently they have been used to power the advance of pistons, anvils, and the like in high-pressure devices. Researchers have tended to think in terms of uniaxial machines. However the problem just ascribed to uniaxial devices can be overcome in apparatus in which compressing members are thrust toward a central region along several axes. The first such device was the tetrahedral anvil press [5]. This was the first high-temperature, high-pressure device capable of synthesizing diamonds that was revealed to the world. It came into being as a matter of necessity, and the reader might be interested in a brief account of these circumstances.

After leaving the General Electric Company in the fall of 1955 to become Director of Research and Professor of Chemistry at Brigham Young University (BYU), I was besieged with requests from scientists everywhere to reveal the details of the belt. Circumstances of company secrecy compounded by a government secrecy order, however, prevented disclosure. In fact, I could not even use the apparatus for research at BYU. It soon became apparent that this problem would not be resolved for some time. Consequently I found myself being encouraged by many persons and institutions to try to develop another apparatus that could be made available to scientists everywhere. Financial support for such an undertaking came first from the Carnegie Institution of Washington. Shortly thereafter, additional financial aid was received from various government agencies, notably the National Science Foundation.

Within two years I had tested several ideas in which pressure was generated by moving "anvils" along several axes. Some of these devices utilized a uniaxial hydraulic press to cause tapered wedges sliding in a cone-shaped ring to move toward each other and thus compress a pyrophyllite cylinder radially inward (Fig. 2.23). Another device was called the "Black Hawk Special." This machine utilizes a number of flat wedges placed between two flat plates. The plates are maintained at a fixed distance corresponding to the thickness of the wedges. Alternate wedges advance or retract, as shown in Fig. 2.24, to reduce the volume

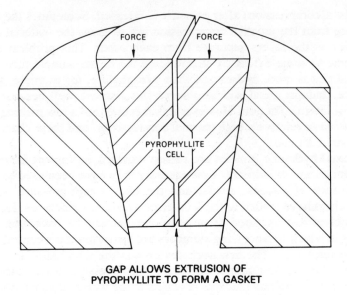

Fig. 2.23. Tapered wedge press.

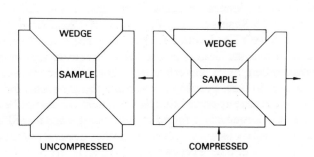

Fig. 2.24. Black Hawk Special.

and thus increase the pressure. The movement of the wedges is exaggerated in the figure to better illustrate the principle. I also built and tested a variety of uniaxial devices. The most successful device, however, used hydraulic rams built into a special press frame to afford symmetrical action in three-dimensional space. I considered tetrahedral presses, hexahedral (cubic), and higher-order multianvil presses, but calculations and intuition showed that the tetrahedral press would be capable of generating the highest pressures. The original tetrahedral press is shown in Fig. 2.25, and a cluster of tetrahedral anvils is shown in Fig. 2.26.

Fig. 2.25. The original tetrahedral press.

During the Christmas holidays of 1957 diamonds were made in this apparatus—a test of the device that seemed important at the time. Patents were obtained on it through BYU and Research Corporation, and all claims were quickly granted on the first office action.

The device was described at the Spring 1958 Meeting of the American Chemical Society in San Francisco and was published in the April 1958 *Review of Scientific Instruments*, and many scientists from all over the world visited Provo to see it. Regardless, the U.S. Government (contrary to previous agreement) placed a secrecy order on the tetrahedral press on January 15, 1959. Note that thousands of copies of the *Review of Sci-*

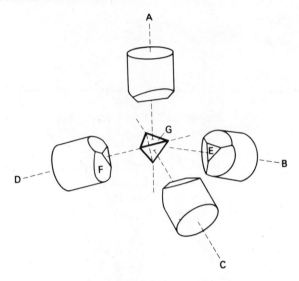

Fig. 2.26. Cluster of tetrahedral anvils.

entific Instruments, which fully described it, had been distributed all over the world nine months before this. Fortunately this secrecy was short-lived, having been rescinded within six months.

A tetrahedral press might be considered a three-dimensional extension of the two-dimensional Bridgman anvil concept. The principle of massive support is still at work in the tetrahedral press and other multianvil presses, but to a lesser extent than in Bridgman anvils. In a regular series of multianvil presses, the solid angle subtended by each anvil decreases as the number of anvils increases. This solid angle is largest in the tetrahedral press and is the primary reason why higher pressures can be obtained in this press than are achieved in cubic and higher-order presses. Some articles in the literature claim that the greater the number of anvils in a multianvil device, the higher the pressure that can be achieved. This runs contrary to my experience.

In multianvil devices the anvils support each other through load on the gaskets, and this tends to compensate for the reduction in massive support over that available in opposed anvils. The four anvils of the tetrahedral press (Fig. 2.26) are driven toward a central point by hydraulic rams whose axes lie along lines normal to the triangular anvil faces. The hydraulic rams and collinear anvil axes (A, B, C, D) intersect at the tetrahedral angle of 109.47° at the center of the regular tetrahedral volume enclosed by the anvil faces.

Tungsten carbide is usually used as the material of construction for the anvils. Each anvil is surrounded by a single press-fit steel binding ring which, as usual, absorbs the tensile loads developed within the tungsten carbide. The sloping shoulders of the binding ring break away at a 2° steeper angle on each side than do the corresponding 45° shoulders of the anvils. This provides a widening gap moving outward from the center of the press and decreases the likelihood of binding rings touching each other in case of slight misalignment. Such touching would constitute an electrical short circuit. The tapering gap also lessens the likelihood of pinching off thermocouple wires or other electrical leads that may be inserted into the sample. The cell in which the pressure is generated consists of a regular pyrophyllite tetrahedron, as is shown in Fig. 2.27.

If the anvils (Fig. 2.26) are advanced symmetrically toward the center of the press until their sloping shoulders F touch, their triangular faces E will define a regular tetrahedron of a given size. The tetrahedral cell G is made larger than this size so that a gasket will be automatically formed by

TETRAHEDRAL CELL

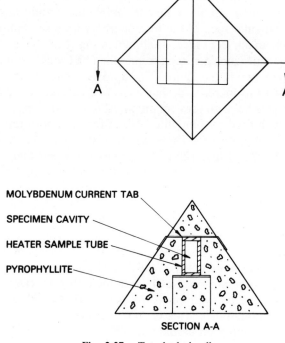

SECTION A-A

Fig. 2.27. Tetrahedral cell.

extrusion of excess pyrophyllite as the anvil faces impinge on this larger cell. In practice, the edges of the tetrahedral cell are made about 25% longer than the corresponding legs on the triangular anvil faces. The pyrophyllite within the tetrahedron transmits pressure to the sample, provides thermal and electrical insulation, and provides the necessary compressing gasket. This cell contains considerably fewer and simpler parts than the corresponding cell and preformed gaskets of the belt. The cell is machined from pyrophyllite by standard machine tools such as lathes and milling machines. Either high-speed steel or tungsten carbide cutting tools may be used. Rough shapes of pyrophyllite are sawed with a bandsaw from commercial blocks which are available up to a 30-cm cube [30]. The material is mined in South Africa.

Referring again to Fig. 2.27, the diagonally disposed heater sample tube is usually made of a high-melting metal, such as molybdenum. The tab is also of molybdenum, and the pyrophyllite prisms provide thermal insulation at the ends. Thermocouple leads may be brought out through edges of the tetrahedron. Four electrically insulated anvils provide four electrical connections to the cell's interior. If only half a tab is used for heating current, two anvils are free for other connections. There are other ways that the heater sample tube and other components may be arranged within the tetrahedron. For example, a heater sample tube could make contact through one triangular face of the tetrahedron with a second tab making contact through an adjacent face, as shown in Fig. 2.28. Several electrical leads for various measuring purposes may then be brought out through an apex as is also shown in the figure. This apex is usually fired in a furnace at 750°C for 12 hr to increase its hardness and strength over that of the natural stone. This helps to prevent pinch-off of the wires passing through the prism. If such fired pyrophyllite is used for the entire tetrahedron, a suitable gasket will not form. There will be crunching sounds and expulsion of large chunks of material as the anvils come together. A smooth, noiseless extrusion of material is needed for proper gasket formation.

Pressure calibration of the tetrahedral press is made in a manner similar to that described for the piston–cylinder apparatus. A typical edge length for the anvil triangular face for use in research is 2 cm. The corresponding length of the triangular edge of the tetrahedral cell would be 2.5 cm. The highest pressure observed in a tetrahedral press at Brigham Young University (with 6 mm on edge anvil faces) is 120,000 atm, based on X-ray diffraction measurements on NaCl contained therein (Decker's NaCl scale). The everyday working pressure used at BYU with 2 cm on edge anvil faces in tetrahedral presses is 70,000 atm. At this pressure and at room temperature, tungsten carbide anvils of 8% cobalt content will last indefinitely. High temperature simultaneously used with high pressure,

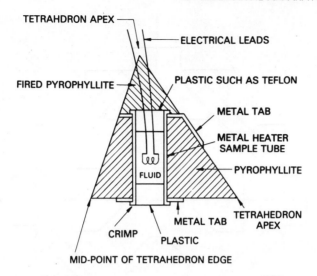

TETRAHDRON APEX

ELECTRICAL LEADS

FIRED PYROPHYLLITE

PLASTIC SUCH AS TEFLON

METAL TAB

METAL HEATER
SAMPLE TUBE

PYROPHYLLITE

FLUID

TETRAHEDRON
APEX

METAL TAB

CRIMP

PLASTIC

MID-POINT OF TETRAHEDRON EDGE

Fig. 2.28. Another version of the tetrahedral cell.

however, causes some breakage. Simultaneous 1500°C at 70,000 atm conditions with the same anvils gives an anvil lifetime of about 2000 runs. This varies with the size of the sample within the tetrahedron and the current flowing through the anvils. To increase lifetime, the sample and the heating current should be small.

The design of the press in multianvil apparatus is important. In the original tetrahedral press, the bases, tie bars, and other appurtenances were not machined to any particular precision. Consequently turnbuckle arrangements as shown in Fig. 2.25 were used to align the tetrahedral axes of the rams. Each of the four hydraulic rams were independently valved. To operate the device, the lower three rams were advanced until the anvils almost touched to form a "nest." Then the pyrophyllite sample cell was placed apex down in the nest. Next, the upper ram was advanced until the triangular face of its anvil touched the upward-facing triangular base of the pyrophyllite tetrahedron. Each anvil was then individually advanced approximately 0.1 mm at a time by admitting hydraulic oil under pressure sequentially to the first, second, third, and fourth rams and repeating the process until the desired pressure was built up within the sample. This procedure required skill, and it was difficult and time consuming to teach others to use the press. Consequently I sought means for synchronizing the motion of the anvils so that at any instant the anvil faces would be equidistant from the center of the tetrahedral frame as defined by the tie bars which hold the press together.

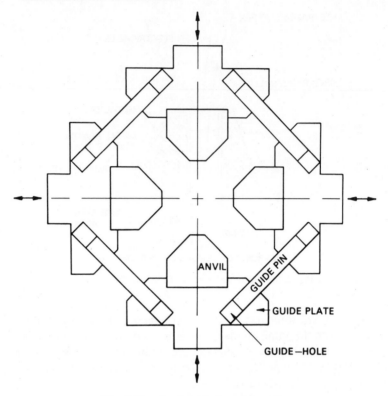

Fig. 2.29. Anvil guide for cubic press.

Consideration was given to mechanisms that would control the flow of oil to the rams or that would operate from a position-indicating transducer. Such devices were complicated and did not satisfactorily solve the problem. The final solution was simple but took longer to conceive than anticipated. The solution is a mechanical device that I have called an anvil guide [31]. Figure 2.29 shows a cross section of an anvil guide device as applied to a cubic press. It consists of 12 guide rods or guide pins and six guide plates. A guide plate is fastened to the moving end of each hydraulic ram. The anvil, with its various supporting structures, is centrally mounted on the guide plate. Each guide plate contains four guide holes symmetrically arranged at 90° angles to each other. The axis of each hole makes an angle of 45° with respect to the ram axis. When assembled, the guide rods are positioned within the guide holes in which they are free to slide as the rams advance or retract. All rams, being interconnected by the guide rods and plates, are forced to move synchronously together as

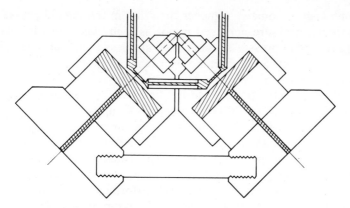

Fig. 2.30. Inverted ram cubic press with anvil-guide retraction system and right/left-hand thread tie bars.

hydraulic oil is simultaneously applied to all the rams. In the tetrahedral press, there are six guide rods and four guide plates. The three holes in each guide plate are disposed at 120° angles about the ram axis, and each makes an angle of 35.26° to the ram axis.

The most advanced cubic (or tetrahedral) press built to date [32] incorporates a hollow guide-pin system for retraction of the hydraulic rams, which saves weight and machining costs and improves mechanical proportions (Fig. 2.30). This system unclutters the anvil region (from water-cooling lines) and provides better access. Oil circulating through the guide pin system, when it is not being used to retract the press, provides cooling. The use of inverted hydraulic rams and right/left-hand thread tie bars and bases reduces weight and cost further. Retraction is accomplished by admitting pressurized hydraulic oil to the anvil guide assembly. Le Chatelier's principle requires the interconnected system to increase in volume, and this is accomplished if the cylinders move (retract) toward the bases. Anvil guide mechanisms can also be applied to any other variety of polyhedral press.

The guide pins and holes need not be symmetrically disposed as indicated in Fig. 2.29 but need only to be parallel to the tie bar axes. They may thus be offset to facilitate the admission of beams of matter or radiation such as neutrons and X-rays into the sample. Alternatively one or more pins may be eliminated without adversely affecting the operation of the anvil guide. Two smaller-diameter guide pins that straddle the normal symmetrical position of a single pin have also been used to give access.

The cubic press is the next member in the series of polyhedral presses.

Anvil breakage is somewhat higher than that in tetrahedral presses, and the maximum pressure achieved is not as high. The everyday working pressure for cubic presses is about 5000 atm less than that of tetrahedral presses. Excellent anvil lifetime is obtained in high-pressure, high-temperature runs at 65,000 atm. There are some secrets to obtaining long anvil lifetime in cubic presses. In my first cubic press there were occasions on which all six anvils would break for no apparent reason. This was very puzzling until I finally realized that the anvils were breaking every time that I first used the press after returning from a meeting or speaking engagement away from the university. This caused me to wonder if there could be such a thing as too long a rest period—a period of nonuse of the anvils. So deliberate experiments were performed to test this idea. Two things were learned. First, the anvils should not remain idle for more than 60 hr if they are being routinely used at 65,000 atm. Second, if the anvils are unused for periods longer than 60 hr, they should be "stress relieved" at a pressure of about 25,000 atm (at room temperature) for a period of 12 hr before resuming routine use at 65,000 atm. This breakage phenomenon does not exist for the tetrahedral press, and no such precautions are required.

An advantage of the cubic press is its rectangular coordinate geometry. A typical cubic cell is shown in Fig. 2.31, and a typical cubic press is shown in Fig. 2.32. It is possible with the cubic press to have six independent contacts emanate from within the high-pressure cell and connect to the outside through the six electrically isolated anvil assemblies.

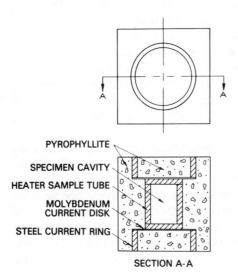

PYROPHYLLITE

SPECIMEN CAVITY

HEATER SAMPLE TUBE

MOLYBDENUM
CURRENT DISK

STEEL CURRENT RING

SECTION A-A

Fig. 2.31. Typical cubic cell.

Fig. 2.32.   Typical cubic press.

Multianvil presses need not be regular as are the tetrahedral and cubic presses so far described. For example, it is possible to have a tetragonal press. Opposite ends of the cell have square faces, and the four sides are rectangular. The anvils for such a press would enclose a tetragon, and the cell assembly would be oversize in order to allow for extrusion of a gasket as is the case with the presses already described.

As can be seen, in multianvil press systems the entire system (hydraulic rams, anvils, guide pins, etc.) is engineered as a unit. This contrasts with most situations for belts, Bridgman anvils, and piston–cylinder devices, where the hydraulic press is a separate, standard item of manufacture that is modified to accept the aforementioned devices. Even with uniaxial systems, however, the best apparatus is a totally engineered, integrated unit, press and all.

10   INSTRUMENTATION

The basic high-pressure apparatus that have been described have been instrumented in numerous ways to study a great variety of phenomena. It is not the purpose of this chapter to delve into the details of such instrumentation. Considerable ingenuity has been used by many researchers, and a few references are listed below to sample briefly what has been done.

Differential thermal analysis (DTA)	[33]
Mössbauer effect	[34]
Diffusion	[13]
Manganin gauge pressure measurement	[35]
Neutron diffraction (time of flight)	[36]
Optical measurements	[37]
X-ray diffraction	[38]
Cryogenic techniques	[39]
Nuclear magnetic resonance	[40]
Electron spin resonance	[41]
Thermal conductivity	[42]

Publications of the author dealing with high-pressure apparatus that have not been referenced are cited in Refs. 43 through 53.

11   THE SYNTHESIS OF DIAMOND: AN EXAMPLE OF THE USE OF HIGH-PRESSURE, HIGH-TEMPERATURE APPARATUS

A cubic press with six 20.3 cm-diameter (320 cm²-area) hydraulic rams and six 1.27 cm on edge square-faced anvils (1.61 cm² area) is used. The hydraulic rams are suspended in an octahedral frame consisting of ram bases and tie bars. This frame is mounted with one triangular face of the octahedron parallel to the floor. I prefer this mounting because debris from the cell (crumbled gasket, etc.) falls between the three lower anvils as the press opens and does not, for the most part, get into the sliding mechanisms of the press. Cubic presses are sometimes mounted with a line through opposite apices of the octahedron perpendicular to the floor. This locates one anvil facing upward with its face parallel to the floor, which is convenient for placing the cell. Gravity holds the cell in place as it rides upward while the anvils advance toward the center of the press. However pyrophyllite debris will fall onto the sliding surfaces of four guide pins unless protective shields are used.

The cubic pyrophyllite sample cell is 1.59 cm on edge. It has a central bore of 6.35 mm and a counterbore on each end to accommodate the current ring and its central pyrophyllite disk. The current rings (two each) are made of AISI 1020 cold-finished steel and are 11.07 mm O.D. by 8.46 mm I.D. by 4.22 mm thick. A pyrophyllite disk 8.46 mm in diameter and 4.22 mm thick is placed inside each current ring and provides thermal insulation at each end of the heater sample tube. Two molybdenum disks 10.90 mm in diameter and 0.25 mm thick are used to carry heating current from the current rings to the sample. The sample consists of alternating disks of graphite and nickel with graphite disks touching the molybdenum disks at each end. The nickel disks are 6.22 mm in diameter and 0.36 mm thick, and the graphite disks are 6.22 mm in diameter and 0.76 mm thick. This stack of alternating nickel and graphite disks will serve as its own heating element.

A paper tab about 5 mm wide and 10 cm long is glued to an edge of the assembled cell with a minimum amount of white glue. The external pyrophyllite portions of the cell are painted with a water suspension of red iron oxide (Fe_2O_3) powder. The cell is then dried at 120°C for $\frac{1}{2}$ hr. With the anvils in retracted position, the cell is roughly centered on one of the lower anvils and held in place by placing a piece of adhesive tape across the paper tab onto the face of an anvil binding ring. The current rings should be placed so that they touch (upon anvil closure) the opposed pair of anvils that carry the heating current. Oil pressure is applied to the hydraulic rams until the anvils impinge on the cell and the pressure (room temperature calibration) is built up to about 60,000 atm. This will occur at an oil pressure of about 476 atm.

After the cell is pressurized, heating current is applied. The characteristics of the electrical system will come into play at this point, and a little experimentation may be required in order to determine the correct setting of the electrical controls. Assume that variable autotransformers are used to control the voltage impressed across the sample. They should be set so that the initial current passing through the sample is about 800 A. If the setting is correct for making diamonds from the graphite, the current will start to decrease within 20 sec of its being applied, and as the current falls, the voltage rises. This is the signal that diamonds are forming! The contents of the cell are at a temperature near 1500°C, and at this temperature graphite dissolves in the molten nickel which acts as a solvent–catalyst for changing graphite to diamond. Diamond is precipitated from the liquid nickel–carbon system, and since diamond is an electrical insulator (it probably conducts some at 1500°C) it impedes the flow of electricity as it is formed. Thus the current falls and the voltage rises.

I noticed this phenomenon in the first synthesis of diamond; in fact, I have a patent on it (U.S. pat. 2,947,608 assigned to General Electric). One can easily find the heating power required to make diamond by this means without actually knowing the temperature inside the cell.

The current may fall to about 200 A as the diamond forms. After 60 sec of heating, the power is disconnected. The sample will cool to near room temperature in about 5 sec. About 20 sec after the heating power is disconnected, the pressure may be released and the anvils retracted. This brief delay is intended to allow adjustments to take place in the cell and anvils following the drastic change in temperature of the sample. This helps to preserve the life of the anvils.

The cell is removed from the press and split open with a mallet blow to a sharp knife set to bisect a square face of the pyrophyllite cube parallel to the sample axis. One or two blows will usually expose the core, which now consists of tiny diamond crystals intermingled with nickel and graphite. The entire mass tends to be a monolithic unit but can usually be cleaved with a knife and mallet along the original nickel–graphite boundaries. The diamond crystals at this point will be covered with a film of nickel metal.

The nickel and graphite may be separated from the diamond by first treating the sample very carefully with hot concentrated H_2SO_4 + $NaNO_3$. The sample is placed in a beaker to which is added the concentrated H_2SO_4 and then about 10% by weight $NaNO_3$. The mixture is heated strongly on a hot plate. Clouds of graphite particles will be seen to rise from the sample. Heating is continued for perhaps an hour until most of the graphite is oxidized and disappears. After the liquid is decanted, the residue is treated with concentrated HCl but is heated only lightly on the hot plate or let stand overnight at room temperature. The diamonds will be reasonably clean at this stage, and their morphological features can be readily observed at 10× magnification. For further cleaning, the H_2SO_4 + $NaNO_3$ and the HCl steps are repeated as many times as necessary to remove all the graphite and nickel. There will still be some residue from the pyrophyllite. This can only be removed by a long-term treatment with concentrated HF (perhaps a week) or by a fusion with NH_4F. The yield of diamond should be about 1 carat (0.20 g).

The diamond obtained is called diamond grit in the trade and could be used in some grades of resinoid-bonded diamond grinding wheels. The diamonds may be black. Transparent, slightly greenish or yellowish diamonds can be formed by lowering the temperature and pressure to a point where diamond is still the thermodynamically stable phase (with respect to graphite) but where the diamonds are grown more slowly. Under these conditions the yield will decrease.

For information concerning the thermodynamics and Kinetics of diamond synthesis, see the *Journal of Chemical Education* article of Ref. (29).

References

1. H. T. Hall, *Chemist,* **47,** 276 (1970).
2. H. T. Hall, "Chemistry at High Temperature and Pressure," in *High*

Temperature—A Tool for the Future, Stanford Research Institute, Menlo Park, California, 1956, pp. 161 and 214.

3. J. C. Jamieson and P. S. DeCarli, *Science*, **133**, 1821 (1961).

4. G. R. Cowan, B. W. Dunnington, A. H. Holtzman, U.S. Pat. 3,401,019 (1968); J. M. Kruse, U.S. Pat. 3,348,918 (1967).

5. H. T. Hall, *Rev. Sci. Instrum.*, **29**, 267 (1968). A good treatise on piston–cylinder use with fluid pressure media is S. E. Babb, Jr., "Techniques of High Pressure Experimentation" in *Technique of Inorganic Chemistry*, Vol. 6, A. Weissberger, Ed., Interscience, New York, 1966, p. 83.

6. H. T. Hall, *Science*, **169**, 868 (1970). Commercial materials are available from Megadiamond Industries, 589 Fifth Avenue, New York City, and from General Electric Specialty Materials Division, Worthington, Ohio.

7. P. W. Bridgman, *Proc. Am. Acad. Arts Sci.*, **74**, 425 (1942).

8. C. A. Parsons, *Proc. Roy. Soc.* (London), **44**, 320 (1880); *Trans. Roy. Soc.* (London), **A220**, 67 (1920). Also see anon. report on Richard Threlfall's discourse at the Royal Institution, *Engineering*, **87**, 425 (1909).

9. L. L. Coes, Jr., *Science*, **118**, 131 (1953).

10. S. M. Stishov and S. V. Popova, *Geokhimiya*, **10**, 837 (1961).

11. L. L. Coes, Jr., "Synthesis of Minerals at High Pressures," in *Modern Very High Pressure Techniques*, R. J. Wentorf, Jr., Ed., Butterworth, London, 1962, p. 137.

12. P. W. Bridgman, *The Physics of High Pressure*, G. Bell, London, 1958, p. 32.

13. H. R. Curtin, D. L. Decker, and H. B. Vanfleet, *Phys. Rev.*, **139**, A1552 (1965).

14. G. J. Piermarini, S. Block, and J. D. Barnett, *J. Appl. Phys.*, **44**, 5377 (1973).

15. P. W. Bridgman, *The Physics of High Pressure*, G. Bell, London, 1958, p. 408.

16. H. T. Hall in E. C. Lloyd, Eds., *Accurate Characterization of the High Pressure Environment*, National Bureau of Standards (U.S.) Special Publ. 326, 1971, pp. 303 and 313.

17. J. D. Barnett, R. B. Bennion, and H. T. Hall, *Science*, **141**, 534 (1963); J. D. Barnett, V. E. Bean, and H. T. Hall, *J. Appl. Phys.*, **37**, 875 (1966).

18. D. L. Decker, *J. Appl. Phys.*, **42**, 3239 (1971).

19. P. W. Bridgman, *Proc. Roy Soc.* (London), **A203**, p. 8 (1950).

20. E. W. Comings, *High Pressure Technology*, McGraw-Hill, New York, 1956, Chapt. 6.

21. P. W. Bridgman, *Proc. Am. Acad. Arts Sci.*, **81**, 167 (1952).

22. C. E. Weir, E. R. Lippincott, A. Van Valkenburg, and E. N. Bunting, *J. Res. Natl. Bur. Stand.* (U.S.), **A63**, 55 (1959); G. J. Piermarini and C. E. Weir, *J. Res. Natl. Bur. Stand.* (U.S.), **A66**, 325 (1962).

23. L. Ming and W. A. Bassett, *Rev. Sci. Instrum.*, **45**, 1115 (1974).

24. J. D. Barnett, S. Block, and G. J. Piermarini, *Rev. Sci. Instrum.*, **44**, 1 (1973).

25. W. A. Bassett, T. Takahashi, and P. W. Stook, *Rev. Sci. Instrum.*, **38**, 37 (1967).

26. L. Merrill and W. A. Bassett, *Rev. Sci. Instrum.*, **45**, 290 (1974).

27. A. Van Valkenburg, "Visual Observations of Single Crystal Transitions under

True Hydrostatic Pressures up to 40 Kilobar,'' in *Conference Internationale sur les Hautes Pressions (1e)*, Creusot, Saone-et-Loire, France, 1965.

28. H. T. Hall, *Rev. Sci. Instrum.*, **31**, 125 (1960).
29. H. T. Hall, "Diamonds," *Proc. 3rd Conf. Carbon, University of Buffalo, Buffalo, New York, June 1957*, Pergamon, London, 1957, pp. 75–84; *J. Chem. Educ.*, **38**, 484 (1961); *Chemist*, **47**, 276 (1970).
30. There is only one commercial source of supply in the United States: The Tennessee Lava Division of the 3M Co., Chattanooga, Tennessee.
31. H. T. Hall, *Rev. Sci. Instrum.*, **33**, 1278 (1962).
32. H. T. Hall, *Rev. Sci. Instrum.*, **46**, 436 (1975).
33. G. C. Kennedy and R. Newton, in *Solids Under Pressure*, D. Warschauer and W. Paul, Eds., McGraw-Hill, New York, 1963. p. 163.
34. W. H. Southwell, D. L. Decker, and H. B. Vanfleet, *Phys. Rev.*, **171**, 354 (1968).
35. R. J. Zeto and H. B. Vanfleet, *J. Appl. Phys.*, **40**, 22 (1969).
36. R. B. Bennion, H. G. Miller, W. R. Myers, and H. T. Hall, *Acta Cryst.*, **25A**, S71 (1969).
37. R. A. Fitch, T. E. Slykhouse, and H. G. Drickamer, *J. Opt. Soc. Am.*, **47**, 1015 (1957).
38. J. D. Barnett and H. T. Hall, *Rev. Sci. Instrum.*, **35**, 175 (1964).
39. P. F. Chester and G. O. Jones, *Phil. Mag.*, **44**, 1281 (1953).
40. G. Benedek, *Sci. Am.* **212**, 102 (1965).
41. J. H. Gardner, M. W. Johansen, C. Larsen, W. Murri, and M. Nelson, *Rev. Sci. Instrum.*, **34**, 1043 (1963).
42. H. H. Schloessin and V. Z. Dvorak, *Geophys. J. R. Astr. Soc.*, **27**, 499 (1972).
43. H. T. Hall, *J. Phys. Chem.*, **59**, 1114 (1955).
44. H. T. Hall, B. Brown, B. Nelson, and L. A. Compton, *J. Phys. Chem.*, **62**, 346 (1958).
45. H. T. Hall, *Science*, **128**, 445 (1958).
46. H. T. Hall, *Sci. Am.*, **201**, 61 (1959).
47. H. T. Hall, "High Pressure Methods," in *Proc. Int. Symposium on High Temperature Technology*, Stanford Research Institute, McGraw-Hill, New York, 1960, pp. 145, 335.
48. H. T. Hall, "High Pressure Apparatus," in *Progress in Very High Pressure Research*, Bundy, Hibbard, and Strong, Eds., Wiley, New York, 1961, p. 1.
49. H. T. Hall, "High Pressure/Temperature Apparatus," in *Metallurgy at High Pressures and Temperatures*, Gschneidner, Hepworth, and Parlee, Eds., Gordon and Breach, New York, 1964, p. 133.
50. J. D. Barnett and H. T. Hall, *Rev. Sci. Instrum.*, **35**, 175 (1964).
51. H. T. Hall, *Rev. Sci. Instrum.*, **37**, 568 (1966).
52. H. T. Hall, *Rev. Phys. Chem.* (Japan), **37**, 63 (1967).
53. H. T. Hall, *Rev. Sci. Instrum.*, **46**, 436 (1975).

Chapter **III**

DETERMINATION OF THE PROPERTIES OF SINGLE-ATOM AND MULTIPLE-ATOM CLUSTERS

J. F. Hamilton

1   INTRODUCTION

One of the most captivating of the emerging areas of science is that dealing with the properties of small metal clusters, intermediate between the isolated atoms on the one hand and bulk metals on the other. Much of the current interest in this subject stems from the common use of finely divided metals as industrial catalysts and from the budding optimism that modern scientific tools and techniques are finally capable of stripping away the veils of black art from the field. From catalytic studies has come the evidence that the activity of very small clusters may be markedly different from either bulk or atomic species and that the changes as a function of size present some interesting, if unpredictable, patterns.

A number of distinctions unique to small clusters are immediately apparent; crystallographically, thermodynamically, coordinatively, and electronically they obviously cannot be regarded simply as a small unit of the bulk. There is currently an intense effort to understand how these basic differences are interrelated and how each affects the catalytic activity. The understanding has not yet been attained, but the concentrated effort, apparent in a rapidly expanding literature on the subject, represents a strong beginning in this direction.

Solid particles or atom clusters in the size range of interest are certainly formed in almost any method of preparing the bulk solid material: solidification from the melt, vapor condensation, heterogeneous reactions, and so on. Hence the existence per se of such disperse solid particles cannot be considered an *extreme condition*. However, these very small clusters are frequently only unstable transient species, present in ill-defined concentrations for short times. Thus it is true that the production and study of well-defined, stable collections of small-sized atom clusters represent an extreme condition in the sense of the definition used in this book.

The "extreme" size range is an intermediate one and may be approached experimentally or theoretically by extension from either limit. Isolated atoms are for many purposes rather easily produced, and their properties can be rationalized, at least, from first principles. On the other hand, bulk solids are also comparatively well understood, and theoretical approaches toward some of the initial changes caused by continued subdivision can be made by suitable modifications of the continuum treatment.

Experimentally, once a large enough sample of fine particles or clusters has been formed, collected, stabilized, and characterized, many of the standard analytical techniques are directly applicable for determination of physical and chemical properties. Results of a number of studies are reviewed here, using techniques such as the normal crystallographic methods; calorimetry; infrared, visible, ultraviolet, and X-ray absorption spectroscopy; magnetic resonance; and the potent new tools of surface analysis. The instrumentation and experimental procedures are for the most part straightforward and thoroughly described in the published literature. These experimental methods are not detailed in this chapter. The specific problems in applying them to small metal particles have to do largely with specimen preparation and manipulation, and more attention is given to those experimental aspects. On the other hand, a review of the results that have been reported gives an index of the techniques that have been proved most suitable up to this time for studies of materials in ultrafinely divided forms.

The problems involve not only the stabilization of the highly dispersed state but also characterization of size or size distribution. Thus the small size limit of experimental studies has frequently been set by the resolution limit of the electron microscopy used in this characterization. The last few years have produced improvements in microscopes and microscopic techniques that have resulted even in a number of reported observations of single atoms of heavy metals. The shorter focal-length lenses now in common use in most commercial high-resolution instruments give them the inherent capability of resolution in this range, and the use of a pointed-filament source gives the required brightness while minimizing thermal drifts and charging problems. The problem of characterization of very small particle sizes is not so much one of resolution as of contrast against the inherent structure of the substrate and of interpretation of the image in terms of a physical size. To improve contrast, dark-field imaging was used by Henkelman and Ottensmeyer [1], Whiting, Ottensmeyer, and Nachod [2], and Hashimoto and his co-workers [3]. Kanaka and his associates [4] and a group working with Parsons [5] made use of phase contrast effects. The structure of the substrate, specifically of vacuum-deposited carbon films, is imaged in phase contrast and, depending upon precise focus conditions, can completely obscure the images of atoms or small clusters. To avoid this problem, a number of single-crystal support films have been recommended [3, 6]. As another approach, Nagata, Matsuda, Komoda, and Hama [7] advocated using a less coherent incident beam to minimize phase-contrast imaging of substrate structure. Prestridge and Yates [8] optimized the objective aperture of their instrument for selectively removing electrons elastically scattered by heavy rhodium clusters while passing those inelastically scattered by the low atomic-numbered elements of their silica support. From a supported rhodium catalyst they obtained the micrograph of Fig. 3.1, presumably showing images of single atoms.

The scanning transmission instrument constructed by Crew and his co-workers at the University of Chicago [9] is capable of atomic resolution and, because of the selective detection of elastic and inelastic scattering, has high discrimination for heavier metal atoms. Striking motion pictures directly revealing Brownian motion and diffusion of single atoms have been made. A similar instrument has been built at Johns Hopkins University [10], and commercial versions are reportedly being developed [11].

These developments in electron microscopy, together with other advances in experimental and theoretical approaches, make small clusters of metal atoms far more accessible to study in the immediate future.

2   PREPARATION METHODS AND EXPERIMENTAL TECHNIQUES

Vapor-Phase Species

The free-atom species of metals exist in the vapor over the molten material under vacuum and in their various ionized forms in flames or arcs. Data on electronic transitions, ionization potentials, and electron affinities of single atoms are derived principally from these sources. Minor fractions of diatomic species are also detectable in metal vapors, and a large number of studies of their properties have been made in this way. The published literature is very extensive, and review articles such as those by Drowart [12], Siegel [13], and Gingerich [14] may be recommended for inclusive references. The source of the vapor is either the Langmuir type, consisting of the free surface of the solid or molten metal, or a Knudsen cell, in which the source metal is enclosed in a furnace and the equilibrated vapor emanates through an orifice. Visible and ultraviolet absorption spectra of the vapor reveal the vibrational and rotational bands of the diatomic molecular species in addition to the line spectra of the monomers. Complete analyses of these spectra give the electronic structures and the equilibrium internuclear distances of the dimeric species. Mass-spectroscopic analyses of the vapor composition are also useful. Except for residual uncertainties about relative ionization cross sections, the data from such a determination give the dimer/monomer ratio as a function of temperature. Data are then analyzed in one of two ways to give a value for the dissociation energy $D_0(x_2)$. In the first method, based

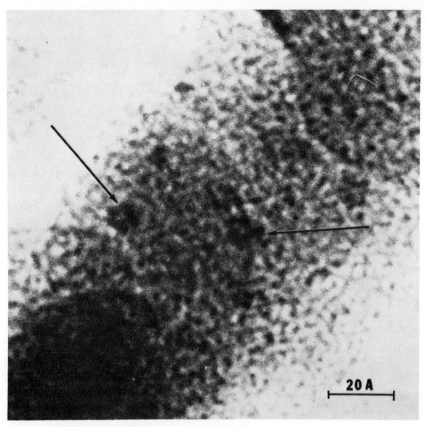

Fig. 3.1. Electron micrograph of clusters of rhodium atoms (arrows) on silica support. From Prestridge and Yates [8].

on the second law, the heat of vaporization $L_T(x_n)$ of a species x_n at temperature T is taken directly as the slope of the plot of ionized current from that species versus $1/T$. Then, to an acceptable approximation

$$D_0(x_2) \sim 2L_T(x_1) - L_T(x_2) \tag{3.1}$$

This method has only limited accuracy and requires that data be collected over a considerable temperature range.

The second method relies on the third law for determination of the absolute entropy and requires an independent knowledge of certain thermodynamic values for the dimer and monomer species. The expression employed is

$$\frac{D_0(x_2)}{2.303RT} = -\log P(x_1) + \log \frac{P(x_2)}{P(x_1)}$$

$$+ \log \frac{T^{3/2}M^{1/2}}{\gamma^2} + \log \frac{Q^2(x_1)}{Q(x_2)} \quad (3.2)$$

where $P(x_n)$ is the equilibrium pressure of (x_n) in the source cell, M is the atomic mass of the monomer, γ is the interatomic distance of the dimer, and $Q(x_n)$ is the partition function of (x_n). Agreement between the two methods has generally been reasonable when both have been used.

Experimental determinations of dissociation enthalpy based on mass-spectrometric or spectroscopic data have been collected for homonuclear dimers of many of the elements, and extensive references may be found in summary articles on the subject. Values taken from the reviews by Siegel [13] and Gingerich [14] are reproduced in Table 3.1. Group IA elements, the alkali metals, form simple but rather weak sigma bonds from the overlapping s orbitals of the atoms. The small binding energies evidently involve repulsive interactions with the valence electrons, for the molecular ion species M_2^+ are found to be more stable than the corresponding neutral molecules within this group.

Elements of groups IIA and IIB have filled s^2 outer shells, and in the dimeric species bonding and antibonding orbitals are equally occupied. The dissociation energies are very weak, arising essentially from van der Waals interactions.

Dimers of the IIIA elements are bonded partly by overlap of the valence p orbitals but have nevertheless comparatively low dissociation energies. The bond energy decreases with increasing molecular weight through the series B_2, Al_2, Ga_2, In_2, Tl_2 of this group, as in fact it does in most other groups as well. The IVA elements, C, Si, Ge, Sn, and Pb have p^2 valence shell configurations and are considerably more strongly bonded than are the IIIA elements. The diatomic molecules N_2, P_2, As_2, Sb_2, and Bi_2 of group VA have singlet sigma ground states, indicating a triple bond. Dissociation energies are highest for this group. The group VI chalcogens form dimers with moderately strong bonds, but the vapor phase is characterized by a strong tendency to form larger aggregates.

The bond energies of dimers of the group IB noble metals Cu, Ag, and Au have been measured by several authors, and the results are in reasonable agreement. This group provides an exception to the general trend of decreasing bond energy with increasing molecular weight, for Au_2 is more strongly bonded than either Cu_2 or Ag_2. The group III transition metals Sc_2, Y_2, and La_2 are also contrary to this trend, with dissociation energy increasing in that order. Dimers of the first-row transition metals have

Table 3.1 Dissociation Energies (kcal/mole) of Symmetrical Diatomic Molecules[a]

Li 26	Be 16											B 66 (66)	C 144	N 225	O 118
Na 18	Mg 7.2											Al 39 (40)	Si 75 (74)	P 116	S 101
K	Ca	Sc (38)	Ti (32)	V (57)	Cr (36)	Mn 3.3 (10)	Fe 37 (29)	Co (39)	Ni 55 (53)	Cu 46 (45)	Zn 6	Ga 33 (33)	Ge 64 (65)	As 91	Se 65
Rb 11	Sr	Y (37)	Zr	Nb	Mo	Tc	Ru	Rh	Pd (25)	Ag 38 (38)	Cd 2	In 22 (24)	Sn 46 (46)	Sb 72	Te 53
Cs 10	Ba	La (58)	Hf	Ta	W	Re	Os	Ir	Pt	Au 52 (53)	Hg 1.4	Te 14	Pb 23 (20)	Bi 47 (47)	Po

[a]Data extracted from the critical summaries of Siegel [13] and Gingerich [14] (in parentheses). Original references may be found in these articles.

been successfully studied, notably by Kant, Strauss, and Lin [15], but except for Pd, the vapor pressures of other transition metals are too low to permit measurements of this type at accessible temperatures.

Larger polymeric aggregates exist in the equilibrium vapor of the group IVA, VA, and VIA elements. Dissociation energies have been determined for a number of these, including Sb_3 and Sb_4 [16], Bi_3 and Bi_4 [17, 18], and Pb_3 and Pb_4 [19]. In Sn [20], C [21], Si [22], and Ge [22, 23], species containing as many as seven atoms have been found at equilibrium. For each of these elements, the energy for the dissociation $M_3 \rightarrow M_2 + M$ is greater than for $M_2 \rightarrow 2M$ or $M_4 \rightarrow M_3 + M$, making the triatomic species a particularly abundant one [13, 24]. For carbon, at least, this trend for odd-sized aggregates to be more stable than even-sized ones has been verified experimentally for larger sizes (through C_9) as well [25]. The tetramers P_4, As_4, and Sb_4 of the group VA elements are the most abundant species in the vapors of these elements [13]. These clusters presumably have tetrahedral geometry. Both S [26] and Se [27, 28] of group VIA vaporize with high concentrations of aggregates with ring structures. S_8 has been found to be the dominant species.

Alkali metal polyatomic clusters are also detectable in the vapor [29]. A report of a high concentration of Ag_3 and Ag_4 species in equilibrium silver vapor [30] has been questioned [13].

Larger clusters form much more readily when the metal vapor expands greatly as it passes through a very fine orifice between a high-pressure vapor source, within which the metal vapor pressure may be as high as 1 atm, and the high-vacuum chamber. The requirement for these conditions is that the mean free path in the vapor be much less than the orifice diameter. Then the gas expands nearly adiabatically, and there is rapid cooling to a state of supersaturation, causing the formation of larger clusters. An orifice of perhaps a few μm is required, and the resulting beam is sometimes called a supersonic or nozzle beam. The process has been studied with the rare gases and simple gas molecules [31–35] and has been employed to produce for mass-spectrometric studies sodium clusters up to eight atoms in size [36, 37]. Many of the available data on polyatomic species were obtained in this way.

Diatomic and larger polyatomic gas-phase species are readily subjected to determinations of ionization potential. The photoionization threshold or the appearance potential for ionization by an electron beam can be determined mass spectrometrically. Although sharp thresholds are not always apparent, the data can be treated to obtain ionization potentials of the various species in the vapor. Results of such studies will be mentioned subsequently.

In some scattered reports of mass-spectrometric studies connected with sputtering processes, the presence of dimeric and even trimeric positive secondary ions was indicated in the sputtered vapor. Subsequently, Krohn [38] reported comparatively high concentrations of polynuclear negative ions emitted from metal targets bombarded by positive cesium ions. The technique was further exploited by Hortig and Müller [39], who used a cesium-coated silver target bombarded with krypton ions of 11 and 15 keV energy. They reported concentrations of negatively charged ionic species containing as many as 40 silver atoms. Throughout the range, the odd-sized clusters were more concentrated than the even-sized ones, indicating to them that odd sizes were more stable. Their results are shown in Fig. 3.2, as replotted by Leleyter [40].

Matrix Isolation

Matrix isolation techniques have been exploited widely in recent years principally for spectroscopic studies on isolated atomic or molecular species [41–45]. The atom in question is evaporated usually from a Knudsen cell and cocondensed from the vapor with a large excess of a matrix constituent. Normally it is desirable that the matrix be totally inert, and frequently this is a rare gas, bled into the chamber through a controlled leak valve and condensed on a substrate held at a temperature of only a few degrees Kelvin. In studying metal vapors, it has been possible by careful control of the metal/matrix gas ratio and the temperature during and after deposition to form dimeric and larger clusters for spectroscopic investigation. The metal atoms diffuse on the transient interfaces that exist during condensation or through the solid phase in the final deposit to form small clusters. Identification of absorptions due to dimeric and trimeric species is made by the concentration dependence of the signal; signals from M_2 increase with the second power of [M], M_3 with their third

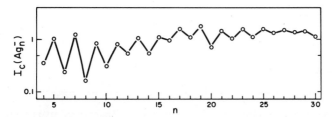

Fig. 3.2. Ionic current I_c of species Ag_n^- excited by 8 keV Cs^+ ions. Data of Hortig and Müller [39] as shown by Leleyter [40].

power, and so on. Ozin and Moskovits and their collaborators [44] have studied dimers of most of the first-row transition metals plus copper and silver. Their absorption measurements for Cr and Cr_2 in an argon matrix are shown in Fig. 3.3. Schulze and Becker [45] reported a study of silver which included Ag_2, Ag_3, and some transitions they attributed probably to Ag_4 and Ag_5.

Hulse and Moskovits [46] have reported studies including also a low concentration of an active gas in the inert matrix along with the metal atoms. They observed spectra corresponding to the adsorption of CO on diatomics and larger clusters of nickel and copper.

Aggregated metallic clusters can be formed from other initially homogeneously disperse mixed systems. Clusters of three and four silver atoms in a frozen sulfuric acid matrix have been studied with ESR [47], and noble metal particles formed by irradiation of solid [48, 49] or frozen [50]

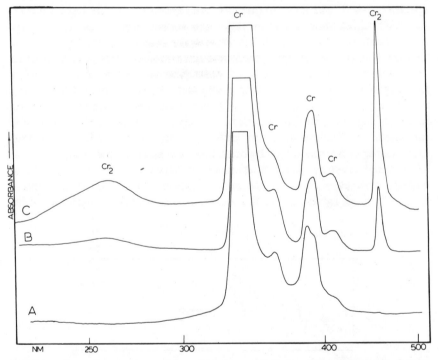

Fig. 3.3. Absorption spectrum of chromium atoms and clusters in an argon matrix. Curves A, B, and C are from samples with increasing chromium concentration. From Kündig, Moskovits, and Ozin [44b].

ionic compounds or doped alkali halide crystals [51–53] or of photosensitive glasses [52–58] have been used for some fine-particle studies. Finely dispersed alkali metal particles can be formed by high-energy irradiation of the corresponding alkali halide [58–62] or alkali azide [63–67], and these have been used in some studies. The problems associated with size characterization in such samples are formidable.

Sols and Other Solution-Derived Forms

Metallic sols with a highly uniform particle size can be made by reduction of the metal ions from solution [68], normally in the presence of some peptizing agent. These can normally only be stabilized in the upper part of the size range of principal interest for this discussion, having mean linear dimensions of 50 to 100 Å and larger. However, they have been used to study some of the size-dependent physical properties expected to be changing in that size range. A novel preparation of palladium nuclei reduced in solution on the surface of suspended microscopic alumina rods was used for catalytic studies [69].

This particular preparation is closely related to the commercially important supported metal catalyst [70, 71]. Supported catalysts are prepared on porous ceramic particles that are infused with one or more metal ions from solution. The metal is reduced to the zero-valence state by high-temperature firing in a hydrogen atmosphere. The metal particle size of such a preparation is commonly evaluated by measuring the ratio of the number of adsorbed gas atoms, presumed equal to the number of surface metal atoms, to the total number of metal atoms. This ratio has been found to be dependent on many factors, particularly the pore size of the support, and in many cases can be made so small that essentially all metal atoms appear to be accessible to adsorbing gases. Dalla Betta and Boudart [72] prepared platinum catalysts on Y zeolite support and concluded, based on the finite cage size of the support, that their particles had sizes of less than six atoms, and Kim and Seff [73] have apparently identified an Ag_6 octahedral cluster in a zeolite cage by single-crystal X-ray diffraction analysis. More definitive characterization of particle size or size distribution of supported catalysts is experimentally difficult. Some of the limiting factors in applying electron microscopy to supported catalysts are discussed by Flynn, Wanke, and Turner [74]. Most notable are the variations in focus owing to the appreciable thickness of such samples and the interference of structural features in the supports. Nevertheless, several size–frequency determinations have been reported [75–78]. Granqvist and Buhrman [79] have pointed out that most of these data fit the log-normal form, a feature they associate with growth by coalescence of existing particles rather than Ostwald ripening by atomic migration. Wanke [80], however, has argued that the evidence is inconclusive.

Prestridge, Via, and Sinfelt [81] published an impressive electron microscopy study of supported catalysts of Ru and Os and mixed clusters of each of these metals with Cu. They gave size–frequency data extending to diameters of only a few angstroms, and presented evidence that many of the particles of Ru and Ru–Cu are raftlike in form, with thickness of a few planes, perhaps only one.

A novel preparation technique was revealed in experiments reported by Smith et al. [82]. They indicated that it was possible to adsorb a $Rh_6(CO)_{16}$ cluster species from chloroform solution to an alumina surface, decarbonylate the cluster, and then reform the original species by CO treatment. These authors speculated that the Rh_6 either retains its integrity or splits into fragments which remain in near proximity and are able to recluster upon carbonylation.

Inert Gas Evaporation

Fine particles of metals can be formed by evaporation of the molten metal in an inert gas atmosphere [83, 84]. An evaporation chamber is first evacuated and then backfilled with, for example, helium, argon, or xenon at a pressure in the range of about one to a few hundred torr. When the evaporation source is heated, a "smoke" forms in the gas over it, and the particles deposit on any unheated surface in the vicinity. Particle sizes have been found to depend upon such variables as the nature and pressure of the gas atmosphere, the metal temperature, and the evaporation rate. In most of the studies reported, the lower limits of the sizes are barely in the range over which any size dependence of physical or chemical properties is expected. However, more recent studies by Granqvist and Buhrman [84] have been successful in producing smaller particles, at least of some metals. Their principal studies were on aluminum, and the methods involved improved, careful temperature control of an oven-type source and controlled surface oxidation of the particles by partial oxygen pressure in the atmosphere, in order to limit coalescence. Mean particle diameters of 30 to 60 Å could be achieved. They also determined a log-normal size distribution for inert gas preparation and suggested that coalescence effects set the lower limit to the size that can be stabilized by this method.

Vacuum Deposition

Vacuum deposition on solid substrates produces a distribution of clusters of various small sizes and in a convenient form for many uses where the small total amount of material is not a serious disadvantage [85–99]. Coagulation effects have been noted [100, 101] over time periods of the order of hours, so that distributions of clusters in the very small size range cannot be taken to be stable except for very limited times. Stability is reported to be extended by preventing contact with air, and many kinds of experiments have been made on specimens of this type.

Such deposits can be made on thin-film substrates, or removed by pseudoreplication from other substrates, in a form ideal for electron microscopy. Thus size distributions are directly determinable in the range greater than a few angstroms. Various proposed mechanisms for the earliest stages of nucleation and growth of such deposits can be analyzed to give predicted size distributions in the range below the direct resolution limit. Variations of a model involving two-dimensional nucleation by mobile adatoms have been treated mathematically by a number of authors [102–112], and the calculated size distributions depend upon the features of the particular model used. For example, Zinsmeister [102] computed a very unsymmetrical size distribution with a sharp large-size cutoff, whereas a treatment by Chakraverty [113], also modified by Robertson [114], predicts a sharp, symmetrical distribution which broadens with increasing deposit thickness. Sacedon and Martin [115] used a method based on the Chakraverty treatment and were able to predict a bimodal distribution at larger mean size, in general agreement with their electron-microscopic observations.

Although the two-dimensional homogeneous nucleation mechanism appears to be operative on some substrates and at elevated temperature on others, Hamilton and Logel [116] found that it was not consistent with their observations of silver, gold, copper, and palladium metal clusters vapor deposited at room temperature on the amorphous substrates carbon and silicon dioxide. They determined that surface diffusion was relatively unimportant and that an incident metal atom was retained on the substrate surface only if it struck within a distance corresponding to at most a few atomic radii of some undetermined type of particularly active site on the surface. Failing to strike within the minimum capture area of an active site, an incident atom suffers some alternate fate, and evidence favors the possibility that it diffuses into the rather porous substrate. A cluster forms and grows by successively capturing other atoms directly from the incident beam. Again, surface diffusion is unimportant.

A confirmation of the general observations leading to this model was reported by Robrieux, Faure, and Desrousseaux [117]. They employed a Kelvin probe set up in the same vacuum chamber used for evaporation to measure the work function change as the thickness of a deposit of silver on carbon increased. They preferred to interpret the possibility that the atoms not included in the visible particles were retained as an adsorbed monolayer on the substrate. Hamilton and Logel, however, maintained that the missing metal constituted considerably more than a monolayer and that this explanation was unlikely. Venables [118], Lewis and Anderson [112], and Anderson and Granqvist [101] also discussed the fate of the missing metal atoms in these experiments.

Under these conditions the growth of the clusters is random, and the particle size distribution would obey Poisson statistics, except for the increase in capture area as the particle grows. With this feature of increasing capture area, exact solutions for the size–frequency distributions are best obtained numerically; but in the approximation of low incident metal coverage, they are given by

$$N_i \cong n_i \cong \frac{n_s^i \sigma_s \sigma_1 \sigma_2 \ldots \sigma_{i-1} m^i}{i!} \tag{3.3}$$

in which n_i is the surface density of clusters of size i, N_i is the surface density of clusters of size i or larger, n_s is the initial surface density of active sites, σ_s is the active-site capture area, σ_j is the capture area of a cluster of j atoms, and m is the total flux of incident metal atoms. A significant feature of the process is that the concentration of clusters of size i or larger increases with the ith power of the incident metal coverage. The parameters of the relationship between N_i and m are n_s and σ_s, the active site density and capture area, respectively, and $\sigma_j(j)$. Experimental data from the range of resolvable particles indicate that $\sigma_j(j)$ is satisfactorily described by the simple geometric relationship $\sigma_j(j) = \pi\epsilon^2(3jv^*/2\pi)^{2/3}$, in which v^* is the volume per atom in the solid metal and ϵ is a flattening factor not greatly different from unity, to account for nonhemispherical shape.

Of the two remaining parameters, n_s can be determined from the saturation nucleus density, and σ_s, by extrapolation of size data to zero coverage.

In Fig. 3.4 are shown a family of curves of N_i versus m for several low values of i, using parameters determined for the experimental case of palladium deposited on amorphous carbon. The low-coverage linear portions of these curves with slope i are apparent.

Partial confirmation of these predicted size distribution curves was accomplished by further vacuum deposition, under conditions favoring continued growth of existing nuclei to easily resolvable sizes but unfavorable to the creation of new nuclei. If homogeneous two-dimensional nucleation were the dominant process, this situation could be achieved by lowering the deposition rate. With the strong influence of active conden-

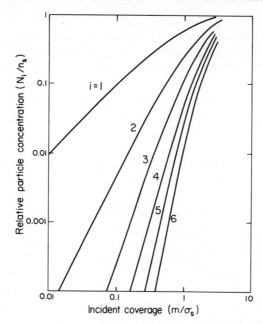

Fig. 3.4. Calculated concentrations N_i of clusters of i or more atoms in vacuum deposits of m incident atoms on amorphous substrates. The substrates contain n_s active sites per unit area, with capture areas of σ_s. According to the model of Hamilton and Logel [116].

sation sites, reduction of the nucleation probability could not be accomplished by lowering the rate. However, the same effect could be produced by changing after the nucleation stage to a second metal whose nucleation probability was lower. The metal used was zinc, whose condensation probability on a clean amorphous carbon substrate is negligible [119]. Chapon, Henry, and Mutaftschiev [120] subsequently reported using cadmium in the same way to reveal subvisible clusters of gold. Hamilton and Logel found that predeposited nuclei of silver, gold, copper, palladium, and most other metals were efficiently grown by the zinc and that the number of visible zinc particles depended on the first power of the coverage of the nucleating metal [89, 116]. This observation was interpreted as showing that a single atom of the nucleating metal, localized at one of the preferential condensation sites, is sufficient to cause the zinc to condense efficiently. The absolute value of the number of nucleating sites compared with the incident metal flux gave an independent measure of the active-site capture area σ_s, which was in reasonable agreement with that determined by extrapolation from data on visible particles.

Deposits such as these provide convenient collections of metal clusters covering the full size range of interest—from single atoms to bulk metal—well characterized with respect to the peak position and the breadth of the size distribution, and measurements of a number of types have been initiated in order to capitalize on these characteristics [95]. Migration and coagulation effects are one of the most serious problems, but these are minimized if the deposits are maintained under vacuum and can be virtually eliminated if the metal deposit can be overcoated with a second vacuum-deposited layer of carbon or silicon dioxide. Obviously some studies require that the bare metal surface be exposed, and in such cases overcoating is not possible. A second limitation is the small total amount of material present. For some studies not involving a surface measurement, this condition may be improved by forming multiple alternate layers of the substrate material and the metal, following the methods used by Zeller and Kuse [90], Ido et al. [91], and Yee and Knight [93].

A technique consisting of vacuum deposition of metal onto the free surface of a low vapor-pressure oil has been reported by Yatsuya, Mihama, and Uyeda [121]. Particles in the 30 to 100 Å size range have been produced. The authors have devised an apparatus employing continuous flow of oil over a rotating disk to collect larger samples.

A related method reported by Wada and Ichikawa [122] uses a frozen solvent as a substrate. The solvent is subsequently melted to obtain a suspension of the particles. Klabunde and collaborators [123] have prepared Cd, Zn, Mg, and Ni clusters and larger particles in this way, which they report to have unusual chemical activity.

It is possible to collect the clusters formed in nozzle beams on solid substrates, if this is desirable. Takagi and collaborators [124] have reported a related method for forming films with unique properties, which they refer to as ionized-cluster beam deposition. The clusters produced in a supersonic nozzle beam are ionized by a crossed electron beam and accelerated toward the substrate by an electrode held at a potential of several kilovolts. Strangely, they find a much higher tendency for surface migration by this technique than by simple deposition with an atomic beam.

3 PROPERTIES OF SMALL METAL CLUSTERS

Because there are complex cause-and-effect interactions between the various aspects of the differences between small clusters and bulk materials, it is not possible to arrive at a completely satisfactory system of classification. Consequently a rather large measure of arbitrariness will be recognized in the organization of the subsequent sections, which review some of the theoretical and experimental efforts to understand the origins of the unique cluster properties.

Geometric and Structural Factors

General

Experimental evidence confirms the theoretical indication that the equilibrium structural form of small clusters of metal atoms may differ markedly from the elements of the crystal lattice of the bulk substance. Ino [125], in 1966, showed electron micrographs of epitaxial gold nuclei in the 100 Å size range, vapor deposited on sodium chloride substrates, having both five- and sixfold symmetry. Contrast effects and anomalous electron diffraction patterns led him to propose that the crystallites were made up of tetrahedron-shaped units of the normal *fcc* lattice, bounded by (111) planes and twin-related to each other. He proposed three specific models. The simplest consists of a (111)-bounded tetrahedral nucleus one face of which rests on the substrate, and the other three bear twin-oriented tetrahedra. Secondary twinned tetrahedra then form on the exposed faces of the primary twins, producing a particle with nearly hexagonal projection. The secondary twins are misaligned from the original nucleus by about 2°, and this change in orientation accounts for both the contrast effects and the diffraction patterns observed.

A second model, illustrated in Fig. 3.5a, consists of five tetrahedral units, each twin-related to its neighbors and all having a single common

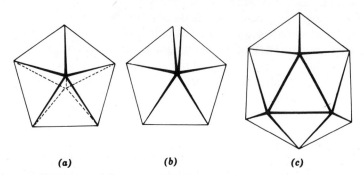

(a)   (b)   (c)

Fig. 3.5. Diagrams of particles composed of tetrahedral subunits. Five tetrahedra (a) form a particle with pentagonal outline, but with perfect tetrahedra, a 7.5° closure gap results (b). An icosahedral particle (c) consists of 20 tetrahedral units and has a hexagonal outline. From Allpress and Sanders [133].

edge. Together these form a pentagonal bipyramid. The geometry of perfect tetrahedra is such that the set of five does not result in complete closure, and a gap of 7.5° results, as shown in Fig. 3.5b. Ino speculated that this would be distributed as a 1.5° misalignment at each of the five boundaries.

A third model required to explain the observations consists of 20 twin-related tetrahedra, all with a common apex, forming together (with similar closure gaps) an icosahedral particle. Such a structure is illustrated in Fig. 3.5c. Ino noted that particles of these forms were common in the early stages of the deposition and concluded that the forms developed as a consequence of the growth process and not by virtue of coalescence effects.

Kimoto and Nishida [126] noted particles with pentagonal symmetry in particles of *fcc* metals formed by vaporization in argon and therefore ruled out strong substrate interactions in the effect. Allpress and Sanders [127] found the multiply twinned particles in vapor deposits of gold, palladium, and nickel on mica and explained their observations in terms of the same models used by Ino. Similar structures have been observed and studied in a variety of particulate metal samples [128].

Komoda [129] used high resolution electron microscopy to resolve the {111} lattice planes in particles of gold with both pentagonal bipyramids and icosahedra. He pointed out that in the former the closure faults, if equally spread among the five interfaces, could be regarded as low-angle boundaries consisting of dislocations with 90 Å spacing. Thus no dislocations would exist in particles with less than 90 Å radius. A micrograph of a pentagonal particle is shown in Fig. 3.6. He observed no dislocations, but the lattice planes were nonparallel, diverging in a slight fanlike manner toward the outer perimeter. Komoda also proposed the first atomistic model for the growth sequence by which such particles form. He suggested that the nucleus of the pentagonal bipyramid is a seven-atom cluster with the same geometry, which can logically be expected to have grown from a tetrahedral grouping of four atoms and which can itself grow by adding successive {111} shells on all faces. In a like manner, a 13-atom cluster, also derived by selected growth from a tetrahedral four-atom precursor, will grow by successive shell addition to an icosahedral form. After addition of the third shell the particle will have reached a size of about 20 Å. Similar growth models were described by Gillet et al. [130].

Sollard, Buffat, and Faes [131] studied gold nuclei formed by vapor deposition on amorphous carbon substrates and then subjected to heat treatment for various times at temperatures over 600°C. This treatment, they expected, would produce stable geometries. They observed proportions of multiply twinned particles approaching 100% for extended ther-

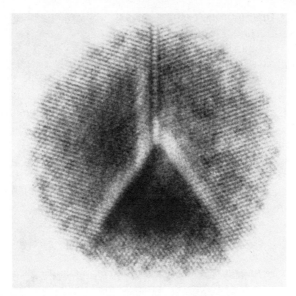

Fig. 3.6.   High-resolution electron micrograph of pentagonal gold particle showing {111} lattice planes. Defects are not detected at the interfaces between tetrahedral subunits, but a dilation of the lattice toward the outer edge can be detected. From Komoda [129].

mal treatment and concluded that these were the most stable forms, with no upper size limit such as had been proposed by Ino [132].

Allpress and Sanders [133] commented on some observations on hydrogen-reduced supported catalysts of nickel on Al_2O_3 by van Harde-veld and van Montfoort [134] and Shephard [135]. They deduced that the metal particles were faulted but not multiply twinned, most particles having the form of normal octahedra. They puzzled over the reason for the structure difference for the two methods of preparation. Avery and Sanders [136] studied nickel, gold, platinum, and palladium on SiO_2 using dark-field electron microscopy and concluded that less than 2% of the metal particles were multiply twinned. Burton [137] maintained that in the micrograph of Prestridge and Yates [8], reproduced as Fig. 3.1., the individual Rh atoms have pentagonal symmetry.

Structural Calculations

Theoretical approaches with ever-increasing degrees of sophistication have been applied in attempting to predict the most stable geometry of small atom clusters. In the most elementary methods, intuitive stability criteria such as the most dense packing or the smallest number of unfilled bonds are used, and various obvious geometric arrangements are compared. Using this approach Romanowski [138] treated lattice elements bounded by different planes from the *fcc, bcc,* and *hcp* crystal structures. Fukano and Wayman [139] made an attempt to account for bonding to the substrate and specifically included the tetrahedrally based forms which had been proposed previously also with a correction for the imperfect closure. They pointed out that the seven-atom pentagonal bipyramid has 32 half-bonds, compared with 30 for the seven-atom cluster formed by triple twinning on a four-atom tetrahedron, and occupies approximately 30% less volume.

A somewhat more realistic approach is to compare the static potential energy of the cluster, as determined by the summation of all pairwise interactions, assuming some specific form for the two-atom potential function. Potential functions normally used are the Mie potential

$$\frac{V}{V_0} = \frac{1}{(n - m)} [nr^{-m} - mr^{-n}] \tag{3.4}$$

in which r is the interatomic separation and m and n are parameters of the model (when $m = 6$ and $n = 12$, this becomes the Lennard–Jones potential), and the Morse potential function

$$\frac{V}{V_0} = \{1 - \exp [a(1 - r)]\}^2 - 1 \tag{3.5}$$

which involves **a** as an adjustable parameter and accounts for somewhat stronger long-range interactions than does the Mie form. The functional dependence of the potentials on the separation distance allows for straightforward investigation of relaxation effects. Allpress and Sanders [133] considered uniform relaxation in closed-shell forms. They found that the reduction in next-nearest neighbor interactions at small cluster size resulted in a dilation of the clusters in this range of as much as 25% for a Morse potential and more than 10% for a Mie potential. In an analysis by Burton [140], each atomic shell was allowed to relax independently, resulting in a dilation of the shells near the surface.

Computerized energy-minimization procedures are feasible even without the geometric constraints, and several strategies have been employed to reach minima expediently and to attempt to distinguish local minima (i.e., metastable isomeric forms) from absolute minima. In general, a starting geometry is chosen, and after calculation of potential gradients at each site the atoms are moved individually in the direction of the force acting upon them. McGinty [141] adopted the procedure of making the displacement of each atom proportional to the magnitude of the force. Successive adjustments are made iteratively until a configuration presenting a stable minimum is reached. Instances of rather severe rear-

rangement are found by this procedure, and Hoare and Pal [142] point out
the importance of an inspired starting configuration in arriving at a stable
minimum within reasonable computation time. Even then, for larger
clusters, the identification of a minimum as the absolute minimum is virtually impossible, and the final configuration may depend upon the starting
conditions.

Hoare and Pal [142–144] and Hoare and McInnes [145] investigated all
metastable isomers of clusters up to nine atoms, pointing out that there
are three polytetrahedral isomers for clusters of seven atoms, four for
eight atoms, nine for nine atoms, and such rapidly increasing numbers of
isomers for larger sizes that even complete listings have not been at-
tempted. In general, in the size range below 50 atoms, the most stable
clusters are never lattice crystallites of *fcc* or *hcp* type but are polytetrahedral structures consisting of face-to-face combinations of four-atom
tetrahedra, distorted as necessary for closure. Examples of these are
shown in Fig. 3.7.

Several authors have considered also the contribution of entropy in
determining the most stable cluster configuration at finite temperature.
The earliest treatments [141, 144, 146–149] employed a harmonic approximation in determining the normal modes from the force-constant

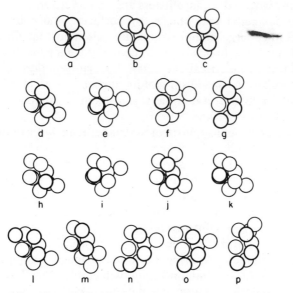

Fig. 3.7. Some of the polytetrahedral isomers of seven-atom (*a* to *c*), eight-atom (*d* to *g*),
and nine-atom (*h* to *p*) clusters. From Hoare and Pal [142]. (An additional eight-atom
isomer and two additional nine-atom ones were subsequently discussed.)

matrix (see Section 3.2). Kristensen, Jensen, and Colterill [150] used the Monte Carlo method originated by Metropolis et al. [151], and this technique has also been employed by Etters and Kaelberer [152]. McGinty [153] and Briant and Burton [154] made calculations using a molecular dynamics approach. Most recently, Etters, Kanney, Gillis, and Kaelberer [155] have made use of a self-consistent phonon calculational procedure, a quantum-mechanical approach which does therefore account for the zero-point energy. All of these treatments deal with pairwise interactions, using classical Lennard–Jones or Morse potential functions. They are in essential agreement in the finding that for such potentials the entropy contributions have minor effects on the most stable configurations, which are determined essentially by energy minimization and are almost always polytetrahedral forms.

Hoare and Pal point out that in the size range of a few atoms, some qualitatively new geometric feature emerges with the addition of each successive atom. Clusters of two, three, and four atoms represent the transition from one- to three-dimensional geometry. The energetic minimum for the five-atom cluster is the first with nonnearest neighbors. Six atoms present the first case with a loss of axial symmetry. Seven atoms result in the pentagonal symmetry. And the 12-atom cluster is the first one with a totally enclosed, nonsurface atom. For the potential functions investigated, some clusters are characterized by surfaces whose bonds are elongated and others, by surfaces whose bonds are compressed. Rather large compressions are tolerated, but only very slight stretching of surface bonds beyond the pair distance results in rearrangement to a less symmetrical geometry. Special attention is drawn to the so-called Mackay icosahedron [156] composed of 55 atoms, which is a particularly stable cluster and the smallest stable one having vestigial *fcc* {111} planes at its surface.

The very elegant and sophisticated structural analyses of these classical models notwithstanding, their applicability remains uncertain. The authors reporting the studies normally begin with the disclaimer that the classical Lennard–Jones or Morse potentials apply only to rare gas aggregates and are not very good approximations even there. Then, understandably impressed by the results that can be obtained, they almost invariably reach the point of comparing their conclusions with experimental results on metal nuclei. Semiempirical quantum-mechanical treatments, although requiring far more extensive computation for a similar degree of energy minimization, will certainly lead to more reliable conclusions. Although these calculations (treated in Section 4 of this chapter) have not yet been extensive enough for a detailed comparison with the predictions from the classical treatments, it does seem possible that some quantum-mechanical differences in stable symmetry might be expected, purely on the basis of orbital parity considerations.

However, it is clear even at this point that structures other than elements of the bulk lattice may exist in small clusters and that these may be the more stable forms. The experimental evidence from metals clearly indicates the importance of the pentagonal geometry and of other forms made up of tetrahedral units, as the classical analyses predict. It is also quite apparent from the classical treatments that a great many stable isomeric forms exist in the small-size range and that the number of isomers with near-minimum energy increases rapidly as size increases.

Experimental Observations

X-Ray and electron diffraction data can be analyzed to give information about the geometry of small metal clusters. Boswell [157], for example, used a carefully calibrated electron diffraction camera to determine the lattice spacings of crystallites of gold and bismuth as a function of size as well as those of several alkali halides. He showed that previous results taken to indicate that alkali halide crystals expanded as the particle size decreased were in error because they used fine gold particles as the calibration standard. He demonstrated, on the contrary, that the lattice spacings of 20 Å gold particles were decreased by almost 2% from the bulk X-ray value. Similarly, 35 Å bismuth particles were contracted by 2%. The general observation was confirmed by Mays, Vermaak, and Kuhlmann-Wilsdorf [158], who took account of a possible temperature rise caused by the measuring electron beam. They interpreted their result in terms of a surface stress. Schroeer and collaborators [159] made similar observations using X-ray diffraction on gold hydrosols and found a corresponding isomer shift in the Mössbauer spectrum from the same samples. A contraction in small silver particles was reported by de Planta and collaborators [160], and Grigson [161], using continuous direct-recording electron diffraction, also observed a contraction in 80/20 NiFe. Vook and Otooni [162], on the contrary, using X-ray diffraction on evaporated gold films, found that the lattice constant normal to the film plane increased with decreasing film thickness, in agreement with an earlier experiment of the same type on tin. They interpreted their results in terms of a suggested increased vibrational amplitude of surface atoms.

The majority of the experimental diffraction data suggest a lattice contraction in small clusters, whereas the indications from the classical cluster model calculations are for a dilation. Briant and Burton [163, 154] have proposed that this apparent discrepancy can be attributed to the clusters of the diffraction experiment having polytetrahedral structures, which have been mistakenly indexed as bulk *fcc* forms. They have calculated radial distributions and interference functions from polytetrahedral clusters and other equilibrium forms resulting from molecular dynamics calculations. They find scattering functions that are distinctly different from those of *fcc* clusters of the same size, but not so different that they could not be incorrectly indexed. They show that if this were done, the conclusion would be that the lattice was contracted, whereas their clusters are actually expanded.

The distinctive features of their calculated interference functions are remarkably similar to those observed by Farges and his collaborators [34, 35] in electron diffraction studies of clusters in a supersonic argon nozzle beam. Similar features also appear in diffraction patterns from so-called "amorphous" metals, which Briant and Burton suggest are composed of randomly oriented 13-atom icosahedral clusters. Some experimental and calculated scattering functions are shown in Fig. 3.8.

In the early 1950s, Ino [164] reported the incorporation of a rotating sector in an electron diffraction camera which compensated for diffuse background scattering and allowed increased accuracy in the measurement of integrated intensity of a larger number of diffraction rings. This change permitted the calculation of radial distribution functions, even on deposits with particle sizes too small to give sharp rings. The method was subsequently used by Morimoto [165] to study vacuum deposits of gold, silver, and aluminum, and by Morimoto and Sakata [166] on nickel and cobalt. In gold, silver, aluminum, and nickel, the nearest-neighbor distances in particles between 10 and 20 Å in size agreed well with bulk values, as did the vibrational amplitude. Except in aluminum, for which the stacking-fault energy is very large, anomalous peaks appeared in the radial distribution functions, which could be explained as resulting from stacking faults. The results on cobalt showed greater anomalies, which Morimoto and Sakata interpreted as indicating vibrational amplitudes twice that of the bulk, and a mixture of *hcp* and diamond structures transforming at high temperatures to *fcc*. Some years later, Watanabe and Miida [167] calculated the radial distribution functions for polytetrahedral structures between 5 and 15 Å in size and found that they could match the experimental ones for cobalt. These results are also reproduced in Fig. 3.8. In view of this interpretation, the absence of these anomalous results for silver, gold, and nickel would seem to indicate that normal *fcc* structures existed for these metals.

Kimoto et al. [168] obtained an electron diffraction pattern from chromium particles formed by inert gas evaporation which could not be interpreted in terms of the normal *bcc* structure. Following further studies, Kimoto, Nishida, and Ueda [169] proposed a new modification of chromium called δ-Cr, having the A-15 structure. Forssell and Persson

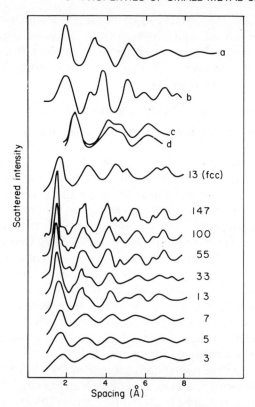

Fig. 3.8. Experimental and calculated X-ray scattering functions for small clusters. Curves *a* and *b*, taken from Farges [35], give the experimental results from argon clusters in a nozzle beam and the calculated function for a *fcc* cluster. Curves *c* and *d* are the experimental results of Morimoto and Sakata [166] on vapor-deposited cobalt particles and the curve calculated by Watanabe and Miida [167] for a polytetrahedral geometry. The remaining curves are those calculated by Briant and Burton [154] for equilibrium geometrics of clusters containing the numbers of atoms indicated by the figures beside the curves.

[170] and also Nishida, Sahashi, and Kimoto [171] observed this same structure in thin vacuum-deposited films and found that its abundance depended upon the incidence angle of the vapor. More recently, Kimoto and Nishida [172] found the same structure in small iron particles.

A novel technique involving multireflection dark-field imaging in a conventional transmission electron microscope has been recommended by Krakow and Siegel [173] for obtaining electron diffraction data from individual metal particles of very small size. The undiffracted beam is masked, and the reflected beams form images displaced about the normal image by the combined effects of defocusing and spherical aberration. Lattice orientations of individual particles may be determined. The method minimizes the background signals from support films, and these authors were able to obtain diffraction patterns from individual gold particles as small as 18 Å in diameter.

Vibrational and Other Thermodynamic Properties

Lattice dynamics will obviously be size dependent in the small cluster range, over and above any possible changes caused by differences in structural form. Methods for calculating vibrational properties of small clusters were mentioned above in connection with the entropy effect in determining the most stable form. By using the harmonic method, the vibrational properties of a small cluster may be determined as an extension of the computerized potential energy minimization techniques discussed in the previous section. Given a geometry and an interatomic potential function, the force-constant matrix—essentially the matrix of second derivatives of the potential functions—is determined. The partition function is then obtained, and from that the entropy, the resonant mode density, and the lattice specific heat may be numerically calculated. Nearest-neighbor interactions give sufficient accuracy for most purposes.

In the molecular dynamics approach [153, 154], a starting configuration and temperature are assumed for an ensemble of atoms, and the classical equations of motion are numerically integrated for a large number of time intervals, the positions and velocities of each atom being recorded at each time. Average values can be analyzed for thermodynamic quantities. The vibrational frequency spectrum is obtained as the Fourier transform of the velocity autocorrelation function.

In the self-consistent phonon method of Etters et al. [155], the resonant modes are derived as eigenvalues from solution of a Hamiltonian taken to represent the assembly of atoms. Frequency distributions for clusters of three through 15 atoms, calculated by Etters et al. [155], are shown in Fig. 3.9. From this and other studies of similar type [141, 144, 147–149], several generalizations may be made. The number of vibrational modes is shown to increase with particle size, as does also the frequency spread. In particular, an increase in low-frequency modes occurs as acoustic-type modes involving cooperative movements of large numbers of atoms become possible. Smaller clusters show only relatively high frequencies associated with breathing modes.

The entropy is found to increase sharply with cluster size in the range up to about 20 atoms, above which size it becomes almost constant. The Helmholtz free energy per atom is found to decrease monotonically with increasing cluster size and to show the same preference for the polytetrahedral structures.

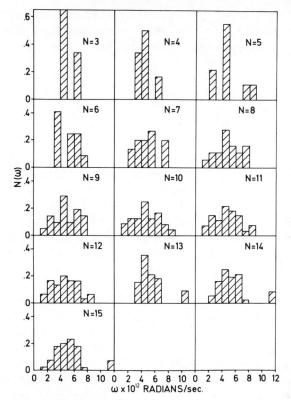

Fig. 3.9. Calculated phonon frequency distributions for clusters with sizes between three and 15 atoms. From Etters et al. [155].

Both the Monte Carlo and the molecular dynamics approaches give dynamic properties as a function of temperature. Results [150, 152, 154] show that the bond energy and the fluctuation of bond lengths exhibit sharp discontinuities with increasing temperatures which are interpreted as melting transitions. The melting temperature is depressed as the cluster size decreases. Kaelberer, Etters, and Raich [174] computed the energetics of four isomers composed of seven atoms as a function of temperature and determined the temperatures of phase transitions from the metastable states to the most stable form.

There is in addition a configurational component to the entropy, arising from the increasing number of stable or metastable isomers as the cluster size increases. This results in an additional stabilizing effect for larger sizes and may be a significant factor in considerations of critical nucleus size, for example, in condensation from the vapor phase [143, 175].

From the vibrational partition function, values can also be computed for the lattice heat capacity C_v. At very low temperatures some structure is seen in the graphs of C_v versus size, but above about 60°K the curves are smooth and close to the classical value of $(3N - 6)k$ per atom [144, 146, 147].

Baltes and Hilf [176] employed a rigorous Bose–Einstein summation of the contributions of all 183 vibrational modes of a 184-atom model and obtained a relation between C_v and temperature in remarkably good quantitative agreement with experimental results. They explained a small residual difference at low temperature in terms of an electronic contribution. Nonnenmacher [177] used a simpler asymptotic expansion, found agreement over part of the temperature range, and attributed a greater effect to the quantum size perturbation of the electronic part of the specific heat.

The small magnitude of the expected differences, coupled with the uncertainty of matrix or substrate effects, makes the measurement of thermodynamic effects attributable to particle size very difficult. For meaningful results, the particles must be constrained not to vibrate or otherwise move as a whole while at the same time the normal modes of the surface atoms must not be affected. Obviously these two conditions can never be perfectly met. In addition, a correction for any supporting medium must be applied. Bogomolov and collaborators [178, 179] forced gallium and mercury into the ~40 Å pores of a leached sodium borosilicate glass and compared several thermodynamic properties of these samples with those of the glass alone and of an equivalent mass of bulk metal. Over the temperature range studied, the expected difference in the lattice specific heat was about equal to the experimental error, and they were not able to detect a deviation from bulk behavior. They did however observe a depression of the melting point and a reduction of 40% for gallium and 24% for mercury in the heat of fusion. In another report [180] a group from the same laboratory used neutron scattering to study the phonon spectrum of the gallium samples, as shown in Fig. 3.10. They demonstrated the absence of low-frequency modes, as compared with the spectrum of bulk gallium, and an enhancement of certain high-frequency modes.

Novotny, Meincke, and Watson [181, 182] studied indium and lead particles, also in matrices of porous glass, whose pore size they varied from 22 to 60 Å. They found in each case an enhancement of the lattice specific heat, which they attributed to low-frequency surface phonon modes. In addition, there was a low-temperature cutoff of the enhancement, which occurred at higher temperatures as the particle size decreased. This effect was attributed to a size limitation on the maximum phonon wavelength.

Novotny and Meincke [182] found that the superconducting critical

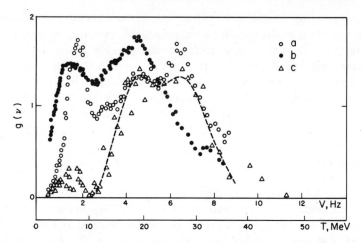

Fig. 3.10. Experimental phonon frequency spectrum for gallium in the solid (*a*), liquid (*b*), and finely dispersed (*c*) states. From Bogomolov et al. [180].

temperature of their particles was increased compared with the bulk value and attributed this effect to changes in the phonon spectrum. Similar size effects on critical phenomena, manifested as an increase of the Curie temperature at small particle size, have been verified by Watson [183], Bogomolov et al. [184], and Lutz and co-workers [185].

Mossbauer spectroscopy has been used to study the lattice dynamics of several samples of finely dispersed metals. Gold particles in the form of hydrosols bound in gelatin were investigated by Marshall and Wilenzick [186] and also by Viegers and Trooster [187]. Tin particles formed by inert gas evaporation were studied by Suzdalev, Gen, Gol'danskii, and Makarov [188] and discontinuous evaporated tin layers, by Akselrod, Pasternak, and Bukshpan [189]. Bogomolov and Klushin [190] worked with tin and gallium in porous glass, and Roth and Hörl [191] examined tungsten particles formed by pyrolysis of tungsten hexacarbonyl and collected by sedimentation from an organic solvent. Results are interpreted in terms of an absence of low-frequency modes and an increased vibrational amplitude for surface atoms, though complications resulting from motion of the particles as a whole prevent a straightforward analysis. Analyses of X-ray and electron diffraction patterns from very small particles also reveal that the mean-square vibrational displacement of surface atoms is several times that of bulk atoms [192].

A depression of the melting point with decreasing size is well documented. The changes are quite large, often 100°C or more, and therefore easily detected in vacuum deposits by temperature-dependent electron microscopy or electron diffraction. By observing the electron diffraction pattern as the specimen temperature is raised, one may judge the melting point by the change from a comparatively sharp ring pattern to one showing only diffuse bands. This is correlated with mean particle size, determined normally by electron microscopy. In the particle size range over which this method is possible ($d > 100$ Å), results are generally found to agree with the macroscopic Gibbs–Thomson–Frenkel concept of a surface energy, and from the data, values of surface energy may be determined. Among the investigations that may be cited are those of Takagi [193] on lead, tin, and bismuth; of Blackman and Curzon [194] on tin; of Gladkich, Niedermayer, and Spiegel [195] on silver and copper; of Wronski [196] on tin; of Boiko, Pugachev, and Bratsykhin [197] on indium; of Coombs [198] on lead and indium; of Bogomolov et al. [199] on Hg, Ga, Sn, Pb, In, and Cd clusters in zeolite cavities; and of Buffat [200] on gold. The results of this last study are shown in Fig. 3.11. Differences between melting and solidification behavior were noted by Takagi [193], Blackman and Curzon [194], and also by Pocza, Barna, and Barna [201] using electron microscopy. These are presumably attributable to shape differences between the solid and molten states. The appearance of the rings in a diffraction pattern is a collective property of the ensemble of particles, with their distribution of sizes. To avoid this uncertainty, Berman and Curzon [202] used high-resolution dark-field electron microscopy, which allowed them to observe melting points of individual particles of tin of different size within the same field. The dark-field image of a particle is only visible when a diffracted beam passes through a selectively

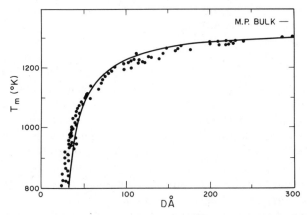

Fig. 3.11. Variation of the melting temperature T_m of gold with crystal diameter D. Experimental points are due to Buffat [200], who used a theoretical model proposed by Sambles [204] to calculate the solid curve.

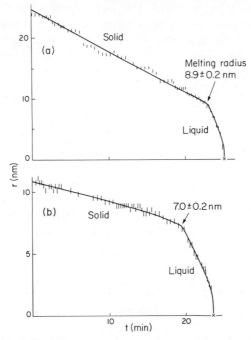

Fig. 3.12. The decreasing radii of two gold particles as they evaporate at 1257°K (*a*) and 1227°K (*b*) in an electron microscope. A marked increase in evaporation rate occurs at the melting radius, marked on each curve by the arrow. From Sambles [204].

positioned objective lens aperture. They used a theory derived by Wronski [196] and found good agreement with data.

In a technique developed by Sambles and collaborators [203, 204], a specimen is observed by electron microscopy at a fixed temperature, and sizes of individual particles are measured as a function of time. Two examples of results [204] are shown in Fig. 3.12. At temperatures near the melting point, the diameter decreases owing to sublimation until the size-dependent melting point reaches that at which the experiment is conducted. At this point the particle melts and a sharp increase in the rate of vaporization is observed. From the data, values of surface energy of 1200 erg/cm² for silver at 1005°K and 1350 erg/cm² for gold at 1160°K were determined.

Thermodynamic parameters play a major role in many of the so-called capillarity or classical theoretical treatments of vapor-phase nucleation or of condensation on a substrate [205]. Some success has been achieved in interpreting data taken from electron microscopy of vapor-deposited nuclei in terms of these concepts. However, under most conditions the critical nucleus size for condensation is so much smaller than the minimum observable size that strong dependence must be placed on theoretical treatments of the stage of growth between these sizes, among which there is considerable difference of opinion. It is not clear that the results add much strength to calculated values at this point.

A recent study by Freund and Bauer [206] uses laser schlieren techniques to measure light scattering from gas-phase condensates of iron in inert gas. Condensation takes place from supersaturated vapors produced by shock heating an unstable compound such as $Fe(CO)_5$. The data are used to determine the binding energy of clusters as a function of size. Only about half the bulk binding energy is reached at a cluster size of 100 atoms.

Konstantinov, Panov, and Malinowski [88] have used electron microscopy for studying the size dependence of the growth versus disappearance of silver nuclei between 20 and 100 Å in solutions of oxidation–reduction couples of various strengths and silver ions (see also the section on chemical and catalytic activity below). They determined a stable critical size as a function of the chemical potential of the solution, as shown in Fig. 3.13. The data are consistent with the concept of a surface energy, for

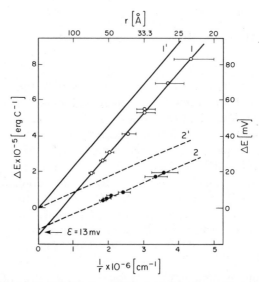

Fig. 3.13.   Chemical potential ΔE of small silver particles in equilibrium with a solution of silver ions as a function of reciprocal radius. The data on line 1 are from uncoated samples and those on line 2, from samples overcoated with gelatin. Lines 1′ and 2′ are shifted by ~13 mV to correct the zero to the classically predicted value. From Konstantinov and Malinowski [88].

which they find a value of 920 erg/cm² on a carbon substrate and 1040 erg/cm² on silver bromide. A coating of gelatin lowers the surface energy on either substrate to about 400 erg/cm². These authors interpret the size criterion of photographic development in terms of the same thermodynamic concepts (see section on chemical and catalytic activity).

It is not uncommon to find the macroscopic and classical surface free-energy approach extended into the cluster size range of a few atoms, where there are obvious conceptual difficulties. Some attempts to rationalize this approach have been made [207], and a surprising numerical aggreement is found between this method and the more atomistic treatments if classical potential functions are used. The agreement, however, seems more fortuitous than fundamental, and the practice must be regarded with some skepticism. Hoare and Pal [144], in fact, refer to it as an absurdity.

Surface-Structural Effects

Atoms located at corners, edges, and other surface singularities are differently coordinated from those in low-index surface planes. For example, the coordination numbers of atoms in the bulk, in the surface, on an edge, and on a corner of an octahedral crystal of an *fcc* metal are 12, 9, 7, and 4, respectively. The larger number of unsaturated bonds can easily be imagined to cause the atoms with lower coordination to be particularly active sites for adsorption and catalysis. Several authors have evaluated the degree of coordination of surface atoms as a function of crystallite size.

Among the first were Poltorak and Boronin [208], who concentrated primarily on *fcc* octahedra as representative of finely divided platinum catalysts. For perfect octahedra of increasing size, the mean coordination number of the surface atoms increases from 4 to 8.64 over the size range from six to 3894 atoms (5.5 to 49.5 Å edge length). They pointed out that the crystallite size range over which most of the change occurs is between 8 and 40 Å linear dimension. The perfect octahedra include only certain select clusters (6, 19, 44, 85, etc., atoms) and have sites only with the particular coordination numbers listed above. For intermediate sizes, Poltorak and Boronin used crystal models to find the configurations with the maximum number of nearest-neighbor bonds and categorized the surface sites on them. Coordination sites with all numbers between three and 12 appear at certain sizes.

Instead of the coordination numbers of existing surface atoms, van Hardeveld and van Montfoort [134] focused attention on high coordination potential adsorption sites. They observed that an infrared-active form of adsorbed nitrogen on nickel, palladium, or platinum was found only on crystallites with diameters between 15 and 70 Å and postulated that it was

Fig. 3.14. B_5 sites at a step on a (110) surface. The atoms forming the nearest B_5 site are marked with dots. From Bond [209].

specifically associated with the so-called B_5 sites, as illustrated in Fig. 3.14, from a paper by Bond [209], where the adsorbate could form bonds with five substrate atoms. They pointed out that such sites do not exist on perfect octahedra or cubo-octahedra. However, the existence of a partial extra plane on either an octahedral or cubic face, omitting only the line of atoms along the edges of the crystallite, produces a maximum number of B_5 sites along the boundary of the extra plane. They determined the maximum ratio of B_5 sites to atoms in the crystallite for modified octahedra and cubo-octahedra as a function of edge length. For both shapes, this ratio rises sharply from zero at a size of about 15 Å, peaks around 20 Å, and thereafter drops again to small values, in general agreement with the experimentally observed active nitrogen form.

Subsequently van Hardeveld and Hartog [210] published a more comprehensive compilation of the numbers of sites of a wide variety of types on lattice elements of many shapes from *fcc, bcc,* and *hcp* crystals and on the same containing the partial planes that produce the B_5 sites.

Apparently no similar attempts have been made to treat surface sites on nonlattice elements, such as the tetrahedrally based clusters, which seem to be stable at small sizes.

Electronic Properties

Finite Level Spacing: The Quantum Size Effect
THEORY

As the number of atoms in a solid decreases, the mean energy separation δ between the individual electronic levels increases. In the energy range of concern, the level spacing is given by the inverse of the single-spin density of states at the Fermi energy, and is inversely proportional to

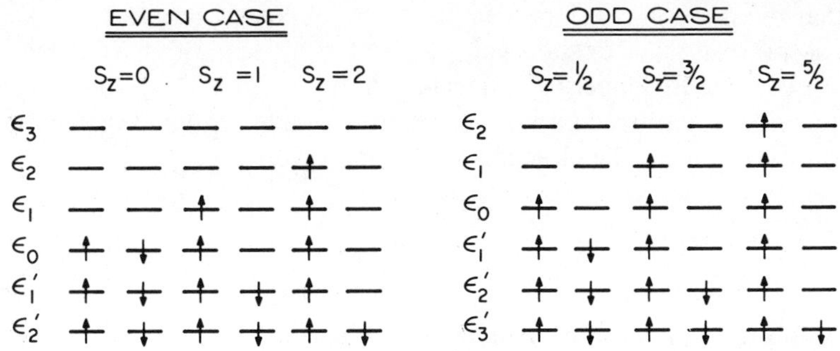

Fig. 3.15. The energy level manifolds of the ground and first two excited states of clusters with even and odd numbers of electrons. From Denton, Mühlschlegel, and Scalapino [214].

the particle volume. Rather than being negligibly small, as in macroscopic solids, for small particles δ becomes comparable with the thermal energy kT, the Zeeman energy $g\mu H$, and electromagnetic energy $\hbar\omega$. Changes in electronic properties are to be expected as a result. Owing to the difference in occupancy of the uppermost level, particles with an odd number of electrons should behave differently from those with an even number (see Fig. 3.15). Using appropriate approximations, partition functions and free energies may be calculated for both odd and even sizes, and these must be averaged over the particle size distribution in order to predict observable changes.

The effect has been dealt with by several authors whose approaches have differed principally in the form of the energy level distribution about the Fermi level, a factor that has a significant effect on the predicted results. Fröhlich [211], apparently the first to discuss this small-particle effect, assumed uniformly spaced levels. Kubo [212] treated the problem in more detail and was the first to attract attention to the most significant of the consequences. He proposed a random statistical level distribution given by

$$P(\Delta, d\Delta) = \exp\left[\frac{-\Delta}{\delta}\right]\frac{d\Delta}{\delta} \tag{3.6}$$

in which Δ is the energy difference between two successive levels and δ is the average value. Kubo calculated that the electronic heat capacity of fine particles would be about two thirds that of the bulk, with values for the odd-sized particles somewhat lower than those for even sizes. He determined that the spin paramagnetism should be higher in odd-sized particles, which should demonstrate Curie-like behavior. Its average value should exceed that of the bulk by half a Bohr magneton. He pointed out that even greater differences are to be expected in dynamic effects such as relaxation of excited spins.

Gor'kov and Éliashberg [213] made the point that scattering of electrons from surface irregularities would randomize the level distribution. They took the mean level separation to be given by

$$\Delta = \frac{2\pi^2}{Vm^*p_0} \tag{3.7}$$

in which V is the particle volume, m^* is the effective electron mass, and p_0 is the limiting momentum of the Fermi surface.

These treatments were extended by Denton, Mühlschlegel, and Scalapino [214], who compared the various possible alternatives extensively. Kubo had concluded that electrostatic energy involved in exchange of an electron was so large that only electrostatically neutral particles need be considered. Denton et al. pointed out that this condition requires the use of the canonical partition function rather than the more familiar grand-canonical one. They cited the similarity of the problem of the distribution function to that of nuclear-level statistics treated by Wigner [215], among others, by whom the distribution function

$$P = \frac{\pi\Delta}{2\delta^2} \exp\left[-\frac{\pi}{4}\left(\frac{\Delta}{\delta}\right)^2\right] \tag{3.8}$$

was proposed. By analogy with the nuclear problem, the appropriate ensemble of levels depends upon the relative values of the spin-orbit coupling energy η and the level spacing δ: for $\eta/\delta << 1$, an orthogonal ensemble has been used; for $\eta/\delta >> 1$, a symplectic ensemble; and for $\mu H >> \delta$ and $\eta/\delta >> 1$, a unitary ensemble has been used. Denton, Mühlschlegel, and Scalapino calculate the expected forms of expressions for the low-temperature electronic heat capacity C and spin susceptibility χ for the different ensembles, and their results are given in Table 3.2.

EXPERIMENTAL TESTS

The various theorists [212–214] involved with this question have discussed the possibilities of experimental verification. For detection of the deviations predicted in specific heat or susceptibility on particles in the 10 to 100 Å range, the condition that $\delta \geqslant kT$ requires a temperature range around 1° to 10°K. In principle the ideas are applicable also to smaller particles at higher temperatures, but the difficulties of particle size characterization have precluded such experiments. In the limit of clusters of a very few atoms, the separations between levels become even more significant, but the approximations of the level distributions used in any of the above treatments are totally inadequate, and quantitative application of these theoretical predictions is meaningless.

Table 3.2 Leading Low-Temperature Behavior of the Electronic Heat Capacity and Spin Susceptibility for Different Ensembles, with Even and Odd Electron Number

Ensemble	Even	Odd
	C/k	
Poisson	$5.02kT/\delta$	$3.29kT/\delta$
Orthogonal	$(3.02 \times 10)\ (kT/\delta)^2$	$(1.78 \times 10)\ (kT/\delta)^2$
Symplectic	$(3.18 \times 10^4)\ (kT/\delta)^5$	$(1.64 \times 10^4)\ (kT/\delta)^5$
Unitary	$(5.88 \times 10^2)\ (kT/\delta)^3$	
	χ	
Poisson	$3.04(\tfrac{1}{2}g\mu)^2/\delta$	$(\tfrac{1}{2}g\mu)^2/kT$
Orthogonal	$7.63(\tfrac{1}{2}g\mu)^2kT/\delta^2$	

Heat Capacity.   Accurate measurement of the heat capacity of an ensemble of small particles in the required temperature range presents severe experimental problems, as discussed in the section on vibrational and other thermodynamic properties. Furthermore, the largest contribution, at least at the high end of the temperature range of interest, is that from the vibrational modes of the atomic nuclei. The effects of nonideal particle shapes and of restraints on displacement imposed by any substrate or matrix are so ill defined as to make the quantitative separation of the vibrational and electronic parts of the heat capacity virtually impossible. The measurements of Novotny, Meincke, and Watson [181, 182] on small indium and lead particles were discussed in the previous section. These authors attribute a part of the effect to the level spacing effect on the electronic heat capacity, as do others who have discussed their results. It is not possible, however, to make a quantitative check of their data with the theory. Stewart [216] used an automated small-sample calorimeter to measure the heat capacity of fine platinum particles formed by cosputtering with SiO_2. He found a definite decrease in the electronic component with decreasing size.

Magnetic Properties.   The expected size effects on the magnetic properties include an increase in the spin susceptibility, as discussed above, and the suppression of certain relaxation mechanisms, resulting in increases in relaxation times. These effects may in principle be observed as an increase in the static susceptibility of small particles at low temperature, compared with bulk values, a narrow NMR resonance with no Knight shift, very narrow-conduction electron spin resonance (CESR) lines in light elements such as the alkali metals, and observable CESR signals in heavy metals, whose bulk resonances are normally too broad for detection.

Gor'kov and Éliashberg [213] expressed the opinion that the static magnetic susceptibility would be too sensitive to exchange interactions involving the magnetic moments of surface atoms to allow a definitive test of the level spacing effect. A few sets of measurements of static susceptibility have been reported, but normally these must be made in conjunction with other relaxation measurements in order to identify surface effects. Among the static measurements that are in agreement with theory are the early studies of Taupin [60] on lithium, the measurements of Kobayashi et al. [217, 218] on copper and aluminum, and recent results of Marzke, Glaunsinger, and Bayard [219] on platinum, shown in Fig. 3.16. All of these results show an enhanced low-temperature susceptibility which does not field saturate. A reciprocal relationship between temperature and susceptibility is demonstrated, and the magnitude of the signal can be rationalized with the requirement of one free spin per odd-sized particle. Marzke et al. point out the paradoxical feature that platinum, with its normal bulk electronic structure, should have an even number of electrons even in odd-sized particles. Hybridization of the s-d orbitals, however, is known to perturb this situation even in the bulk, and there is evidence for a change toward the d^9s^1 structure for particles of small size (see later).

The expected reduction or elimination of the Knight shift in the NMR spectrum has been confirmed experimentally by Taupin and Charvolin [61] for thin lithium platelets in irradiated lithium fluoride; by Charles and Harrison [220] for lead in porous glass; by Kobayashi, Takahashi, and Sasaki [217, 218, 221] for both aluminum and copper prepared by inert gas

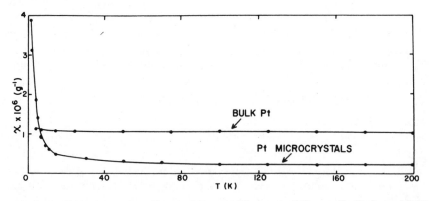

Fig. 3.16. Temperature dependence of the magnetic susceptibility χ of bulk platinum and for finely dispersed platinum particles in a gelatin matrix. From Marzke, Glaunsinger, and Bayard [219].

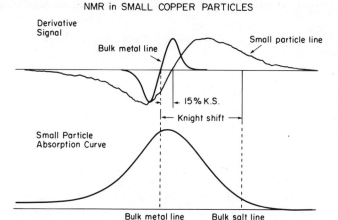

Fig. 3.17. NMR derivative spectra of bulk copper and 100-Å vacuum-deposited copper particles. The small-particle curve is integrated below. From Yee and Knight [93].

evaporation; by Ido, Shibukawa, and Hoshino [91] for copper prepared both by inert gas evaporation and by vacuum deposition on SiO_2; and by Yee and Knight [93] for copper deposited on SiO_2 (Fig. 3.17). Fujita and co-workers [222], on the contrary, found that both the shift and the linewidth in aluminum particles prepared by gas evaporation were as in the bulk, but that the signal strength decreased with decreasing temperature for fine-particle samples. The linewidth was observed to decrease at small sizes in lithium [80], but only a broadening attributed to surface effects was observed for lead [220] and copper [93]. Kobayashi et al. [218] observed only a broadening in their copper particles. This they attributed to the high spin-orbit coupling, which prevents the disappearance of the signal from the even-sized particles. For aluminum [217] they were able to analyze the spectrum in terms of a sharp peak arising from even particles and a broader tail from the odd ones.

The theory of finite level effects on the ESR signal has been developed in detail by Kawabata [223], based on the original formalism of Elliott [224] for spin-orbit coupling. This treatment shows that the manifestations of the finite level separation will be observable when two conditions are met: (1) when the level separation is greater than the Zeeman energy, $\hbar\omega_z/\delta \ll 1$; and (2) when $\hbar/\tau\delta \ll 1$, in which τ, the spin-orbit coupling parameter, is given by $a/V_F \Delta g^2$, a being the particle dimension, V_F the Fermi velocity, and Δg the g shift. In principle these requirements are sufficient even if $kT/\delta > 1$, though in practice the entire resonance peak becomes unobservable except at low temperature, owing to thermal broadening.

Kawabata pointed out the advantage of studying the light elements, that is, alkali metals (except lithium, for which the spin-orbit coupling is too weak). The majority of the early studies [60, 61, 63, 64, 225] did deal with alkali metal particles and are in general agreement with the theory. Sodium and potassium show a narrowing of the resonance line with decreasing temperature and particle size. Lithium particles have in some experiments manifested a size-dependent width (either a narrowing or a broadening), but the absence of a temperature dependence identifies this effect as a surface spin flip. Smithard [66] and Gordon [67] have reported exceptions to the predicted behavior of sodium in irradiated sodium azide. Both explain their results in terms of surface exchange interactions, which mask the presumed level spacing effect. Borel and co-workers [226] studied lithium matrix-isolated in CO_2. They observed a signal from single atoms, which decayed owing to aggregation upon warming. The resonance from the resulting particles exhibited a size-dependent relaxation time and line broadening from the surface. The line shape observed by Berim et al. [227] for lithium colloids formed in LiF crystals by electrolysis and annealing agreed with their calculations based on the quantum size effect.

An early report [48] of a CESR signal from silver particles in irradiated silver halide was contested [49] as arising from an impurity effect, and this situation is still not clear. Meanwhile Monot, Narbel, and Borel [228] have observed signals in silver reduced by hydrogen gas from an aqueous solution of silver oxide and from silver matrix-deposited with CO_2, and Jain, Arora, and Reddy [51] reported positive results from silver colloids in KCl crystals. These results were further discussed by Borel [229] in terms of the Kawabata formalism [223], and additional results in agreement with this theory were given by Chatelain, Millet, and Monot [230] for silver aggregates in frozen gas matrices. Abou-Kais and co-workers [231] recently reported a CESR signal from conventional supported catalysts of silver with a mean size of 53 Å on silica.

Millet and Monot [232] also obtained CESR signals from matrix-condensed aluminum particles. Resonances in gold particles were reported by Dupree, Forwood, and Smith [233] and later by Monot, Chatelain, and Borel [234], though the details of the two investigations were not in agreement. Glaunsinger and Marzke [235] observed a CESR signal in small particles of platinum.

Although totally unrelated to the magnetic effects produced by the finite spacing of energy levels, an additional magnetic property of small ferromagnetic particles—superparamagnetism—may be mentioned at this point [236]. In a ferromagnetic material, due to the finite domain wall energy, there is a minimum or critical domain size below which it is energetically unfavorable for separate domains to exist. Thus particles with sizes below the critical dimensions are single-domain particles. Within them the direction of magnetization cannot change by domain wall motion, as is characteristic of bulk ferromagnetics, but only by a collective reorientation of the spins of all unpaired electrons in the particle (or by a rotation of the particle itself). For a collection of moderately large particles a classical theoretical analysis is appropriate, and if anisotropics are small, the magnetization $\overline{M}$ parallel to an applied magnetic field H is given by

$$\overline{M} = n\mu\left[\coth\left(\frac{\mu H}{kT}\right) - \frac{kT}{\mu H}\right] \tag{3.9}$$

where n is the number of particles and μ is their magnetic moment. This expression simplifies to the low-field and high-field approximations

$$\overline{M}_{LF} \cong n\mu^2\frac{H}{3kT} \tag{3.10}$$

and

$$\overline{M}_{HF} \cong n\mu\left(1 - \frac{kT}{\mu H}\right) \tag{3.11}$$

Having no metastable configurations, such particles exhibit no hysteresis effects; and when magnetization is plotted against H/T, data taken at different temperatures superimpose. These are the characteristics of paramagnetism, and indeed such a particle does behave as a giant paramagnetic atom, but with a magnetic moment μ equal to the collective moments of all its atoms; hence the term superparamagnetism. In practice, accurate fits of experimental data to expressions (10) and (11) are not found, owing to the distributions of particle size and hence of magnetic moment. From measurements of resonance as a function of temperature, and assuming randomness to the directions of easy magnetization of the particles relative to the applied field, a particle size distribution can be deduced. This technique has sometimes been used for determinations of particle size of finely divided ferromagnetic materials [237], but the expected relationships are found to break down for sizes below about 10 Å.

Polarizability. One further proposal of Gor'kov and Éliashberg [213] was the onset of an anomalously high electrostatic polarizability α for a particle of dimension a as the electric field strength is reduced such that $eEa \le \delta$. At low field, α was predicted to be proportional to a^5, compared with the usual a^3 dependence when $eEa \gg \delta$. Polarizability is a significant physical property because dielectric shielding plays an important part in adsorption and desorption phenomena and is involved in experimental determinations of work function.

A criticism of the Gor'kov–Éliashberg analysis was given by Strässler, Rice, and Wyder [238], who suggested that their assumption that the local field equaled the applied field neglected the screening effect of the system of conduction electrons, which they contend would be effective irrespective of whether the energy levels are discrete or continuous. Subsequent to this, Rice, Schneider, and Strässler [239] analyzed the size effect on the screening by a finite number of electrons trapped in an infinite spherical potential wall and satisfying the Poisson equation. They predicted a decrease in polarizability at small sizes owing essentially to the inadequate amount of charge for full shielding in a small particle. Cini and Ascarelli [240] expanded this free-electron analysis, with similar results.

Experimental attempts to detect an anomalously high dielectric constant were made by Meier and Wyder [241] using gold particles in a photosensitive glass and by Dupree and Smithard [242] using silver particles formed either by irradiation of photosensitive glass or by reactive diffusion into nearly molten glass. In spite of care to select proper conditions, no indications of a size effect were found in either study.

Surface Plasmons

In an alternating electromagnetic field, the electron charge cloud near the surface of a particle responds by resonant collective oscillations known as surface plasmons. These surface plasmons dominate the optical absorption of collections of metal particles of moderate size. Experimental measurements are relatively straightforward, and results have been reported for a variety of metals and suspension media. Among those to be noted are the observations of Doremus [55, 58, 243] (Fig. 3.18a), principally on silver and gold particles, but also including several other metals; those of Kreibig et al. [56, 244], also principally on silver and gold particles; of Papavassiliou and Kokkinakis [245], including copper and alloys of silver and gold; of Skillman and Berry on silver [246]; Smithard's studies of silver [57, 247] and of sodium in sodium chloride [62] and in irradiated sodium azide [65]; and other isolated observations many of which are referred to in these papers. Within a year of this writing, papers have appeared by Jarrett and Ward [248] on discontinuous gold films; by Truong and Scott [249] on films of all three noble metals; by Rasigni and Rasigni [250] on lithium deposits; by Eversole and Brodia [251] on silver, copper, and gold particles formed by inert gas evaporation; and by Norman et al. [252] on discontinuous gold films.

Classical theoretical treatments of this effect date back to those of Maxwell-Garnett [253] and Mie [254]. These have generally been found inadequate, and the more recent investigations involve a quantum-mechanical solution of the particle-in-a-box problem, in which the periodic potential in the metal is neglected and the charge is treated as a free-electron gas. Refinements in the theory have included the consideration of scattering of the electron gas at the surface, by Kawabata and Kubo [255] and later by Smithard and co-workers [256], and the concept of a diffuse surface barrier rather than the abrupt change in potential, as in the analyses of Eguiluz, Ying, and Quinn [257] and Ascarelli and Cini [258]. References to earlier theoretical treatments are given in these papers. When surface scattering effects are included in the theory, predictions of the dependence of peak width on particle size are in good agreement with experiment (Fig. 3.18b), showing first a narrowing with decreasing size, and then a minimum and a subsequent broadening as scattering effects become more important. The calculated peak position (Fig. 3.18c) shows only a short-wavelength shift with decreasing size for the model of an abrupt potential barrier, but Ascarelli and Cini [258] showed that by use of a diffuse barrier they could account for a long-wavelength shift at very small sizes, in agreement with observations.

The problem has also been treated [259, 260] by an alternative, effective-medium approach, which becomes essentially equivalent to the Maxwell-Garnett analysis in dilute systems but is claimed to be more appropriate when the concentration of metal particles is high. Gittleman and Abeles [261], however, conclude that this treatment is less accurate in describing experimental results than the Maxwell-Garnett theory.

Genzel, Martin, and Kreibig [262], in solving the particle-in-a-box problem for small numbers of electrons, point out that for very small clusters a fine structure should appear on the plasma absorption peak, corresponding to transitions between discrete levels in the conduction band. A possible observation of such an effect is described in the section on correlation with experiments.

The surface plasmon resonance can also be detected and studied by energy loss spectroscopy of transmitted electron beams. Studies of this type have been reported by Kunz [263], Creuzburg [264], Fujimoto et al. [265], Carillon [266], and Kreibig and Zacharias [244]. Kokkinakis and Alexopoulos [267] have also reported on an X-ray-excited luminescence from small silver particles, which they ascribe to a radiative plasmon decay. Subsequently a similar effect was observed in copper particles by Kokkinakis and Papavassiliou [268]. Borziak and co-workers [269] reported an emission as a result of passing a current through a discontinuous film. They proposed that the luminescence resulted from the decay of

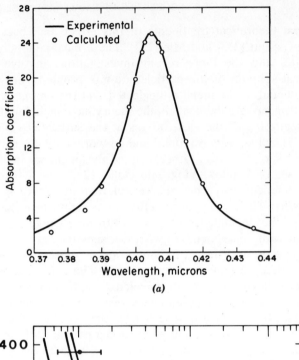

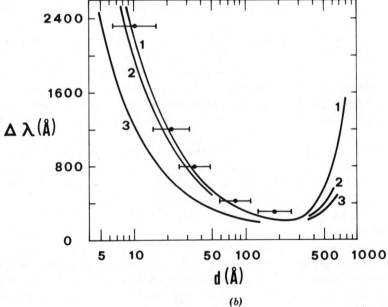

Fig. 3.18. (*a*) Measured and calculated spectral absorption of finely dispersed silver particles, from Doremus [55]. (*b*) Experimental data of Smithard [247] on the width of the absorption peak $\Delta\lambda$ as a function of particle diameter, compared with predicted curves from three different models. (*c*) Similar data and calculations for absorption peak position λp.

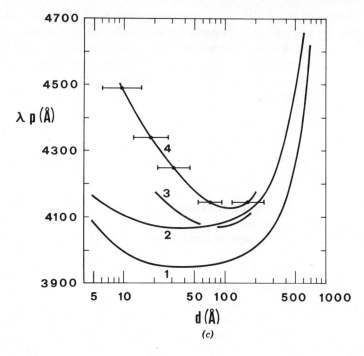

Figure 3.18 *continued*

plasmons excited by electrons accelerated across the gaps between the individual particles in the deposit.

Surface Electronic Levels

Very small metal particles, with their relatively large specific surface, have allowed energy levels which differ from those in the bulk simply because of the potential energy discontinuity at the surface. Theoretical approaches to the solution of the Schrödinger equation across such a boundary have been pursued with increased activity in the past decade or so, with notable success. As with the case of an infinite crystal, the simplifications necessary for a solution have been principally of one of two types. In the approximation of nearly free electrons (NFE), the periodic potential within the metal is taken to be small compared with the total kinetic energy of the electrons, so that perturbation theory can be used in determining the wave functions. In the tight-binding (TB) approximation, on the other hand, the extent of the atomic potential functions is taken to be small compared with the interatomic spacing, and atomic orbitals with only nearest-neighbor interactions are used as the basis potential functions.

The NFE approach involves obtaining first a Bloch-wave solution within the crystal and an exponential function for the region outside the surface, and then testing for a match of the wave functions and their first derivatives at the interface. The relaxation of infinite periodicity allows for Bloch solutions having complex wave vectors at the surface, and near a band gap this possibility can give rise to surface states within the gap. For a gap at a Brillouin zone boundary, the matching of the wave functions at the surface, and hence the existence of a surface state, is possible only if the periodic potential is negative at the matching plane, or, in other terms, if the highest state of the lower band has s-type symmetry. The dependence on symmetry was emphasized by Shockley [270] and is sometimes referred to as Shockley's theorem. Forstman [271] showed that for a band gap within a Brillouin zone, arising from the crossing of two bands, a surface state will always exist.

The NFE formalisms used by various authors differ in the exact form of the periodic potential function and the mathematics of solving the Schrödinger equation. Reviews discussing the merits of various approaches are available [272].

The NFE approach has been most extensively applied to the elemental semiconductors with diamond-type lattices, for which the technological importance of surface states in the gap is obvious. It has also been used for analyses of the noble metals and for the nonlocalized s- and p-type bands in the transition elements. For d bands, the tight-binding approximation has found more extensive use. For this simplification, orbital overlap beyond nearest neighbors is neglected, and the basis of the Hamiltonian is a linear combination of localized atomic states. Several methods for treating the problem have been explored. The "method of moments" has been developed principally by Cyrot-Lackmann and coworkers [273–278]. This technique, involving the calculation of a finite but sizable number of moments of the local density-of-states function, is capable of arriving at such a density of states without relying upon the Bloch theorem or calculating the energy bands. It has been used to calculate local density-of-states functions for low-index planes of a simple cubic metal [274] and the effects of an adsorbate thereon [275]. Subsequent treatments dealt with the d states of the low-index surfaces of *fcc* Ni and *bcc* Cr and αFe [276], with surfaces of Mo and W [277], and with flat and stepped Pt surfaces [278], results of which are shown in Fig. 3.19.

The recursion method of Haydock, Heine, and Kelly [279] for handling the Green-function matrix element also results in a local density-of-states distribution without resorting to the Bloch formalism. Haydock and Kelly [280] have also explored how these states vary through the first few atomic planes near the (100) and (111) surface of a *fcc* d-band metal (Fig.

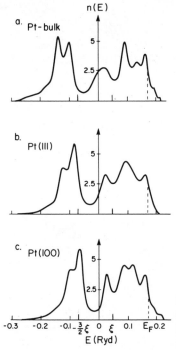

Fig. 3.19. Calculated density of states for bulk platinum (*a*) and the local density of states on a platinum (111) (*b*) and (100) (*c*) surface. From Desjonqueres and Cyrot-Lackmann [278], using the method of moments.

3.20) and have included the effect of lattice dilation near the surface. A recent extension by Kerker [281] dealt with surface and volume *d*-electron states in alloys of transition metals. Dempsey and Kleinman [282] used the recursion method with *s* and *p* as well as *d* orbitals and concluded that hybridization has a significant effect on the surface density of states.

A method introduced by Garcia-Moliner and Rubio [283], based on a Green-function approach, has the advantage of being applicable to an arbitrary interfacial configuration between two media. The technique has been further developed by Büttner and Gerlach [284], Inglesfield [285], and Glasser [286]. Green-function methods have also been used within the tight-binding formulation by Allan and Lenglart [287] and Kalkstein and Soven [288]. Still another formalism, called the effective surface potential method, was introduced by Foo, Thorpe, and Weaire [289].

The treatments by Forstmann and Heine [290] and Forstmann, Pendry, and Gurman [291] are notable in that they deal with the hybridization of an *s* or *sp* band with a tight-binding *d* band. A localized surface state in the hybridized band gap is predicted, and they suggested that anomalous peaks in the photoemission spectra of copper and nickel were manifesta-

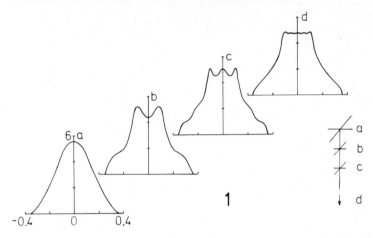

Fig. 3.20. Calculated local density of states at a (100) surface atom (*a*), the first two subsurface atoms (*b* and *c*), and a bulk atom (*d*), using a simple cubic tight-binding *s*-band and the method of Haydock and Kelly [280].

tions of these states. The experimental evidence bearing on the question was subsequently reviewed by Eastman [292], who concluded that these peaks had to be attributed to adsorbed impurities and that at that time there was no experimental evidence for surface states on either copper or nickel. More extensive photoemission studies, employing angular selectivity for both the exciting radiation and the analyzed electrons, have been reported within the last few years. The disagreement over the experimental evidence for a surface state on copper, however, continues. Gartland, Berge, and Slagsvold [293] identified a peak 0.40 eV below the Fermi level as a surface band and reported that it disappears upon exposure to oxygen. It is shown as the peak labeled 2 in Fig. 3.21, from this reference. Lloyd, Quinn, and Richardson [294] and Ilver and Nilsson [295] observed peaks they could not explain by the bulk band structure, and both sets of authors mentioned the possibility that these may arise from surface states. Dietz, Becker, and Gerhardt [296], by contrast, found no indication of surface modification. Nilsson and Ilver [297] observed no surface-state effects in the spectra from gold crystals, but Heimann, Neddermeyer, and Roloff [298] reported evidence for intrinsic surface-state effects in the photoemission spectra from all three noble metals. The existence of surface states on clean tungsten surfaces has been conclusively demonstrated both in field emission spectra by Swanson and Crouser [299] and Plummer and Gadzuk [300] and in photoemission spectra by Waclawski and Plummer [301] and Feuerbacher and associates [302].

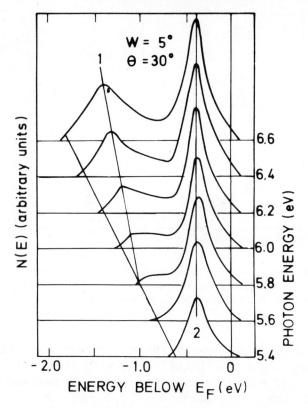

Fig. 3.21. Experimental energy distribution curves for nearly forward photoemission from a copper (111) surface by photons with energies between 5.4 and 6.6 eV. The peak labeled 1 is interpreted as a bulk emission, whereas that labeled 2 is thought to be from a surface electronic level. From Gartland and Slagsvold [293].

Photoemission studies to reveal the occupied states of thin deposits of copper and palladium on silver substrates were reported by Eastman and Grobman [303] and are reproduced in Fig. 3.22. Although they allowed that some islanding of the deposits may have occurred, they cited evidence suggesting that this effect was minimal and that the overlayers behaved essentially as uniform layers. For copper, as the thickness increased above about one monolayer, they observed a shift of the d-state peak toward the Fermi level, a sharpening of the principal d-band peak, and the appearance of two lower-lying shoulders, as in the density of states of bulk copper. At a thickness of approximately four monolayers, all the features of bulk copper were well developed. For the thinnest palladium overlayers, the band of d states was broad, and its upper edge

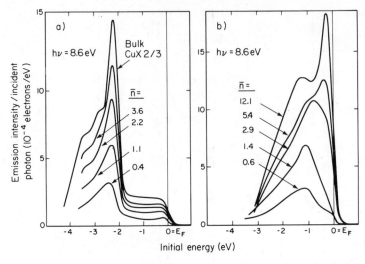

Fig. 3.22. The extra emission due to thin overlayers of copper (*a*) and palladium (*b*) on silver. In (*a*), the bulk copper spectrum is also shown. The values of $\bar{n}$ give the number of monolayers of the overlayer metal. From Eastman and Grobman [303].

was cut off by the Fermi level, indicating vacant *d* states. As the thickness was increased, a strong buildup of states near the Fermi level occurred, and by a thickness of about five monolayers, a strong peak had appeared about 0.3 eV below E_F, characteristic of bulk palladium.

Calculations relevant to these studies were reported by Cooper and Bennett [304], followed later by variations due to Cottey [305] and Pendry and Gurman [306]. Conditions appropriate for *d*-band transition metals and boundary conditions for two symmetrical interfaces separated by one or more atomic layers were chosen. Similarity to the experimental results was demonstrated. Various approaches to the thin-film calculations continue to appear [307–310], and the justifications of the approximations involved are still being argued.

The theoretical methods used in these treatments of the effect of the surface on electronic states explicitly involve a semiinfinite crystal with periodicity at least in the interfacial plane. More than a qualitative application to very small clusters of atoms is therefore not possible. Nevertheless, the qualitative parallel is real, and it is conceptually necessary to consider the simple discontinuity at the surface as one of the factors responsible for the difference of electronic properties between bulk metal and small clusters. To this extent there is a connection between surface properties in general and the specific characteristics of small clusters.

Because of the more localized character of the potential function, results obtained by tight-binding approaches would be expected to be more nearly applicable to clusters.

Cluster Calculations

There are however other theoretical alternatives. Over the past few years several types of semiempirical, quantum-mechanical cluster calculations have been applied to metal particles. These approaches are in marked contrast to those just discussed, which employ essentially boundary-modified bulk-solid formalisms. Cluster methods, on the other hand, begin with Hartree–Fock descriptions of the isolated constituent atoms and deal with the interactions of atomic wave functions as these atoms are combined. In principle, at least, results account for both the boundary effect and that resulting from the finite spacing of levels and can be applied to any arbitrary geometry, requiring no periodicity in any dimension.

The calculations derive the manifold of allowed states and the occupancy rules for these. From them, the total energy may be calculated and, by comparison with the total energy of the corresponding isolated atoms, the binding energy of the assembly. By systematic variation of the cluster size, its structure and separation distance, and the total electronic charge, it is possible to determine the most stable geometry and to obtain quantities such as ionization potential, electron affinity, orbital symmetry, charge distribution, binding energy, bond length, and vibrational constants.

A complete Hartree–Fock self-consistent field calculation using a linear combination of the atomic orbitals to represent each molecular orbital (HF-SCF-LCAO) is impractical on present computers except for very small combinations of very simple atoms. Results have been reported for dimers of some low-numbered elements and for larger clusters of lithium and beryllium [311]. Except in such simple cases, some approximations must be used, and methods of approach have followed those developed by theoretical chemists for treating molecules of moderate size. The molecular-orbital techniques adopt the expedient of neglecting the core electrons and treating explicitly only the interactions of electrons involved in the bonding. Account is taken of the interactions with core electrons by parameterization of calculations on diatomics to fit experimental data.

Two of the molecular-orbital computational schemes first applied to metal clusters were the extended Hückel (EH) and complete neglect of differential overlap (CNDO) methods. Each has its strengths and weaknesses, and these have been discussed by several authors, particularly

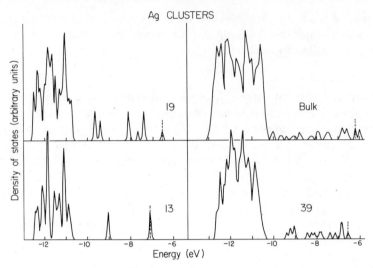

Fig. 3.23. Calculated densities of states for silver clusters of various sizes. From Baetzold, [323] using the CNDO method.

Baetzold [94, 312–323], who has applied them extensively to the problems of metal clusters. The EH method, a nonself-consistent one, is a formalization of a rather simplified intuitive view of bonding in which electron–electron and electron–core interactions are not explicitly calculated but are approximated by evaluating the extent of orbital overlap. The repulsion component is therefore not strictly realistic, with the result that the EH method generally predicts excessive charge transfer and has limited accuracy for determining the total energy of a cluster.

In the CNDO method, charge densities and electronic and nuclear repulsions are explicitly calculated and iterative techniques are employed to reach self-consistency, but rather severe approximations are involved. The diagonal matrix elements of the Hamiltonian are parameterized, and the off-diagonal elements involve the evaluation of an overlap integral. Baetzold reports that CNDO generally predicts bond lengths too large by about 50%, and that there are inaccuracies in the calculated energy level of the lowest unoccupied orbital due to self-interaction considerations inherent to the Hartree–Fock approach.

Johnson, Messmer, and co-workers [324–327] have criticized both of these molecular orbital techniques. They have used a different type of self-consistent approach, employing the multiple-scattering method equivalent to that of the Koringa–Kohn–Rostoker technique for infinite crystals and Slater's statistical approximation for the exchange interaction involving the parameter α, the self-consistent field Xα scattered wave

(SCF-Xα-SW) method. The potential is represented by a muffin tin model, the one-electron wave equation is solved to find the spin orbitals, an occupancy criterion is applied, the charge density and potential are determined and averaged as required for the muffin tin approximation, and the procedure is repeated to convergence. A claim is made for computational economy and for better agreement with experimental data than the MO approaches, but Baetzold [322] has challenged the latter claim.

A common criticism of the SCF-Xα-SW method is the strong reliance on spherical symmetry of potential functions in the muffin tin model. Clearly, this seems questionably appropriate, at least for some cluster geometries.

Johnson and Messmer have emphasized the rapid convergence of Xα-calculated properties toward bulk values with increasing cluster size, whereas Baetzold pointed out that with EH and CNDO some calculated properties do not approach experimental bulk values, even for clusters of dozens of atoms. Both cite experimental results with which agreement is claimed.

Until recently Baetzold's strongest arguments have had to do with the binding energy per atom, which is definitely size dependent to quite a large size. However, even on the basis of classical theory, this is the expected result and not a strong argument in favor of either calculation method.

Johnson and Messmer have emphasized general similarities between their Xα-calculated density-of-state functions and the calculated bulk band structure. Baetzold [323] has recently reported a method of cyclic manipulation of the boundary conditions of a CNDO calculation so as to arrive at the limit of the calculated properties for essentially an infinite solid. Some of his calculated density-of-state functions for clusters of various sizes are shown in Fig. 3.23. These curves are produced by a Gaussian broadening technique from the discrete levels calculated in this way. It can be seen that there are large and apparent changes over the range of the first few atoms but that significant differences still exist between a 39-atom cluster and the infinite solid.

In view of the controversies still existing, there is merit in the opinion expressed by Blyholder [328] that all types of approximate calculations be compared and, in the absence of direct experimental confirmation, greatest confidence be placed on those results for which agreement is obtained with several approximate methods.

CALCULATED CLUSTER PROPERTIES

Studying isolated clusters of various metals, Baetzold [313, 315] concluded from his calculations that the most stable geometry is three-dimensional for Pd and Ni but linear for Ag, Cd, and Na up to sizes of several tens of atoms. He drew attention to the experimental ESR results of Eachus and Symons [47] on silver radicals frozen in aqueous sulfuric acid. They reported an ESR signal which they identified as arising from an Ag$_4^+$ species. The symmetry of the signal indicates two inequivalent pairs of silver nuclei consistent with a linear configuration. A reinterpretation of the signal has recently been given by Forbes and Symons [329]. The preference for linear geometry is contrary to such intuitive stability criteria as total bond energy, total occupied volume, and surface energy, and Baetzold rationalized the result by an argument involving the parity of antibonding orbitals. The calculated bond length indicates a contraction at small sizes, in agreement with most experimental observations and contrary to classical predictions (see section on structural calculations above). For monovalent elements there is a marked oscillation of the energy levels between odd and even sizes as the occupancy of the uppermost occupied level alternates between one and two. Superimposed on the oscillation is a general trend for the ionization potential to decrease and the electron affinity to increase with increasing cluster size, converging toward a value taken as the calculated bulk Fermi level. Predictions are that the electronic bandwidths in palladium clusters approach those of the bulk at smaller sizes than do those of silver.

The type of results obtained is illustrated by Fig. 3.24, which gives calculated ionization potentials for sodium clusters with sizes between

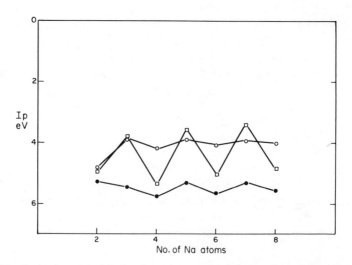

Fig. 3.24. Ionization potential of sodium clusters as a function of cluster size: (□) CNDO calculation by Baetzold and Mack [317]; (●) extended Hückel calculation by the same authors; (○) experimental determination of Robbins, Leckenby, and Willis [36].

Table 3.3 Calculated d-Band Holes in Palladium Clusters (Extended Hückel Method [271])

Size (No. of Atoms)	Geometry	Numbers of d Holes/Atom
2	—	1.0
3	Triangle	0.67
4	Tetrahedron	0.52
5	Trigonal bipyramid	0.41
6	Octahedron	0.35
7	Pentagonal bipyramid	0.33
8	Cube	0.35

two and eight atoms, using both EH and CNDO methods [317], compared with the experimental results of Robbins, Leckenby, and Willis [36]. This series represents one of the very few for which experimental data exist over this large a size range. This figure shows that the absolute values are better for CNDO than for EH but that the odd–even oscillations are much larger than the experimental ones. With either method the oscillations are larger than any overall trend and introduce some uncertainty about the size dependence. It appears, however, that in this size range the EH calculations predict almost no overall change in ionization potential but that a slight increase is indicated by CNDO.

Because of the directionality of d orbitals, their overlap with the orbitals on neighboring species can be high. They are expected to be important in adsorption and catalysis and have been emphasized in the molecular calculations. Baetzold [313] found that empty d states occur in isolated palladium aggregates with as few as two atoms, as indicated in Table 3.3. For silver, the $4d$ levels lie so far in energy below the $5s$ levels that they remain filled and have only a minor effect on the bonding.

Comparing the calculated properties of fcc and icosahedral clusters, Baetzold [321] reported that for palladium the icosahedral structure is the more stable but that the reverse is true for silver. The calculated ionization potentials are lower for icosahedral particles but only very slightly so (0.2 eV for the noble metals), and the occupancy of d states of palladium clusters is calculated to be the same for both forms.

Blyholder [330] calculated the properties of nickel clusters up to 13 atoms in size using the CNDO method. He found the stability of various geometries to fall in the order three-dimensional > planar > linear. He arbitrarily parameterized to make Ni_6 properties match those of the bulk and observed the trends with size of the orbital structure. The highest occupied level remained relatively fixed, whereas the lowest unoccupied level was monotonically lowered with increasing size.

Baetzold also calculated the effects of simplified substrate models on the electronic properties of Pd aggregates. The results depend strongly on the ionization potential of the substrate material. For carbon [315], silver aggregates are found to transfer electrons to the substrate, but palladium aggregates accept electrons, causing the otherwise empty d states to be filled in all atoms in contact with the carbon. Only if the palladium aggregate assumes three-dimensional geometry do d-band holes appear. In a two-dimensional palladium nucleus, the criterion for empty d states is that the Pd $5s$ orbitals must be lower in energy than the highest occupied levels of the substrate.

Isolated Ni clusters have geometric and electronic properties similar to those of palladium but have smaller fractional occupancy of the d states [315]. In a Ni–Pd mixed aggregate, electrons transfer from the nickel to the palladium, and all vacant d states have the character of the nickel levels. Baetzold and Mack [320] found that in a mixed aggregate of Ni–Cu, the copper donates electrons to the nickel, causing a decrease in the number of d-band holes as the copper fraction increases. The ionization potential is constant for lower copper fractions but then decreases by about 1 eV with higher copper content. Alloy clusters of silver and palladium behave similarly. Each silver atom contributes 0.3 to 0.4 electron to the filling of palladium d-band holes. A similar drop in ionization potential occurs at higher silver fractions.

Davidson and Fain [331] applied EH calculations to clusters of 13, 43, 79, and 135 atoms (closed fcc shells) of the noble metals and their alloys. They found that the size dependence of the ionization potential depended upon the basis set chosen for the calculation. In Ag–Au and Cu–Au clusters, the ionization potential increases with gold content but shows deviations from a linear compositional dependence which are affected by the inclusion of d orbitals in the basis set.

The most extensive results of applying the SCF-Xα-SW approach to metal clusters are those reported by Messmer, Knudsen, Johnson, Diamond, and Yang [326]. They determined the electronic structures of eight-atom cubes and 13-atom cubo-octahedra of copper, nickel, palladium, and platinum. Their results for nickel are compared with band-structure calculations in Fig. 3.25. In the copper models, the d band is completely filled and lies well below the Fermi level; but in nickel, palladium, and platinum, the Fermi level intersects the d band, leaving in each case the uppermost d states unfilled. The orbital degeneracy at the Fermi level in nickel clusters results in unpaired spins, leading to a net

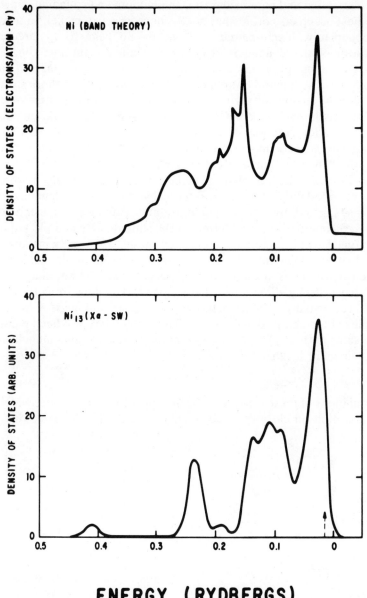

ENERGY (RYDBERGS)

Fig. 3.25. Density of states for nickel using a bulk band structure calculation and a 13-atom cluster calculation, using the SCF-X α-SW method. From Messmer, Knudsen, Johnson, Diamond, and Yang [326].

spin polarization and paramagnetism. In platinum and palladium clusters, the highest occupied d orbital has not the degeneracy of nickel, so that it is fully occupied with spin-paired electrons and no magnetism is predicted. The width of the d band increases systematically through the series nickel, palladium, platinum, and in each case there are localized d states split off from the top and bottom of the band, which the authors speak of as the cluster analogs of surface states. The general similarities of these properties to those of bulk metals are emphasized, although the calculated higher ionization potential of the metal clusters (~7 eV vs. ~5 eV) and reduced paramagnetic magneton number of Ni clusters are both attributed to the small size and large effective surface.

Thirteen-atom clusters of iron, nickel, and copper arranged so as to represent a model that compares a stepped surface with a plane one have been investigated by Jones, Jennings, and Painter [332] using SCF-$X\alpha$-SW calculations. They determined the state density and plotted orbital charge density contours for each of these cases, showing that the stepped surface exhibits bonding orbitals not present on the flat surface. The density of states at the Fermi level is greatest for iron, and iron and nickel have strong directional charge lobes extending from the steps, which are virtually absent in the copper clusters. Figure 3.26 shows examples of the charge density maps. The method has also been modified by the same authors [333] to apply to manganese and vanadium clusters, with their more localized d orbitals.

ADSORPTION TO CLUSTERS

Calculations of these types have been applied somewhat more widely to gas adsorption on clusters, though some of these studies have directed

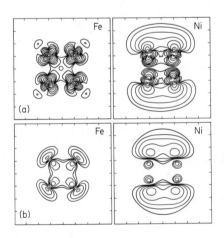

Fig. 3.26. Orbital charge densities for states near the Fermi energy for a model consisting of a four-atom terrace on the cubic face of iron and nickel surfaces: (a) the plane through the four terrace atoms; (b) a plane above those atoms by half of the step height. From Painter, Jennings, and Jones [322].

almost no critical attention to the calculated properties of the metal aggregates themselves. In a series of papers published in 1971, Bennett, McCarroll, Messmer, and Singal [334–337] used both EH and CNDO techniques to investigate the adsorption of a number of gas atoms to a graphite surface. Initially they reported [334] the results of EH methods, which predicted that the preferred adsorption site for atomic H is directly over a carbon atom. The indications were that only a very low barrier exists for the motion of the adsorbed atom along the line connecting nearest-neighbor carbons but that the center of the hexagonal graphite cell is energetically the least-favored site. Subsequently an improvement in the graphite model was made, and the CNDO method was used [335–337] to investigate not only hydrogen adsorption but also carbon, nitrogen, oxygen, and fluorine. Results indicated that the strength of the binding increases in the order H, F, O, N, C. Preferred adsorption sites were over the center of the hexagon for carbon and nitrogen but in a bridged configuration between two nearest-neighbor carbons for hydrogen, fluorine, and oxygen. The revised stable position for the hydrogen atom was attributed to the change in the manner of treating the graphite model rather than to the calculation method. The work function was predicted to be decreased by the adsorption of hydrogen or nitrogen but increased by carbon, oxygen, or fluorine.

The adsorption of atomic hydrogen on nickel and copper surfaces was explored by Fassaert, Verbeek, and van der Avoird [338] using EH techniques. They found that an adsorption site directly over a metal atom was more favorable than bridged or pyramidal arrangements. The adsorption bond was even stronger for copper than for nickel, and bond order considerations confirmed that, contrary to much speculation, the adsorption involves principally the $4s$ orbitals and that the d orbitals of nickel are relatively unimportant. Anders, Hansen, and Bartell [339] used similar techniques to study hydrogen on a tungsten(100) surface and also found the site directly over a substrate atom to be most favorable.

Carbon monoxide adsorption on nickel was investigated by Robertson and Wilmsen [340] using EH calculations. They found that a bridged arrangement with the carbon atom nearest the surface was the most energetically favorable but that a location directly over a nickel atom, with the adsorbed molecule in the same orientation, was nearly as advantageous. Surprisingly the energy minimum in the bridged configuration does not occur at the midpoint between nickel atoms but at a site slightly off center. Blyholder [328, 341] used CNDO to study the same system: carbon monoxide on nickel. In contrast, he found the most stable bonds to be either a bridge bond with the carbon atom directly between two nickel atoms or a multicenter bond with the carbon over a hole formed by three or four nickel atoms. He found that the bonding occurs principally through the nickel s and p orbitals and that the d orbitals are relatively unimportant.

Baetzold [318] investigated the dissociative adsorption of the H_2 molecule by clusters of a variety of metals. He found that there is a strong adsorptive bond and a small activation barrier for dissociation on palladium but no tendency to adsorb on silver or the other noble metals. He established that the criterion for adsorption has to do with the degree of occupancy of the diffuse s orbitals of the metal cluster, which become destabilized upon adsorption of H_2. The dissociative activation barrier results as the $1s$ molecular orbital of the H_2 destabilizes with increasing separation. When the separation becomes great enough, an energetically favored transfer of electrons to the antibonding orbital of the adsorbate occurs, forming the hydride species. Baetzold reports little effect on results by a change in the metal cluster geometry or size, in the range from two to nine atoms.

Itoh [342] used EH calculations to investigate the adsorption of hydrogen on nickel and copper clusters. He concluded that the $4s\,4p$ orbitals control the adsorption on both metals and that dissociation on nickel involves an interaction with the $3d$ orbitals.

Johnson and Messmer [324] used the SCF-Xα-SW method to investigate the adsorption of oxygen on nickel at high coverages. From models of linear NiO, bridged Ni_2O, and pyramidal Ni_4O they obtained state densities which they compared with experimental UPS data of Eastman and Cashion [343]. None of these agreed, and they were prompted to assume that at high coverage the oxygen penetrates the nickel lattice, forming an oxide layer. They found good agreement between the principal calculated features of a $NiO_6{}^{10-}$ cluster and the experimental data.

CORRELATION WITH EXPERIMENT

Experimental data suitable for direct comparison with cluster calculations are not extensive. The gas-phase determinations of dissociation energies of homonuclear dimers have been discussed in Section 2 of this chapter. Ionization potentials and excitation energies of many dimeric species have also been measured, either in the gas phase or in an inert matrix. Although these data are fairly extensive, they are used collectively for parameterization in the calculations and do not provide an independent check, except for internal consistency of the parameters. Calculated values for properties of dimers as given by Baetzold [312] using both EH and CNDO methods and by Cooper, Clarke, and Hare [344] using EH are given in Table 3.4. The experimental values used by these authors for comparison are also listed in the table. Specific references to the original experimental papers appear in these two articles.

The binding energy determinations of Freund [206] for larger particles of iron condensing from the gas phase and of Konstantinov, Panov, and Malinowski [88] for silver particles in equilibrium with an ionic solution have also been reported in an earlier section. Neither of these experimental techniques is capable of extension into the few-atom size range, where quantum-mechanical interactions are likely to cause significant deviation from the classical Gibbs–Thomson–Frenkel thermodynamic concepts.

The odd–even oscillations in the abundance of polynuclear secondary ions produced by ion bombardment (see Section 2 and Fig. 3.2) reflect oscillating binding energies. Although such an effect is qualitatively predicted by the quantum-mechanical cluster calculations, it is also rather simply rationalized on the basis of qualitative orbital occupancy concepts. It is not apparent that a more quantitative confirmation of specific calculations can be obtained from such experiments.

Existing data on the electronic structure of small clusters include the measurements [36] of ionization potential for sodium, mentioned previously and shown in Fig. 3.25, and similar data on selenium clusters by Hoareau et al. [345]. The selenium measurements include determinations of ionization potential and heats of sublimation of the various species, and these have been compared by the same authors [346] with calculated values. Extended Hückel methods were found to give poor agreement with experiment, but CNDO calculations were much more satisfactory, provided d orbitals were included along with the s and p orbitals in the basis set. Experimental and calculated results are given in Fig. 3.27. These authors also take a novel approach to the presentation of data by giving the total energy divided by one less than the number of atoms in the cluster. The rationale for this viewpoint is that this quantity more nearly represents the energy per bond, which may have more fundamental significance than the binding energy per atom. It is interesting to note that this quantity decreases with increasing size, as opposed to the binding energy per atom, and approaches a bulk value at smaller cluster size.

The photoemission studies of Eastman and Grobman [303] using thin copper deposits on palladium and silver substrates were discussed in a previous section. These represent one of the earliest investigations of the density of states of metal samples of small dimensions. More recently Mason, Baetzold, Gerenser, and Lee [94, 99, 347] have reported occupied-state density determinations from X-ray photoelectron spectra (XPS) of vapor-deposited clusters of silver, gold, platinum, and palladium. The cluster size distributions of the samples were determined, and the measurements cover a range from that consisting of essentially single atoms to others dominated by particles tens of angstroms in size. They noted how the threshold emission energy, the energy of the core

Table 3.4 Calculated and Experimental Properties of Homonuclear Diatomics

Dimer	Method	Bond Distance (Å)	Binding Energy (eV)	Ionization Potential (eV)	Excitation Energy (eV)	Vibrational Frequency (cm^{-1})	Ref.
Ag_2	EH	3.2	2.60	9.86	2.98	313	312
	CNDO	3.0	2.60	7.23	3.80	500	
	Exptl.	2.5	1.63	< 7.6	2.7	192	
Cu_2	EH	2.1	4.14	9.21	0.24	390	312
	CNDO	3.0	2.17	7.21	3.07	510	
	Exptl.	2.22	1.98	< 6.7	2.70	265	
Au_2	EH	2.75	3.24	10.68	3.65	328	312
	CNDO	3.25	2.46	7.67	2.87	500	
	Exptl.	2.47	2.24	< 9.2	2.94	142	
Pd_2	EH	2.00	1.25	7.94	0.37	264	312
	CNDO	2.50	1.40	9.22	0.10	410	
	Exptl.	2.57	1.10	< 8.3	—	344	
Na_2	EH	3.6	1.79	6.04	1.83	—	312
	CNDO	3.5	1.40	4.92	2.08	—	
	Exptl.	3.07	0.73	4.9	1.85	—	
Cd_2	EH			—————Unstable—————			312
	CNDO	3.0	0.22	9.37	7.70	—	
	Exptl.	2.46	0.09	—	—	—	

Ca₂	EH	4.0	0.29	7.29	5.60	—	312
	CNDO	3.48	0.22	—	—	—	
	Exptl.						
Sc₂	EH	2.20	1.25	5.50	0.34	250	344
	Exptl.	2.50	1.13	5.7	—	230	
Ti₂	EH	2.30	1.88	5.3	0.51	250	344
	Exptl.	2.50	1.40	5.8	—	288	
V₂	EH	2.10	3.05	5.6	—	300	344
	Exptl.	2.32	2.48	5.7	—	325	
Cr₂	EH	1.90	2.6	7.9	0.40	300	344
	Exptl.	2.22	1.90	5.8	—	400	
Mn₂	EH	1.90	1.22	8.0	0.25	325	344
	Exptl.						

Ca_2, Sc_2, Ti_2, V_2, Cr_2, Mn_2 — Unstable

Fe₂	EH	1.90	3.10	8.9	—	—	344
	Exptl.	2.22	1.0	5.9	—	365	
Co₂	EH	2.30	1.63	9.2	3.29	370	344
	Exptl.	2.31	1.72	6.8	—	335	
Ni₂	EH	2.21	2.45	8.9	2.25	370	344
	Exptl.	2.30	2.37	6.6	—	325	
Zn₂	EH	2.6	0.36	7.1	1.83	100	344
	Exptl.	2.5	0.29	8.4	2.79	—	

Fe_2, Co_2, Ni_2, Zn_2 — van der Waals Molecule

135

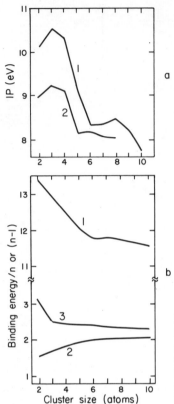

Fig. 3.27. Calculated (CNDO) and experimental values of ionization potential (*a*) and binding energy (*b*) for selenium clusters. Curves 1 show experimental data, and curves 2 and 3, calculated values. In (*b*), curve 2 is the binding energy per atom, whereas curves 1 and 3 give the energy per bond. Taken from the results of Horeau et al. [345, 346].

levels, and the state density in the *d* bands varied with particle size. The spectra of the lowest coverage were in excellent agreement with predictions based on gas-phase spectroscopic data of the atomic species, as shown in Fig. 3.28 [347]. As the particle size increased, the spectrum changed as illustrated in Fig. 3.29 [99]. The Fermi level shifted by about 1 eV toward the vacuum level, the core levels were displaced much less in the same direction, and a splitting appeared in the *d* levels, attributable to the combined effects of spin-orbit and crystal-field interactions. The transitions occurred over the cluster size range of one to ten atoms. These results were compared with CNDO calculations [94], which show some of the same features.

XPS studies of thin gold deposits on substrates of aluminum, sapphire, and SiO_2 were reported by Liang, Salaneck, and Aksay [348]. Their results are similar to those of Mason and Baetzold [94]. Some splitting of the *d* levels is apparent even at the lowest detected gold levels, but as the

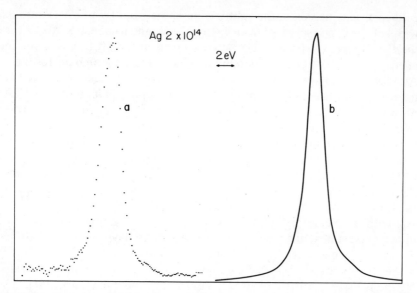

Fig. 3.28. Experimental XPS data (*a*) from a vacuum-deposited silver sample consisting principally of isolated atoms, and the calculated spectrum (*b*), based on atomic spectroscopic data. From Mason [347].

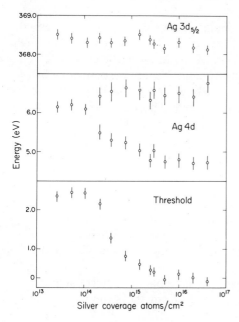

Fig. 3.29. Changes in the XPS features of silver with deposit thickness (and therefore particle size). Shown are the $3d_{5/2}$ core level, the splitting of the 4*d* level, and the shift of the emission threshold. From Mason and Baetzold [94].

137

gold coverage increases, the splitting increases from 1.9 to 2.5 eV and the width of the band increases from 4.4 to 5.7 eV, and its center shifts toward the Fermi level by 1.5 eV. The $4f$ core levels also shift to lower binding energy by 1.1 eV. On the aluminum substrate, they compare results to AuAl alloys and speak of the formation of $AuAl_2$ species at intermediate deposit thicknesses. These authors mention the possibility that level shifts observed by Kim and Winograd [349] for gold implanted in SiO_2, ascribed by them to a matrix perturbation, might actually be a size effect.

Wehking, Beckermann, and Niedermayer [350] used ultraviolet photoelectron spectroscopy to study the state density of very thin silver layers formed by vapor deposition on a (111) silicon substrate. They measured the mean thickness of the silver deposit and concluded that it existed in the form of isolated clusters but made no study of the increase in cluster size with the mean thickness. They reported the evolution of a full d band of occupied states, at a mean thickness of 15 monolayers, from the atomic d states at a mean thickness of about 0.2 monolayers.

Measurements of the XPS valence state density in finely divided platinum were reported by Ross, Kinoshita, and Stonehart [351]. The samples were prepared after the manner of industrial catalysts by high-temperature hydrogen reduction of a platinum salt solution on a silica support. Compared with the spectrum from platinum foil, an accumulation of states more tightly bound than the d levels was observed. These authors attributed them to either surface states or levels arising from bonding to the support.

Mason et al. [99] have reported an interesting comparison between the XPS data for palladium and platinum on carbon substrates. The high-energy edge of the d-band spectrum of platinum particles smaller than about 10 Å is less steep than the instrument resolution, indicating incomplete filling of this band. However, at about that size a sharpening of the edge is observed as the states become filled. These results, illustrated in Fig. 3.30, are evidence that in this configuration, at least, the electronic structure of platinum changes from something like $5d^9 6s^1$ at small sizes toward $5d^{10} 6s^0$ as the size increases. A shift in energy of the core electronic levels is interpreted in the same way. The observed transition in electronic structure occurs at precisely the same size that the platinum loses its catalytic activity for nickel electroless plating. Palladium particles, which are catalytic for this reaction regardless of size, also have only fractional occupancy of the highest d level over the full size range. The correlation is in agreement with claims of the importance of d-band holes in catalysis.

A report by Lytle [352] describes a method for determining d-state occupancy in noble and transition metals from the structure of the L_{III}

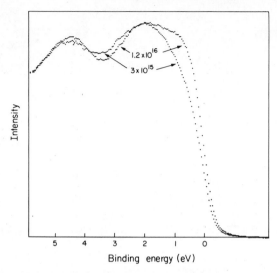

Fig. 3.30. XPS data from vacuum-deposited platinum particles of two sizes. The change in steepness of the d-band edge indicates a difference in occupancy of these levels. From Mason, Gerenser, and Lee [99].

X-ray absorption edge. He points out that this type of structure is distinct from the extended X-ray absorption fine structure (EXAFS) used for structure determinations, although it appears as a bonus in collecting EXAFS data. The absorption probability at the edge reflects the density of unfilled states for $2p$-to-$5d$ transitions. Using the EXAFS spectrometer at the SSRP facility, Lytle obtained data which show that a peak in the bulk metal series Au, Pt, Ir, Ta correlates well with the band structure calculations for unfilled d states of 0, 0.8, 1.9, and 5.8, respectively. Results are also given comparing bulk platinum, α-PtO$_2$, and a supported platinum catalyst sample before and after treatment for 2 hr in flowing hydrogen at 500°C. The reducing treatment decreased the vacant d-state density for reasons about which Lytle could only speculate.

Tanner, Sievers, and Buhrman [353] measured absorption spectra of fine particles of copper, aluminum, tin, and lead in the far-infrared (5 to 50 cm^{-1}) spectral region, where transitions between electronic levels just around the Fermi level should occur. They observed a structure such as that shown in Fig. 3.31, which they compared with the statistical level distribution schemes of the nonquantum-mechanical treatments. They derived mean level spacings which they considered to be in satisfactory agreement with the theory. It is not clear whether or not more specific

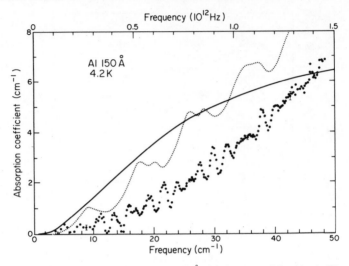

Fig. 3.31. Infrared absorption spectrum of 150-Å aluminum particles (dots). The solid and broken curves are calculated using two approximations to the ensemble of energy levels. From Tanner, Sievers, and Buhrman [353].

conclusions about state densities can be obtained from such measurements.

Chemical and Catalytic Activity

In view of the many differences between the fundamental properties of small clusters of metal atoms and those of the corresponding bulk metals, it is no wonder that there are size-dependent chemical and catalytic properties as well. On the contrary, the surprising thing is the large number of documented reports of research efforts to study such a size dependence in which none is found, beyond the trivial proportionality to reactive surface area. Perhaps part of the reason is the lack of capability for producing and characterizing particles in the sensitive size range in a form suitable for such studies. With recent advances in this area, results of future studies of this type will be of increasing interest.

Straightforward changes in bond energy with size affect the chemical reactivity. Examples of experimental results dealing with such effects are mostly from the range of sizes large enough that the concept of surface energy has been applied. Measurements of changes in reduction and oxidation potential have been noted previously [88]. More recently, matrix isolation studies with noninert carriers have documented differences in the reactivity of gas-phase metal atoms and bulk metals, and an entire new practice of low-pressure gas-phase synthesis has arisen [41–44, 354, 355]. Single atoms and also metal–atom pairs have been found to react to form chemical species that cannot be produced by more conventional synthetic methods.

The size effects of metallic particles used as catalysts have been studied extensively, and several recent review articles are devoted exclusively to this subject [70, 71, 209, 356–362]. Complete references to the early work in the field can be found in these summaries. More recent studies include those of Dalla Betta and Boudart [72], Ostermaier, Katzer, and Manogue [363], Kobayashi and Shirasaki [364], Nakamura, Yamada, and Amano [78], Vannice [365], Corolleur and co-workers [366], and Dartigues, Chambellan, and Gault [367].

Original controversies about the existence of size effects arose because only certain catalyzed reactions are affected by the size-dependent properties of catalysts, whereas others are not, at least in the size range normally studied. Boudart and co-workers [368] first recognized the distinction, and they coined the terms *demanding* and *facile,* respectively, for the two classes, also frequently designated as structure sensitive and structure insensitive. Further disagreement in results can be found, however, and there are obvious questions concerning the validity of the size characterizations and about the influence of other variables assumed to be unimportant. Some studies rely on comparisons between supported catalysts on different supports or even with unsupported metals, and in others the particle size is changed by sintering at different temperatures. Changes in factors other than size can be imagined. When direct electron microscopy is used for size data, the lower size limit rarely goes below 15 Å. A few studies have been extended to smaller sizes, as deduced from measurements of the property which catalyst chemists call *dispersion,* namely, the fraction of the atoms available at the surface for gas adsorption. This practice depends upon the gas adsorption itself being structure insensitive, and considerable attention has been directed to this question [369]. It is generally agreed that platinum dispersion is accurately measured by hydrogen adsorption, and although substrate effects and sample history can affect the results of oxygen adsorption studies, the titration of adsorbed oxygen with hydrogen is reported to be reliable and to have increased sensitivity over simple hydrogen adsorption.

Many catalyzed chemical reactions studied over ranges of catalyst particle sizes from sometimes 10 or 20 Å to bulk have been reported to demonstrate a strict proportionality between reaction rate and surface area and hence are structure insensitive. For catalysts of the platinum metal group, these include hydrogenation of 1-hexene, cyclohexene, benzene, and allyl alcohol, dehydrogenation of cyclohexane and isopropanol, hydrogenolysis of cyclopropane and cyclopentane, and hydrogen–deu-. terium exchange. In these reactions, apparently every surface atom on the catalyst has equal activity.

Over the same size range, hydrogenolysis reactions involving the breaking of carbon–carbon bonds and certain skeletal isomerization reactions have been reported to be structure sensitive, the specific activity (i.e., activity per unit surface area) of platinum, nickel, and rhodium catalysts decreasing markedly with increasing particle size. In many studies it happens that a structure-sensitive and a structure-insensitive reaction are operating in parallel, with the result that the selectivity of a catalyst for a certain product is a strong function of particle size. A recent example of such effects is shown in Fig. 3.32, taken from the results of Dartigues, Chambellan, and Gault [367]. Using a series of seven Pt–Al_2O_3 catalysts with particle sizes varying between 8.5 and 170 Å, as measured by H_2 adsorption, they studied the selectivity of the hydrogenolysis of methylcyclopentane, and used ^{13}C labeling to investigate the isomerization of 2-methylpentane into 3-methylpentane and the isomerization of n-hexane into 3-methylpentane. The sharp break in the region of the 20 Å particle size is evident from their data.

Interpretations of data of this type have relied heavily on surface geometric features, emphasizing the size-dependent concentrations of corners, edges, and other sites with coordination different from those on a plane surface. The B_5 sites of van Hardeveld and co-workers [210, 357]

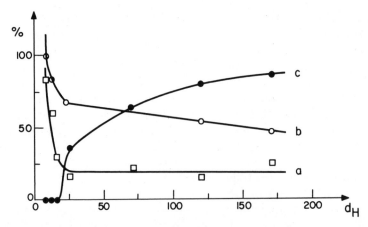

Fig. 3.32. Changes in selectivity of platinum catalysts as a function of particle size. The reactions are the isomerization of 2-methylpentane to 3-methylpentane (*a*), cyclic isomerization of *n*-hexane to 3-methylpentane (*b*), and selective hydrogenolysis of methylcyclopentane (*c*). From Dartigues, Chambellan, and Gault [367].

(see section on surface-structural effects above) are frequently invoked. Wells [361] has proposed the generalization that most structure-sensitive reactions are those that require multiple bonding between an adsorbed species and the substrate.

The importance of selective surface sites is confirmed by the extensive studies of Somorjai and his co-workers [370] on stepped surfaces of platinum and other metals. Plane surfaces are cut and polished on macroscopic metal crystals at controlled low angular inclinations from low-index crystallographic planes. LEED measurements show that such surfaces are composed of well-ordered arrays of monatomic surface steps and kinks whose density is predictably determined by the rotations of the surface out of the low-index plane. Comparative experiments [371] indicate that for many reactions the stepped surfaces have far higher activity than do the corresponding nonstepped planes.

The general similarity of these results to particle size effects is evident, though it is not yet completely clear that surface features are the whole story of size-dependent effects. Some disagreements, in fact, are found. For example, Poltorak and Boronin [372] found H_2-D_2 exchange to be structure insensitive on platinum catalyst particles over the size range from 10 to 50 Å, certainly covering that for which surface defects are expected to change, whereas Bernasek and Somorjai [371] reported an increase in rate of 10^3 to 10^4 on a stepped surface compared with a (111) surface (argued by Wachs and Madix [373] to indicate an enhancement exchange probability of about 20 times, in agreement with results of Lu and Rye [374]).

Dartigues, Chambellan, and Gault [367] argue that the size-associated changes in catalytic properties revealed in their data are too abrupt to be accounted for by differences in surface defect concentrations. They speculated about a possible structural change involving, for example, icosahedral forms. In connection with the abruptness of the change in properties, Luss [375] has considered the statistical effects of the size distribution on observed results. The size-dependent change in occupancy of the d levels of platinum particles, observed by Mason et al. [99], was discussed in the section on correlation with experiment. This change does occur in just the size range over which most of the reported catalytic effects are observed. If this electronic effect proves more general, it must certainly be considered as likely as the simple geometric factors in controlling the catalytic activity. Although calculations [278, 332, 333] have indicated a change in the directionality of d orbitals at models of kinks, there is no indication that the occupancy could be itself controlled by the concentration of surface defects. Partial charge exchange with the support, however, cannot be ruled out.

There are a few documented reports of structure sensitivity of the opposite sense, that is, with decreasing activity for small particles. Burton and co-workers [376] reported such behavior for the electroreduction of oxygen on platinum (though they allowed for the possibility of a substrate effect), and Pusateri, Katzer, and Manogue [377] observed this type of size dependence for the platinum-catalyzed reduction of nitric oxide by ammonia in the presence of oxygen, as did Ostermaier, Katzer, and Manogue [363] for oxygen reduction of ammonia. Selective oxygen poisoning of highly dispersed platinum was reported by Boudart and his associates [368], and poisoning effects must be suspected in all of the above results.

Hill and Selwood [378] found an indication of a minimum effective size of nickel catalysts for benzene hydrogenation, though their results are subject to question. Yates and Sinfelt [379] reported a loss of activity of extremely finely divided rhodium for ethane hydrogenolysis and discussed the results in terms of a "loss of metallic character" at small sizes.

Photographic development is a common catalyzed chemical reaction for which there is abundant evidence indicating a minimum cluster size for the silver catalyst, the latent image. Because of the practical importance of this phenomenon, a number of model experiments have been devised to confirm and investigate it. In the early 1930s Reinders and co-workers [380] approached the question using silver clusters formed by vacuum deposition and a solution of a reducing agent and a soluble silver salt, a formulation termed a physical developer. After a suitable reaction time, they counted the number of clusters that the deposited silver metal had made visible by microscopy and statistically analyzed the dependence of their counts on the amount of silver in the vacuum deposits. They arrived at the conclusion that the minimum size for a cluster to possess catalytic activity was four atoms. However, their analysis was viewed skeptically for it depended upon a number of unverified assumptions about the nucleation of the vacuum deposits. The general approach has been revived in recent years, now taking advantage of electron microscopy and a thorough characterization of the size distribution in the vacuum deposits. The experiments of Konstantinov, Panov, and Malinowski [88] were discussed in Section 3 of this chapter. They used silver nuclei in the size range of 10 Å and larger and compared size–frequency data before and after a time in the reacting solution. The data revealed a critical size below which silver from the nuclei was oxidized, passing into solution, and above which silver ions from the solution were catalytically reduced on the existing particles. The reduction–oxidation potential of the solution was varied (always much less strongly reducing than in practical photographic development), and the dependence of the critical size on the

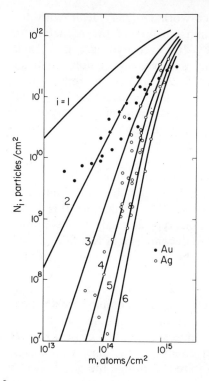

Fig. 3.33. Particle count data used for the determination of the minimum catalyst size for catalytic deposition of silver ion from solution on silver and gold catalysts. The minimum size for silver is approximately four atoms, for gold, two atoms. From Hamilton [95].

potential was analyzed to determine a surface energy. Similar but less extensive experiments were reported by Galashin, Senchenkov, and Chibisov [381].

Hamilton and Logel [92, 95] used a more active reducing solution on subvisible vacuum deposits of silver and gold which had been size-characterized as discussed in the section on vacuum deposition. They counted enlarged centers and compared the data with their calculated size distributions to determine a minimum active cluster size, as illustrated in Fig. 3.33. For their conditions they concluded that the critical size for silver clusters is approximately four atoms, whereas that for gold is distinctly smaller, probably two atoms.

Konstantinov, Panov, and Malinowski [88] analyzed their results purely in terms of the thermodynamic stability of the particles as affected by the surface energy, and Hillson [382] also considered the process using these concepts. Other proposals dealing more with kinetic factors determined by the changing electronic structure have been made [383], however, and the data are too incomplete for decisive conclusions at this time.

A related study by Kirstein and co-workers [384] dealt with the catalysis by vacuum-deposited palladium clusters of the reduction of palladium ion from aqueous solution by dissolved hydrogen gas. Their conclusion was that three or four atoms were required before a cluster became catalytic. Hamilton and Logel [89, 95] studied the reduction of nickel ion from solution (the electroless plating process) on silver and palladium vacuum deposits and found that the minimum catalyst size depended upon the reducing agent. They determined that with a sufficiently active reducing agent, single silver atoms were catalytic. These methods all depend upon reaction to produce an insoluble product that deposits upon the original catalyst nucleus, enlarging it to visible size so that microscopic counts can be made. They are therefore limited to autocatalytic reactions, essentially the reduction of metal ions. It is possible, however, to use vacuum deposits of this type as catalysts in solution- or gas-phase reactions and to analyze the dependence of reaction rate on metal deposit coverage to determine the minimum cluster size for catalysis.

There is reason for the optimistic viewpoint that more extensive catalytic data in this range of very small cluster size, compared with measurements and calculations of the thermodynamic and electronic properties of the same clusters, may aid in understanding the critical factors involved in heterogeneous catalysis of chemical reactions. Certainly a focus on this range of sizes will be required to bridge the gap between heterogeneous and homogeneous catalysis by single metal atoms and their * complexes and by the organometallic cluster compounds, the catalytic properties of which are just beginning to be exploited.

References

1. R. M. Henkelman and F. P. Ottensmeyer, *Proc. Natl. Acad. Sci. USA*, **68,** 3000 (1971).
2. R. F. Whiting, F. P. Ottensmeyer, and F. C. Nachod, *Angew. Chem. Int. Ed.*, **13,** 536 (1974).
3. H. Hashimoto, A. Kumao, K. Hino, H. Yatsumoto, and A. Ono, *Jap. J. Appl. Phys.*, **10,** 1115 (1971); *J. Electron Microsc.*, **22,** 123 (1973); H. Hashimoto, *ibid.*, **24,** 56 (1975).
4. K. Kanaka, T. Oikawa, S. Ono, K. Hojou, and K. Adachi, *J. Electron Microsc.*, **24,** 218 (1975).
5. J. R. Parsons, H. M. Johnson, C. W. Hoelke, and R. R. Hosbons, *Phil. Mag.*, **27,** 1359 (1973).
6. K. Mihama, A. Horata, and R. Uyeda, *Jap. J. Appl. Phys.*, **13,** 377 (1974);

K. Mihama, S. Shima, and R. Uyeda, *J. Crystal Growth*, **24/25**, 323 (1974); K. Mihama, *J. Electron Microsc.*, **24**, 57 (1975); N. Tanaka, K. Mihama, and R. Uyeda, *ibid.*, **24**, 213 (1975); K. Mihama and N. Tanaka, *ibid.*, **25**, 65 (1976).

7. F. Nagata, T. Matsuda, and T. Komoda; *Jap. J. Appl. Phys.*, **14**, 1815 (1975); F. Nagata, T. Matsuda, T. Komoda, and K. Hama, *J. Electron Microsc.*, **25**, 237 (1976).

8. E. B. Prestridge and D. J. C. Yates, *Nature*, **234**, 345 (1971).

9. A. Crew, *J. Microsc.*, **100**, 247 (1974); M. Retsky, *Optik*, **41**, 127 (1974); M. S. Isaacson, D. Kopf, N. W. Parker, and M. Utlaut, in *Proc. 34th Annual EMSA Meeting, Miami*, 1976, pp. 498, 584; M. S. Isaacson and N. W. Parker, in *Scanning Electron Microscopy/1976*, IIT Research Institute, Chicago, 1976.

10. J. W. Wiggins, M. Beer, S. D. Rose, M. Cole, A. A. Waldrop, J. Zubin, J. W. Platner, L. Marzilli, C. H. Chang, and L. Kapili, in *Scanning Electron Microscopy/1976*, IIT Research Institute, Chicago, 1976.

11. H. Todokoro, S. Nomura, and T. Komoda, *J. Electron Microsc.*, **25**, 191 (1976).

12. J. Drowart, in *Condensation and Evaporation of Solids*, E. Rutner, P. Goldinger, and J. P. Hirth, Eds., Gordon and Breach, New York, 1965, p. 255.

13. B. Siegel, *Q. Rev.*, **19**, 77 (1965).

14. K. A. Gingerich, *J. Cryst. Growth*, **9**, 31 (1971).

15. A. Kant and B. Strauss, *J. Chem. Phys.*, **41**, 3806 (1964); A. Kant, *ibid.*, **41**, 1872 (1964); A. Kant and B. Strauss, *ibid.*, **45**, 3161 (1966); S.-S. Lin and A. Kant, *J. Phys. Chem.*, **73**, 2450 (1969); A. Kant, S.-S. Lin, and B. Strauss, *J. Chem. Phys.*, **49**, 1983 (1968); A. Kant, *ibid.*, **48**, 523 (1968); S.-S. Lin, B. Strauss, and A. Kant, *ibid.*, **51**, 2282 (1969).

16. J. Kordis and K. A. Gingerich, *J. Chem. Phys.*, **58**, 5141 (1973).

17. L. Rovner, A. Drowart, and J. Drowart, *Trans. Faraday Soc.*, **63**, 2906 (1967).

18. F. J. Kohl, O. M. Uy, and K. D. Carlson, *J. Chem. Phys.*, **47**, 2667 (1967).

19. K. A. Gingerich, D. L. Cocke, and F. Miller, *J. Chem. Phys.*, **64**, 4027 (1976).

20. K. A. Gingerich, A. Desideri, and D. L. Cocke, *J. Chem. Phys.*, **62**, 731 (1975).

21. J. Drowart, R. P. Burns, G. DeMaria, and M. G. Ingram, *J. Chem. Phys.*, **31**, 1131 (1959).

22. R. E. Honig, *J. Chem. Phys.*, **22**, 1610 (1954).

23. J. Drowart, G. DeMaria, A. J. H. Boerboom, and M. G. Ingram, *J. Chem. Phys.*, **30**, 308 (1959).

24. E. Clementi, *J. Am. Chem. Soc.*, **83**, 4501 (1959).

25. J. Franzen and H. Hintenberger, *Z. Naturforsch.*, **A16**, 535 (1961).

26. J. Berkowitz and J. R. Marquart, *J. Chem. Phys.*, **39**, 275 (1963).

27. J. Berkowitz and W. Chupka, *J. Chem. Phys.*, **45**, 4289 (1966).

28. A. G. Sigai, *J. Vac. Sci. Technol.*, **12**, 958 (1975).

29. R. E. Leckenby and E. J. Robbins, *J. Phys. B.*, **1**, 441 (1968).
30. A. W. Searcy, R. D. Freeman, and M. C. Michel, *J. Am. Chem. Soc.*, **76**, 4050 (1954).
31. R. E. Leckenby and E. J. Robbins, *Proc. Roy. Soc.*, **A291**, 389 (1966).
32. T. A. Milne and F. T. Greene, *J. Chem. Phys.*, **47**, 4095 (1967).
33. O. F. Hagena and W. Obert, *J. Chem. Phys.*, **56**, 1793 (1972).
34. J. Farges, B. Raoult, and G. Torchet, *J. Chem. Phys.*, **59**, 3454 (1973).
35. J. Farges, *J. Cryst. Growth*, **31**, 79 (1975).
36. E. J. Robbins, R. E. Leckenby, and P. Willis, *Adv. Phys.*, **16**, 739 (1967).
37. P. J. Foster, R. E. Leckenby, and E. J. Robbins, *J. Phys. B*, **2**, 478 (1969).
38. V. E. Krohn, Jr., *J. Appl. Phys.*, **33**, 3523 (1962).
39. G. Hortig and M. Müller, *Z. Phys.*, **221**, 119 (1969).
40. M. Leleyter, Thesis, Université de Paris-Sud, Centre d'Orsay, 1975.
41. L. Andrews and G. C. Pimentel, *J. Chem. Phys.*, **44**, 1361 (1966); *ibid.*, **47**, 2905 (1967).
42. L. Brewer, B. A. King, J. L. Wang, B. Meyer, and G. F. Moore, *J. Chem. Phys.*, **49**, 5209 (1968); L. Brewer and B. King, *ibid.*, **53**, 3981 (1970); L. Brewer and J. L. F. Wang, *J. Mol. Spectrosc.*, **40**, 95 (1971); L. Brewer and C.-A. Chang, *J. Chem. Phys.*, **56**, 1728 (1972).
43. A. Bos and A. T. Howe, *J. Chem. Soc. Faraday II*, **70**, 451 (1974).
44. E. P. Kündig, M. Moskovits, and G. A. Ozin, *Angew. Chem. Int. Ed.*, **14**, 292 (1975); E. P. Kündig, M. Moskovits, and G. A. Ozin, *Nature*, **254**, 503 (1975); H. Huber, E. P. Kündig, M. Moskovits, and G. A. Ozin, *J. Am. Chem. Soc.*, **97**, 2097 (1975); T. A. Ford, H. Huber, W. Klotzbücher, E. P. Kündig, M. Moskovits, and G. A. Ozin, *J. Chem. Phys.*, **66**, 524 (1977); G. A. Ozin, *Acc. Chem. Res.*, **10**, 21 (1977); R. Busby, W. Klotzbücher, and G. A. Ozin, *J. Am. Chem. Soc.*, **98**, 4013 (1976); M. Moskovits and J. E. Hulse, *J. Chem. Phys.*, **66**, 3988 (1977).
45. W. Schulze and H. U. Becker, *J. Phys. (Paris)*, **38**, C2-7 (1977).
46. J. E. Hulse and M. Moskovits, *Surface Sci.*, **57**, 125 (1976); M. Moskovits and J. E. Hulse, *ibid.*, **57**, 302 (1976).
47. R. S. Eachus and M. C. R. Symons, *J. Chem. Soc. (A)*, 1329 (1970).
48. M. J. A. Smith and D. J. E. Ingram, *Proc. Phys. Soc.*, **80**, 139 (1962).
49. D. J. Greenslade, *Br. J. Appl. Phys.*, **16**, 1921 (1965).
50. L. Shields and M. C. R. Symons, *Mol. Phys.*, **11**, 57 (1966).
51. S. C. Jain, N. D. Arora, and T. R. Reddy, *Phys. Lett.*, **58A**, 53 (1975).
52. A. E. Hughes and S. C. Jain, *Phys. Lett.*, **58A**, 61 (1976).
53. P. G. Baranov, R. A. Zhitnikov, and V. A. Khramtsov, *Phys. Status Solidi* (b), **76**, K109 (1976).
54. R. D. Maurer, *J. Appl. Phys.*, **29**, 1 (1958).
55. R. H. Doremus, *J. Chem. Phys.*, **42**, 414 (1965).
56. U. Kreibig and C. v. Fragstein, *Z. Phys.*, **224**, 307 (1969); U. Kreibig, *Appl. Phys.*, **10**, 255 (1976).
57. M. A. Smithard and R. Dupree, *Phys. Status Solidi* (a), **11**, 695 (1972).
58. R. H. Doremus and A. M. Turkalo, *J. Mater. Sci.*, **11**, 903 (1976).
59. M. Creusburg, *Z. Phys.*, **194**, 211 (1966).

60. C. Taupin, *J. Phys. Chem. Solids*, **28**, 41 (1967).

61. C. Taupin and J. Charvolin, *Proceedings Colloque Ampere XIV*, R. Blinc, Ed., North-Holland, Amsterdam, 1967, p. 489.

62. M. A. Smithard and M. Q. Tran, *Helv. Phys. Acta*, **46**, 869 (1974).

63. R. C. McMillan, G. J. King, B. S. Miller, and F. F. Carlson, *J. Phys. Chem. Solids*, **23**, 1379 (1962).

64. R. C. McMillan, *J. Phys. Chem. Solids*, **25**, 773 (1964).

65. M. A. Smithard, *Solid State Commun.*, **14**, 407 (1974).

66. M. A. Smithard, *Solid State Commun.*, **14**, 411 (1974).

67. D. A. Gordon, *Phys. Rev. B*, **13**, 3738 (1976).

68. J. Turkevich, *Am. Sci.*, **47**, 97 (1959); J. Turkevich, J. Hillier, and P. C. Stevenson, *Discuss. Faraday Soc.*, **11**, 55 (1951).

69. J. Turkevich and G. Kim, *Science*, **169**, 873 (1970).

70. G. C. Bond, *Catalysis by Metals*, Academic Press, New York, 1962.

71. P. Wynblatt and N. A. Gjostein, *Progress in Solid State Chemistry*, Vol. 9, J. O. McCaldin and G. Somorjai, Eds., Pergamon, Oxford, 1975; *Acta Met.*, **24**, 1165, 1175 (1976).

72. R. A. Dalla Betta and M. Boudart, *Catalysis; Proceedings 5th International Congress, Miami, 1972*, J. W. Hightower, Ed., North-Holland, Amsterdam, 1973, p. 1329.

73. Y. Kim and K. Seff, *J. Am. Chem. Soc.*, **99**, 7055 (1977).

74. P. C. Flynn, S. E. Wanke, and P. S. Turner, *J. Catal.*, **33**, 233 (1974).

75. G. R. Wilson and W. K. Hall, *J. Catal.*, **17**, 190 (1970); *ibid.*, **24**, 306 (1972).

76. D. Pope, W. L. Smith, M. J. Eastlake, and R. L. Moss, *J. Catal.*, **22**, 72 (1971).

77. J. A. Bett, K. Kinoshita, and P. Stonehart, *J. Catal.*, **35**, 307 (1974).

78. M. Nakamura, M. Yamada, and A. Amano, *J. Catal.*, **39**, 125 (1975).

79. C. G. Granqvist and R. A. Buhrman, *J. Catal.*, **42**, 477 (1976); *ibid.*, **46**, 238 (1977).

80. S. E. Wanke, *J. Catal.*, **46**, 234 (1977).

81. E. B. Prestridge, G. H. Via, and J. H. Sinfelt, *J. Catal.*, **50**, 115 (1977).

82. G. C. Smith, T. P. Chojnacki, S. R. Dasgupta, K. Iwatate, and K. L. Waters, *Inorg. Chem.*, **14**, 1419 (1975).

83. S. Yatsuya, S. Kasukabe, and R. Uyeda, *Jap. J. Appl. Phys.*, **12**, 1675 (1973); K. Kasukabe, S. Yatsuya, and R. Uyeda, *ibid.*, **13**, 1714 (1974); T. Ohno, S. Yatsuya, and R. Uyeda, *ibid.*, **15**, 1213 (1976).

84. C. G. Granqvist and R. A. Buhrman, *J. Appl. Phys.*, **47**, 2200 (1976), and references therein; *Solid State Commun.*, **18**, 123 (1976).

85. S. Yamaguchi, *J. Phys. Soc. Japan*, **15**, 1577 (1960).

86. L. Ward, *Br. J. Appl. Phys.*, **2**, 123 (1969).

87. J. R. Anderson and R. J. MacDonald, *J. Catal.*, **19**, 227 (1970).

88. I. Konstantinov, A. Panov, and J. Malinowski, *J. Photogr. Sci.*, **21**, 250 (1973); I. Konstantinov and J. Malinowski, *J. Photogr. Sci.*, **23**, 1, 145 (1975).

89. J. F. Hamilton and P. C. Logel, *J. Catal.*, **29**, 253 (1973).

90. H. R. Zeller and D. Kuse, *J. Appl. Phys.*, **44**, 2763 (1973).

91. M. Ido, A. Shibukawa, and R. Hoshino, *J. Phys. Soc. Japan*, **34**, 556 (1973).
92. J. F. Hamilton and P. C. Logel, *Photogr. Sci. Eng.*, **18**, 507 (1974).
93. P. Yee and W. D. Knight, *Phys. Rev. B*, **11**, 3261 (1975).
94. M. G. Mason and R. C. Baetzold, *J. Chem. Phys.*, **64**, 271 (1976).
95. J. F. Hamilton, *J. Vac. Sci. Technol.*, **13**, 319 (1976).
96. M. Rasigni, J. P. Gasparini, G. Rasigni, and R. Fraisse, *Solid State Commun.*, **18**, 629 (1976).
97. D. N. Jarrett and L. Ward, *J. Phys. D: Appl. Phys.*, **9**, 1515 (1976).
98. K. S. Liang, W. R. Salaneck, and I. A. Aksay, *Solid State Commun.*, **19**, 329 (1976).
99. M. G. Mason, L. J. Gerenser, and S. T. Lee, *Phys. Rev. Lett.*, **39**, 288 (1977).
100. J. G. Scofronick and W. B. Phillips, *J. Appl. Phys.*, **38**, 4791 (1967); Ya. E. Geguin, Yu. S. Kaganovskii, and V. V. Kalinin, *Sov. Phys.—Solid State*, **11**, 203 (1969); H. Jaeger, P. D. Mercler, and R. C. Sherwood, *Surface Sci.*, **13**, 349 (1969); G. G. Paulson and A. L. Friedberg, *Thin Solid Films*, **5**, 47 (1970); E. I. Tochitski and V. E. Obuhov, *Thin Solid Films*, **32**, 205 (1976); J. J. Metois, K. Heinemann, and H. Poppa, *Appl. Phys. Lett.*, **29**, 134 (1976); K. Heinemann and H. Poppa, *Thin Solid Films*, **33**, 273 (1976).
101. T. Anderson and C. G. Granqvist, *J. Appl. Phys.*, **48**, 1673 (1977).
102. G. Zinsmeister, *Vacuum*, **16**, 529 (1966); *Thin Solid Films*, **2**, 497 (1968); *ibid.*, **4**, 363 (1969); *ibid.*, **7**, 51 (1971); *Kristall und Technik*, **5**, 207 (1970); *Vakuum-Technik*, **22**, 85 (1973); *Jap. J. Appl. Phys.*, **Suppl. 2**, 545 (1974).
103. B. Lewis and D. S. Campbell, *J. Vac. Sci. Technol.*, **4**, 209 (1967); B. Lewis, *Surface Sci.*, **21**, 273, 289 (1970); B. Lewis and G. J. Rees, *Phil. Mag.*, **29**, 1253 (1974); B. Lewis and V. Halpern, *J. Cryst. Growth*, **33**, 39 (1976); B. Lewis and S. Fujiwara, *J. Appl. Phys.*, **47**, 1735 (1976).
104. A. C. Zettlemoyer, *Nucleation*, Marcel Dekker, New York, 1969.
105. V. Halpern, *J. Appl. Phys.*, **40**, 4627 (1969).
106. I. Markov, *Thin Solid Films*, **6**, 119, 281 (1970).
107. M. J. Stowell and T. E. Hutchinson, *Thin Solid Films*, **8**, 411 (1971); M. J. Stowell, *Phil. Mag.*, **26**, 349, 361 (1972); *J. Cryst. Growth*, **24/25**, 45 (1974).
108. R. A. Sigsbee, *J. Appl. Phys.*, **42**, 3904 (1971).
109. R. J. H. Voorhoeve, *Surface Sci.*, **28**, 145 (1971).
110. J. A. Venables, *Phil. Mag.*, **27**, 697 (1973).
111. S. Fujiwara, H. Terajima, and S. Ozawa, *J. Appl. Phys.*, **45**, 4242 (1974).
112. B. Lewis and J. C. Anderson, *Nucleation and Growth of Thin Films*, Academic Press, New York, in press.
113. B. K. Chakraverty, *J. Phys. Chem. Solids*, **28**, 2413 (1967).
114. F. Robertson, Ph.D. Thesis, Stanford University, 1972.
115. J. L. Sacedon and C. S. Martin, *Thin Solid Films*, **10**, 99 (1972).
116. J. F. Hamilton and P. C. Logel, *Thin Solid Films*, **16**, 49 (1973); *ibid.*, **23**, 89 (1974); *ibid.*, **29**, L24 (1975).
117. R. Robrieux, R. Faure, and G. Desrousseaux, *Thin Solid Films*, **31**, 311 (1976).
118. J. A. Venables, *Thin Solid Films*, **18**, 511 (1973).
119. A. G. Kaspaul and E. E. Kaspaul, in *Transactions of the Tenth National*

Vacuum Symposium of the American Vacuum Society, G. H. Bancroft, Ed., Macmillan, New York, 1963, p. 422; H. G. Wehe, in *Electron and Ion Beam Science and Technology*, Vol. 2, R. Bakish, Ed., American Inst. of Mining, Metallurgical and Petroleum Engineers, New York, 1966, p. 813.

120. C. Chapon, C. Henry, and B. Mutaftschiev, *J. Cryst. Growth*, **33**, 291 (1976); C. Henry, C. Chapon, and B. Mutaftschiev, *Thin Solid Films*, **33**, L1 (1976).

121. S. Yatsuya, K. Mihama, and R. Uyeda, *Jap. J. Appl. Phys.*, **13**, 749 (1974).

122. N. Wada and M. Ichikawa, *Jap. J. Appl. Phys.*, **15**, 755 (1976).

123. K. J. Klabunde, H. F. Efner, L. Satek, and W. Donley, *J. Organomet. Chem.*, **71**, 309 (1974); K. J. Klabunde, H. F. Efner, T. O. Murdock, and R. Ropple, *J. Am. Chem. Soc.*, **98**, 1021 (1976); T. O. Murdock and K. J. Klabunde, *J. Org. Chem.*, **41**, 1076 (1976).

124. T. Takagi, I. Yamada, and A. Sasaki, *J. Vac. Sci. Technol.*, **12**, 1128 (1975); T. Takagi, I. Yamada, and A. Sasaki, *Thin Solid Films*, **39**, 207 (1976).

125. S. Ino, *J. Phys. Soc. Japan*, **21**, 346 (1966).

126. K. Kimoto and I. Nishida, *J. Phys. Soc. Japan*, **22**, 940 (1967); K. Kimoto and I. Nishida, *Jap. J. Appl. Phys.*, **16**, 941 (1977).

127. J. G. Allpress and J. V. Sanders, *Surface Sci.*, **7**, 1 (1967).

128. S. Ogawa, S. Ino, T. Kato and H. Ota, *J. Phys. Soc. Japan*, **21**, 1963 (1966); S. Ino and S. Ogawa, *ibid.*, **22**, 1365 (1967); B. G. Bagley, *J. Cryst. Growth*, **6**, 323 (1970); K. Reichelt and H. Schreiber, *Surface Sci.*, **43**, 644 (1974); K. Yagi, K. Takayanagi, K. Kobayashi, and G. Honjo, *J. Cryst. Growth*, **28**, 117 (1975).

129. T. Komoda, *Jap. J. Appl. Phys.*, **7**, 27 (1968).

130. E. Gillet and M. Gillet, *J. Cryst. Growth*, **13/14**, 212 (1972); E. Gillet, A. Renou, and M. Gillet, *Thin Solid Films*, **29**, 217 (1975).

131. C. Sollard, P. Buffat, and F. Faes, *J. Cryst. Growth*, **32**, 123 (1976).

132. S. Ino, *J. Phys. Soc. Japan*, **27**, 941 (1969).

133. J. G. Allpress and J. V. Sanders, *Aust. J. Phys.*, **23**, 23 (1970).

134. R. van Hardeveld and A. van Montfoort, *Surface Sci.*, **4**, 396 (1966).

135. F. E. Shephard, *J. Catal.*, **14**, 148 (1969).

136. N. R. Avery and J. V. Sanders, *J. Catal.*, **18**, 129 (1970).

137. J. J. Burton, *Cat. Rev.—Sci. Eng.*, **9**, 209 (1974).

138. W. Romanowski, *Surface Sci.*, **18**, 373 (1969).

139. F. Fukano and C. M. Wayman, *J. Appl. Phys.*, **40**, 1656 (1969).

140. J. J. Burton, *J. Chem. Phys.*, **52**, 345 (1970); *Nature*, **229**, 335 (1971).

141. D. J. McGinty, *J. Chem. Phys.*, **55**, 580 (1971).

142. M. R. Hoare and P. Pal, *Adv. Phys.*, **20**, 161 (1971).

143. M. R. Hoare and P. Pal, *Nature, Phys. Sci.*, **230**, 5 (1971); *ibid.*, **236**, 35 (1972).

144. M. R. Hoare and P. Pal, *J. Cryst. Growth*, **17**, 77 (1972); *Advan. Phys.*, **24**, 645 (1975).

145. M. R. Hoare and J. McInnes, *Faraday Discuss. Chem. Soc.*, **61**, 12 (1976).

146. J. J. Burton, *J. Chem. Phys.*, **52**, 345 (1970).

147. J. J. Burton, *Chem. Phys. Lett.*, **3**, 594 (1969); *ibid.*, **7**, 567 (1970); *J. Chem. Phys.*, **56**, 3133 (1972).

148. J. M. Dickey and A. Paskin, *Phys. Rev. Lett.*, **21**, 1441 (1968); *Phys. Rev. B*, **1**, 851 (1971).
149. K. Nishioka, R. Shawyer, A. Beinenstock, and G. M. Pound, *J. Chem. Phys.*, **55**, 5082 (1971).
150. W. D. Kristensen, E. J. Jensen, and R. M. J. Colterill, *J. Chem. Phys.*, **60**, 4164 (1974).
151. N. Metropolis, A. W. Rosenbluth, M. N. Rosenbluth, A. H. Teller, and E. Teller, *J. Chem. Phys.*, **21**, 1087 (1953).
152. R. D. Etters and J. Kaelberer, *Phys. Rev.*, **A11**, 1068 (1975); J. B. Kaelberer and R. D. Etters, *J. Chem. Phys.*, **66**, 3233 (1977); R. D. Etters and J. Kaelberer, *J. Chem. Phys.*, **66**, 5112 (1977).
153. D. J. McGinty, *J. Chem. Phys.*, **58**, 4733 (1973).
154. C. L. Briant and J. J. Burton, *J. Chem. Phys.*, **63**, 204 (1975); C. L. Briant, *Faraday Discuss. Chem. Soc.*, **61**, 25 (1976).
155. R. D. Etters, L. Kanney, N. S. Gillis, and J. Kaelberer, *Phys. Rev.*, **B15**, 4056 (1977).
156. A. L. Mackay, *Acta Cryst.*, **15**, 916 (1962).
157. F. W. C. Boswell, *Proc. Phys. Soc. (London)*, **A64**, 465 (1951).
158. C. W. Mays, J. S. Vermaak, and D. Kuhlmann-Wilsdorf, *Surface Sci.*, **12**, 134 (1968).
159. D. Schroeer, R. F. Marzke, D. J. Erickson, S. W. Marshall, and R. M. Wilenzick, *Phys. Rev.*, **B11**, 4414 (1970).
160. T. de Planta, R. Ghez, and F. Piuz, *Helv. Phys. Acta*, **37**, 74 (1964).
161. C. W. B. Grigson, *Nature*, **212**, 750 (1966).
162. R. W. Vook and M. A. Otooni, *J. Appl. Phys.*, **39**, 2471 (1968).
163. C. L. Briant and J. J. Burton, *Surface Sci.*, **51**, 345 (1975).
164. T. Ino, *J. Phys. Soc. Japan*, **8**, 92 (1953).
165. H. Morimoto, *J. Phys. Soc. Japan*, **13**, 1015 (1958).
166. H. Morimoto and H. Sakata, *J. Phys. Soc. Japan*, **17**, 136 (1962).
167. D. Watanabe and R. Miida, *Jap. J. Appl. Phys.*, **11**, 296 (1972).
168. K. Kimoto, Y. Kamiya, M. Nonoyama, and R. Uyeda, *Jap. J. Appl. Phys.*, **2**, 702 (1963).
169. K. Kimoto, I. Nishida, and R. Uyeda, *J. Phys. Soc. Japan*, **20**, 1963 (1965); K. Kimoto and I. Nishida, *ibid.*, **22**, 744 (1967).
170. J. Forssell and B. Persson, *J. Phys. Soc. Japan*, **27**, 1368 (1969); *Ibid.*, **29**, 1532 (1970).
171. I. Nishida, T. Sahashi, and K. Kimoto, *Thin Solid Films*, **10**, 265 (1972).
172. K. Kimoto and I. Nishida, *Thin Solid Films*, **17**, 49 (1973).
173. W. Krakow and B. M. Siegel, *J. Appl. Cryst.*, **9**, 325 (1976).
174. J. B. Kaelberer, R. D. Etters, and J. C. Raich, *Chem. Phys. Lett.*, **41**, 580 (1976).
175. J. J. Burton, *Chem. Phys. Lett.*, **17**, 199 (1972).
176. H. P. Baltes and E. R. Hilf, *Solid State Commun.*, **12**, 369 (1973).
177. Th. F. Nonnenmacher, *Phys. Lett.*, **51A**, 213 (1975).
178. V. N. Bogomolov, R. Sh. Malkovich, I. A. Smirnov, V. V. Tikhonov, and F. A. Chudnovskii, *Sov. Phys.—Solid State*, **12**, 938 (1970).

179. V. N. Bogomolov, N. A. Klushin, M. J. Romanova, I. A. Smirnov, and V. V. Tikhonov, *Sov. Phys.—Solid State*, **14**, 2330 (1973).

180. V. N. Bogomolov, N. A. Klushin, N. M. Okuneva, E. L. Plachenova, V. I. Pogrebnoi, and F. A. Chudnovskii, *Sov. Phys.—Solid State*, **13**, 1256 (1971).

181. V. Novotny, P. P. M. Meincke, and J. H. P. Watson, *Phys. Rev. Lett.*, **28**, 901 (1972).

182. V. Novotny and P. P. M. Meincke, *Phys. Rev.*, **B8**, 4186 (1973).

183. J. H. P. Watson, *Phys. Rev.*, **B2**, 1282 (1970).

184. V. N. Bogomolov, R. H. Malkovich, and F. A. Chudnovskii, *Sov. Phys.— Solid State*, **11**, 2483 (1970).

185. H. Lutz, J. D. Gunton, H. K. Schurmann, J. E. Crow, and T. Mihalisin, *Solid State Commun.*, **14**, 1075 (1974).

186. S. W. Marshall and R. M. Wilenzick, *Phys. Rev. Lett.*, **16**, 219 (1966).

187. M. P. A. Viegers and J. M. Trooster, *Phys. Rev.*, **B15**, 72 (1977).

188. I. P. Suzdalev, M. Ya. Gen, V. I. Gol'danskii, and E. F. Makarov, *Sov. Phys. JETP*, **24**, 79 (1967).

189. S. Akselrod, M. Pasternak, and S. Bukshpan, *Phys. Rev.*, **B11**, 1040 (1975).

190. V. N. Bogomolov and N. A. Klushin, *Sov. Phys.—Solid State*, **15**, 375 (1973).

191. S. Roth and E. M. Hörl, *Phys. Lett.*, **25A**, 299 (1967).

192. A. A. Maradudin, E. W. Montroll, and G. M. Weiss, *Solid State Phys., Suppl.* **3**, 595, 1963; Y. Kashiwase, I. Nishida, Y. Kainuma, and K. Kimoto, *J. Phys. Soc. Japan*, **38**, 899 (1975).

193. M. Takagi, *J. Phys. Soc. Japan*, **9**, 359 (1954).

194. M. Blackman and A. E. Curzon, *Structure and Properties of Thin Films*, C. Neugebauer, J. B. Newkirk, and D. A. Vermelyea, Eds., Wiley, New York, 1959, p. 217.

195. N. T. Gladkich, R. Niedermayer, and K. Spiegel, *Phys. Status Solidi*, **15**, 181 (1966).

196. C. R. M. Wronski, *Br. J. Appl. Phys.*, **18**, 1731 (1967).

197. B. T. Boiko, A. T. Pugachev, and V. M. Bratsykhin, *Sov. Phys.—Solid State*, **10**, 2832 (1969).

198. C. J. Coombs, *J. Phys. F: Metal Phys.*, **2**, 441 (1972).

199. V. N. Bogomolov, A. I. Zadorozhnii, A. A. Kapanadze, E. L. Lutsenko, and V. P. Petranovskii, *Sov. Phys.—Solid State*, **18**, 1777 (1977).

200. P.-A. Buffat, *Thin Solid Films*, **32**, 283 (1976).

201. J. F. Pocza, A. Barna, and P. B. Barna, *J. Vac. Sci. Technol.*, **6**, 472 (1969).

202. R. P. Berman and A. E. Curzon, *Can. J. Phys.*, **52**, 923 (1974).

203. M. Blackman and J. R. Sanders, *Nature*, **226**, 938 (1970); J. R. Sambles, L. M. Skinner, and N. D. Lisgarten, *Proc. Roy. Soc. (London)*, **A318**, 507 (1970).

204. J. R. Sambles, *Proc. Roy. Soc. (London)*, **A324**, 339 (1971).

205. J. Lothe and G. M. Pound, *J. Chem. Phys.*, **36**, 2080 (1962); J. P. Hirth and G. M. Pound, *Progress in Material Science*, Vol. XI, B. Chalmers, Ed., Pergamon, Oxford, 1963; S. J. Hruska, J. P. Hirth, and G. M. Pound, *Surfaces and Interfaces*, J. J. Burke, Ed., Syracuse University Press,

Syracuse, N.Y., 1967, p. 305; J. Lothe and G. M. Pound, *Nucleation*, A. C. Zettlemoyer, Ed., Marcel Dekker, New York, 1969, p. 109.

206. H. J. Freund, Ph.D. Thesis, Cornell University, 1975; H. J. Freund and S. H. Bauer, *J. Phys. Chem.*, **81**, 994 (1977).

207. B. Lewis, *Thin Solid Films*, **9**, 305 (1972); A. Bonissent and B. Mutaftschiev, *J. Chem. Phys.*, **58**, 3727 (1973); A. Milchev, S. Stoyanov, and R. Kaischev, *Thin Solid Films*, **22**, 267 (1974).

208. O. M. Poltorak and V. S. Boronin, *Russ. J. Phys. Chem.*, **40**, 1436 (1966).

209. G. C. Bond, *Platinum Metals Rev.*, **19**, 126 (1975).

210. R. van Hardeveld and F. Hartog, *Surface Sci.*, **15**, 189 (1969).

211. H. Fröhlich, *Physica*, **4**, 406 (1937).

212. R. Kubo, *J. Phys. Soc. Japan*, **17**, 975 (1962).

213. L. P. Gor'kov and G. M. Éliashberg, *Sov. Phys. JETP*, **21**, 940 (1965).

214. R. Denton, B. Mühlschlegel, and D. J. Scalapino, *Phys. Rev. Lett.*, **26**, 707 (1971); *Phys. Rev.*, **B7**, 3589 (1973).

215. E. P. Wigner, *Proc. Camb. Phil. Soc.*, **47**, 790 (1951).

216. G. R. Stewart, *Phys. Rev.*, **B15**, 1143 (1977).

217. S. Kobayashi, T. Takahashi, and W. Sasaki, *J. Phys. Soc. Japan*, **31**, 1442 (1971).

218. S. Kobayashi, T. Takahashi, and W. Sasaki, *J. Phys. Soc. Japan*, **3**, 1234 (1972).

219. R. F. Marzke, W. S. Glaunsinger, and M. Bayard, *Solid State Commun.*, **18**, 1025 (1976).

220. R. J. Charles and W. A. Harrison, *Phys. Rev. Lett.*, **11**, 75 (1963).

221. S. Kobayashi, T. Takahashi, and W. Sasaki, *Phys. Lett.*, **33A**, 429 (1970).

222. T. Fujita, K. Ohshima, N. Wada, and T. Sakakibara, *J. Phys. Soc. Japan*, **29**, 797 (1970).

223. A. Kawabata, *J. Phys. Soc. Japan*, **29**, 902 (1970).

224. R. J. Elliott, *Phys. Rev.*, **96**, 266 (1954).

225. H. S. Gutowsky and P. J. Frank, *Phys. Rev.*, **94**, 1067 (1954); R. A. Levy, *ibid.*, **102**, 31 (1956); N. S. Garif'yanov, *Sov. Phys. JETP*, **5**, 111 (1957); M. Ya. Gen and V. I. Petinov, *ibid.*, **21**, 19 (1965); K. Asayama and Y. Oda, *J. Phys. Soc. Japan*, **22**, 937 (1967); K. Saiki, T. Fujita, Y. Shimizu, S. Sakoh, and N. Wada, *ibid.*, **32**, 447 (1972).

226. J.-P. Borel, C. Borel-Narbel, and R. Monot, *Helv. Phys. Acta*, **47**, 537 (1975).

227. G. O. Berim, F. G. Cherkasov, E. G. Kharakhaskyan, and Y. I. Talanov, *Phys. Status Solidi* (a), **40**, K53 (1977).

228. R. Monot, C. Narbel, and J.-P. Borel, *Nuovo Cimento*, **19**, 253 (1974).

229. J.-P. Borel, *Phys. Lett.*, **57A**, 253 (1976).

230. A. Chatelain, J.-L. Millet, and R. Monot, *J. Appl. Phys.*, **47**, 3670 (1976).

231. A. Abou-Kais, M. Jarjoui, J. C. Vedrine, and P. C. Gravelle, *J. Catal.*, **47**, 399 (1977).

232. J.-L. Millet and R. Monot, in *Proceedings, 18th Ampere Congress, Nottingham*, P. S. Allen, E. R. Andrew, and C. A. Bates, Eds., University of Nottingham, Nottingham, 1974, p. 319.

233. R. Dupree, C. T. Forwood, and M. J. A. Smith, *Phys. Status Solidi*, **24**, 525 (1967).
234. R. Monot, A. Chatelain, and J.-P. Borel, *Phys. Lett.*, **34A**, 57 (1971).
235. W. S. Glaunsinger and R. F. Marzke, *Bull. Am. Phys. Soc.*, **(2)20**, 411 (1975).
236. C. P. Bean, *J. Appl. Phys.*, **26**, 1381 (1955); C. P. Bean and J. D. Livingston, *J. Appl. Phys. Suppl.*, **30**, 1205 (1959); P. W. Selwood, *Adsorption and Collection Magnetism*, Academic Press, New York, 1962; A. E. Berkowitz, *Magnets and Metallurgy*, Academic Press, New York, 1969.
237. P. W. Selwood, S. Adler, and T. R. Phillips, *J. Am. Chem. Soc.*, **77**, 1452 (1955); J. L. Carter and J. H. Sinfelt, *J. Catal.*, **10**, 134 (1968); M. Boudart, A. Delbouille, J. A. Dumesic, S. Khammouma, and H. Topsoe, *J. Catal.*, **37**, 486 (1975); R. A. Buhrman and C. G. Granqvist, *J. Appl. Phys.*, **47**, 2220 (1976); J. T. Richardson and P. Desai, *J. Catal.*, **42**, 294 (1976).
238. S. Strässler, M. J. Rice, and P. Wyder, *Phys. Rev.*, **B6**, 2575 (1972).
239. M. J. Rice, W. R. Schneider, and S. Strässler, *Phys. Rev.*, **B8**, 474 (1973).
240. M. Cini and P. Ascarelli, *J. Phys. F: Metal Phys.*, **4**, 1998 (1974).
241. F. Meier and P. Wyder, *Phys. Lett.*, **39A**, 51 (1972).
242. R. Dupree and M. A. Smithard, *J. Phys. C: Solid State Phys.*, **5**, 408 (1972).
243. R. H. Doremus, *J. Appl. Phys.*, **35**, 3456 (1964).
244. U. Kreibig and P. Zacharias, *Z. Phys.*, **231**, 128 (1970); U. Kreibig, *ibid.*, **234**, 307 (1970); *J. Phys. F: Metal Phys.*, **4**, 999 (1974).
245. G. C. Papavassiliou and Th. Kokkinakis, *J. Phys. F: Metal Phys.*, **4**, L67 (1974); G. C. Papavassiliou, *J. Phys. F: Metal Phys.*, **6**, L103 (1976).
246. D. C. Skillman and C. R. Berry, *J. Chem. Phys.*, **48**, 3297 (1968); C. R. Berry and D. C. Skillman, *J. Appl. Phys.*, **42**, 2818 (1971); D. C. Skillman and C. R. Berry, *J. Opt. Soc. Am.*, **63**, 707 (1973).
247. M. A. Smithard, *Solid State Commun.*, **13**, 153 (1973).
248. D. N. Jarrett and L. Ward, *J. Phys. D: Appl. Phys.*, **9**, 1515 (1976).
249. V. V. Truong and G. D. Scott, *J. Opt. Soc. Am.*, **67**, 502 (1977).
250. M. Rasigni and G. Rasigni, *J. Opt. Soc. Am.*, **67**, 510 (1977).
251. J. D. Eversole and H. P. Brodia, *Phys. Rev.*, **B15**, 1644 (1977).
252. S. Norman, T. Andersson, and C. G. Granqvist, *Solid State Commun.*, **23**, 261 (1977).
253. J. C. Maxwell-Garnett, *Phil. Trans.*, **A203**, 385 (1904); *ibid.*, **205**, 237 (1906).
254. G. Mie, *Ann. Phys.*, **25**, 377 (1908).
255. A. Kawabata and R. Kubo, *J. Phys. Soc. Japan*, **21**, 1765 (1966).
256. J.-D. Ganière, R. Rechsteiner, and M. A. Smithard, *Solid State Commun.*, **16**, 113 (1975).
257. A. Eguiluz, S. C. Ying, and J. J. Quinn, *Phys. Rev.*, **B11**, 2118 (1975).
258. P. Ascarelli and M. Cini, *Solid State Commun.*, **18**, 385 (1976).
259. D. Stroud, *Phys. Rev.*, **B12**, 3368 (1975).
260. D. M. Wood and N. W. Ashcroft, *Phil. Mag.*, **35**, 269 (1977).
261. J. I. Gittleman and B. Abeles, *Phys. Rev.*, **B15**, 3272 (1977).
262. L. Genzel, T. P. Martin, and U. Kreibig, *Z. Phys.*, **21**, 339 (1975).
263. C. Kunz, *Phys. Lett.*, **15**, 312 (1965).

264. M. Creuzburg, *Z. Phys.*, **194**, 211 (1966).
265. F. Fujimoto, K. Komaki, and K. Ishida, *J. Phys. Soc. Japan*, **23**, 1186 (1967).
266. A. Carillon, *C.R. Acad. Sci. Paris*, **267**, 332 (1968).
267. Th. Kokkinakis and K. Alexopoulos, *Phys. Rev. Lett.*, **28**, 1632 (1972).
268. Th. Kokkinakis and G. C. Papavassiliou, *Phys. Status Solidi* (b), **77**, K49 (1976).
269. P. Borziak, I. Konovalov, Yu Kulyupin, and K. Pilipchak, *Thin Solid Films*, **35**, L9 (1976).
270. W. Shockley, *Phys. Rev.*, **56**, 317 (1939).
271. F. Forstman, *Z. Phys.*, **235**, 69 (1970).
272. V. Heine, *Proc. Roy. Soc. (London)*, **A331**, 307 (1972); P. Mark and W. R. Bottoms, in *Progress in Solid State Chemistry*, Vol. 6, H. Reiss and J. O. McCaldin, Eds., Pergamon, New York, 1971, Chap. 2; R. O. Jones, in *Surface Physics of Phosphors and Semiconductors*, C. G. Scott and C. E. Reed, Eds., Academic Press, London, 1975, Chap. 2, and references therein.
273. F. Cyrot-Lackmann, *Adv. Phys.*, **16**, 393 (1967); F. Cyrot-Lackmann, *Surface Sci.*, **15**, 535 (1969); J. P. Gaspard and F. Cyrot-Lackmann, *J. Phys. C: Solid State Phys.*, **6**, 3077 (1973).
274. F. Cyrot-Lackmann and M. C. Desjonquères, *Surface Sci.*, **40**, 423 (1973).
275. F. Cyrot-Lackmann, M. C. Desjonquères, and J. P. Gaspard, *J. Phys. C: Solid State Phys.*, **7**, 925 (1974).
276. M. C. Desjonquères and F. Cyrot-Lackmann, *Surface Sci.*, **53**, 429 (1975); *Solid State Commun.*, **20**, 855 (1976).
277. M. C. Desjonquères and F. Cyrot-Lackmann, *J. Phys. F: Metal Phys.*, **6**, 567 (1976).
278. M. C. Desjonquères and F. Cyrot-Lackmann, *Solid State Commun.*, **18**, 1127 (1976).
279. R. Haydock, V. Heine, and M. J. Kelly, *J. Phys. C: Solid State Phys.*, **5**, 2845 (1972); *ibid.*, **8**, 2591 (1975).
280. R. Haydock and M. J. Kelly, *Surface Sci.*, **38**, 139 (1973).
281. G. Kerker, *J. Phys. F: Metal Phys.*, **6**, L113 (1976).
282. D. G. Dempsey and L. Kleinman, *J. Phys. F: Metal Phys.*, **7**, 113 (1977).
283. F. Garcia-Moliner and J. Rubio, *Proc. Roy. Soc. (London)*, **A324**, 257 (1971).
284. H. Büttner and E. Gerlach, *J. Phys. F: Metal Phys.*, **2**, 302 (1972).
285. J. E. Inglesfield, *J. Phys. C: Solid State Phys.*, **4**, L14 (1971); *J. Phys. F: Metal Phys.*, **2**, 63 (1972).
286. M. L. Glasser, *Surface Sci.*, **64**, 141 (1977).
287. G. Allan and P. Lenglart, *Surface Sci.*, **30**, 641 (1972).
288. D. Kalkstein and P. Soven, *Surface Sci.*, **26**, 85 (1971).
289. E-N. Foo, M. F. Thorpe, and D. Weaire, *Surface Sci.*, **57**, 323 (1976).
290. F. Forstmann and V. Heine, *Phys. Rev. Lett.*, **24**, 1419 (1970).
291. F. Forstmann and J. B. Pendry, *Z. Phys.*, **235**, 75 (1970); S. J. Gurman and J. B. Pendry, *Phys. Rev. Lett.*, **31**, 637 (1973).
292. D. E. Eastman, *Phys. Rev. B*, **3**, 1769 (1971).

293. P. O. Gartland, S. Berge, and B. J. Slagsvold, *Phys. Rev. Lett.*, **30**, 916 (1973); P. O. Gartland and B. J. Slagsvold, *Phys. Rev. B*, **12**, 4047 (1975).

294. D. R. Lloyd, C. M. Quinn, and N. V. Richardson, *J. Phys. C: Solid State Phys.*, **8**, L371 (1975).

295. L. Ilver and P. O. Nilsson, *Solid State Commun.*, **18**, 677 (1976).

296. E. Dietz, H. Becker, and U. Gerhardt, *Phys. Rev. Lett.*, **36**, 1397 (1976).

297. P. O. Nilsson and L. Ilver, *Solid State Commun.*, **17**, 667 (1975).

298. H. F. Roloff and H. Neddermeyer, *Solid State Commun.*, **21**, 561 (1977); P. Heimann, H. Neddermeyer, and H. F. Roloff, *J. Phys. C: Solid State Phys.*, **10**, L17 (1977).

299. L. W. Swanson and L. C. Crouser, *Phys. Rev. Lett.*, **16**, 389 (1966); *Phys. Rev.*, **163**, 622 (1967).

300. E. W. Plummer and J. W. Gadzuk, *Phys. Rev. Lett.*, **25**, 1493 (1970).

301. B. J. Waclawski and E. W. Plummer, *Phys. Rev. Lett.*, **29**, 783 (1972).

302. B. Feuerbacher and B. Fitton, *Phys. Rev. Lett.*, **29**, 786 (1972); B. Feuerbacher and N. E. Christensen, *Phys. Rev. B*, **10** (1974).

303. D. E. Eastman and W. D. Grobman, *Phys. Rev. Lett.*, **30**, 177 (1973).

304. B. R. Cooper and A. J. Bennett, *Phys. Rev. B*, **1**, 4654 (1970); B. R. Cooper, *Phys. Rev. Lett.*, **30**, 1316 (1973); B. R. Cooper, *J. Vac. Sci. Technol.*, **10**, 713 (1973).

305. A. A. Cottey, *J. Phys. C: Solid State Phys.*, **5**, 2583 (1972).

306. J. B. Pendry and S. J. Gurman, *Surface Sci.*, **49**, 87 (1975); S. J. Gurman, *J. Phys. F: Metal Phys.*, **5**, L194 (1975).

307. R. V. Kasowski, *Phys. Rev. Lett.*, **33**, 83 (1974); *Solid State Commun.*, **17**, 179 (1975).

308. N. Kar and P. Soven, *Phys. Rev. B*, **11**, 3761 (1975).

309. W. Kohn, *Phys. Rev. B*, **11**, 3756 (1975).

310. L. Kleinman, *Phys. Rev. B*, **13**, 4640 (1976); D. G. Dempsey, L. Kleinman, and E. Caruthers, *ibid.*, **14**, 279 (1976); E. Caruthers, D. G. Dempsey and L. Kleinman, *ibid.*, **14**, 288 (1976).

311. H. F. Schaefer, *The Electronic Structure of Atoms and Molecules*, Addison-Wesley, Reading, Mass., 1972; C. W. Bauschlicher, D. H. Liskow, C. F. Bender, and H. F. Schaefer, *J. Chem. Phys.*, **62**, 4815 (1975); C. E. Dykstra and H. F. Schaefer, *J. Chem. Phys.*, **65**, 5141 (1976); R. F. Marshall, R. J. Blint, and A. B. Kunz, *Phys. Rev. B*, **13**, 3333 (1976).

312. R. C. Baetzold, *J. Chem. Phys.*, **55**, 4355 (1971).

313. R. C. Baetzold, *J. Chem. Phys.*, **55**, 4363 (1971).

314. R. C. Baetzold, *Surface Sci.*, **36**, 123 (1972).

315. R. C. Baetzold, *J. Catal.*, **29**, 129 (1973).

316. R. C. Baetzold, *Comments Solid State Phys.*, **4**, 62 (1972).

317. R. C. Baetzold and R. E. Mack, *Inorg. Chem.*, **14**, 686 (1975).

318. R. C. Baetzold, *Surface Sci.*, **51**, 1 (1975).

319. R. C. Baetzold, *J. Catal.*, **39**, 158 (1975).

320. R. C. Baetzold and R. E. Mack, *J. Chem. Phys.*, **62**, 1513 (1975).

321. R. C. Baetzold, *J. Phys. Chem.*, **80**, 1504 (1976).

322. R. C. Baetzold, *J. Phys. Chem.*, **82**, 738 (1978).

323. R. C. Baetzold, *J. Chem. Phys.*, **68**, 555 (1978).

324. K. H. Johnson and R. P. Messmer, *J. Vac. Sci. Technol.*, **11**, 236 (1974).

325. R. P. Messmer, C. W. Tucker, and K. H. Johnson, *Chem. Phys. Lett.*, **36**, 423 (1975).

326. R. P. Messmer, S. K. Knudsen, K. H. Johnson, J. B. Diamond, and C. Y. Yang, *Phys. Rev. B*, **13**, 1396 (1976).

327. R. P. Messmer, in *The Physical Basis for Heterogeneous Catalysis*, E. Drauglis and R. I. Jaffee, Eds., Plenum, New York, 1975, p. 261.

328. G. Blyholder, *J. Phys. Chem.*, **79**, 756 (1975).

329. C. E. Forbes and M. C. R. Symons, *Mol. Phys.*, **27**, 467 (1974).

330. G. Blyholder, *Surface Sci.*, **42**, 249 (1974).

331. E. R. Davidson and S. C. Fain, Jr., *J. Vac. Sci. Technol.*, **13**, 209 (1976).

332. G. S. Painter, P. J. Jennings, and R. O. Jones, *J. Phys. C: Solid State Phys.*, **8**, L199 (1975); R. O. Jones, P. J. Jennings, and G. S. Painter, *Surface Sci.*, **53**, 409 (1975).

333. P. J. Jennings, G. S. Painter, and R. O. Jones, *Surface Sci.*, **61**, 255 (1976).

334. A. J. Bennett, B. McCarroll, and R. P. Messmer, *Surface Sci.,* **24**, 191 (1971).

335. A. J. Bennett, B. McCarroll, and R. P. Messmer, *Phys. Rev. B*, **3**, 1397 (1971).

336. B. McCarroll and R. P. Messmer, *Surface Sci.*, **27**, 451 (1971).

337. R. P. Messmer, B. McCarroll, and C. M. Singal, *J. Vac. Sci. Technol.*, **9**, 891 (1972).

338. D. J. M. Fassaert, H. Verbeek, and A. van der Avoird, *Surface Sci.*, **29**, 501 (1972).

339. L. W. Anders, R. S. Hansen, and L. S. Bartell, *J. Chem. Phys.*, **59**, 5277 (1973).

340. J. C. Robertson and C. W. Wilmsen, *J. Vac. Sci. Technol.*, **9**, 901 (1972).

341. G. Blyholder, *J. Vac. Sci. Technol.*, **11**, 865 (1974).

342. H. Itoh, *Jap. J. Appl. Phys.*, **15**, 2311 (1976).

343. D. E. Eastman and J. K. Cashion, *Phys. Rev. Lett.*, **27**, 1520 (1971).

344. W. F. Cooper, G. A. Clarke, and C. R. Hare, *J. Phys. Chem.*, **76**, 2268 (1972).

345. A. Hoareau, J. M. Reymond, B. Caband, and R. Uzan, *J. Phys. (Paris)*, **36**, 7573 (1975).

346. A. Hoareau, P. Joyes, and B. Caband, *J. Phys. (Paris)*, **36**, 979 (1975); A. Hoareau, P. Joyes, B. Caband, and R. Uzan, *Surface Sci.*, **57**, 279 (1976).

347. M. G. Mason, paper presented at 1976 Winter Meeting, Upstate New York Chapter, AVS, Corning, N.Y.

348. K. S. Liang, W. R. Salaneck, and I. A. Aksay, *Solid State Commun.*, **19**, 329 (1976).

349. K. S. Kim and N. Winograd, *Chem. Phys. Lett.*, **30**, 91 (1975).

350. P. Wehking, H. Beckermann, and R. Niedermayer, *Thin Solid Films*, **36**, 265 (1976).

351. P. N. Ross, K. Kinoshita, and P. Stonehart, *J. Catal.*, **32**, 163 (1974).

352. F. W. Lytle, *J. Catal.*, **43**, 376 (1976).

353. D. B. Tanner, A. J. Sievers, and R. A. Buhrman, *Phys. Rev. B*, **11**, 1330 (1975).

354. K. J. Klabunde, *Acc. Chem. Res.*, **8**, 393 (1975); *Am. Lab.*, **7**, 35 (1975); K. J. Klabunde and T. O. Murdock, *Chem. Technol.*, **5**, 624 (1975).

355. Symposium on Metal Atoms in Chemical Synthesis, Darmstadt, May 1974. *Angew. Chem. Int. Ed.*, **14**, 193 (1975).

356. M. Boudart, *Adv. Catal. Relat. Subj.*, **20**, 153 (1969).

357. R. van Hardeveld and F. Hartog, *Adv. Catal. Relat. Subj.*, **22**, 75 (1972).

358. P. B. Wells, in *Surface and Defect Properties of Solids*, Vol. 1, The Chemical Society, London, 1972, p. 236.

359. A. D. O. Anneide and J. K. A. Clarke, *Catal. Rev.*, **7**, 213 (1972).

360. T. E. Whyte, *Catal. Rev.*, **8**, 117 (1973).

361. P. B. Wells, in *Proc. Symposium on Electrocatalyses, San Francisco, 1974*, M. W. Breiter, Ed., The Electrochemical Society, Princeton, 1975, p. 1.

362. P. Stonehart, K. Kinoshita, and J. A. S. Bett, in *Proc. Synposium on Electrocatalyses, San Francisco, 1974*, M. W. Breiter, Ed., The Electrochemical Society, Princeton, 1975, p. 275.

363. J. J. Ostermaier, J. R. Katzer, and W. H. Manogue, *J. Catal.*, **33**, 457 (1974).

364. M. Kobayashi and T. Shirasaki, *J. Catal.*, **39**, 148 (1975).

365. M. A. Vannice, *J. Catal.*, **40**, 129 (1975).

366. C. Corolleur, F. G. Gault, D. Juttard, G. Maire, and M. J. Muller, *J. Catal.*, **27**, 466 (1972).

367. J.-M. Dartigues, A. Chambellan, and F. G. Gault, *J. Am. Chem. Soc.*, **98**, 856 (1976).

368. M. Boudart, A. Aldag, J. E. Benson, N. A. Dougharty, and C. G. Harkins, *J. Catal.*, **6**, 92 (1966).

369. J. E. Benson and M. Boudart, *J. Catal.*, **4**, 704 (1965); J. Frul, *ibid.*, **25**, 149 (1972); G. R. Wilson and W. K. Hall, *ibid.*, **17**, 190 (1970); G. R. Wilson and W. K. Hall, *ibid.*, **24**, 306 (1972); N. Kinoshita, J. Lundquist, and P. Stonehart, *ibid.*, **31**, 325 (1973).

370. G. A. Somorjai, *Catal. Rev.*, **7**, 87 (1972); L. L. Kesmodel and G. A. Somorjai, *Acc. Chem. Res.*, **9**, 392 (1976); G. A. Somorjai, *ibid.*, **9**, 248 (1976); *J. Electrochem. Soc.*, **124**, 205 (1977); *Adv. Catal. Relat. Subj.*, **26**, 1 (1977).

371. S. L. Bernasek and G. A. Somorjai, *J. Chem. Phys.*, **62**, 3149 (1975).

372. O. M. Poltorak and V. S. Boronin, *Zh. Fiz. Khim.*, **39**, 1476 (1965).

373. I. E. Wachs and R. J. Madix, *Surface Sci.*, **58**, 590 (1976).

374. K. E. Lu and R. R. Rye, *Surface Sci.*, **45**, 677 (1974).

375. D. Luss, *J. Catal.*, **23**, 119 (1971).

376. K. F. Burton, P. Greenberg, H. G. Oswin, and D. R. Rutt, *J. Electrochem. Soc.*, **119**, 559 (1972).

377. R. J. Pusateri, J. R. Katzer, and W. H. Manogue, *A.I.Ch.E. J.*, **30**, 219 (1974).

378. F. N. Hill and P. W. Selwood, *J. Am. Chem. Soc.*, **71**, 2522 (1949).
379. D. J. C. Yates and J. Sinfelt, *J. Catal.*, **8**, 348 (1967).
380. W. Reinders and L. Hamberger, *Rec. Trav. Chim.*, **50**, 475 (1931); *Z. Wiss. Photogr.*, **31**, 32 (1932); *ibid.*, **31**, 265 (1933); W. Reinders and R. W. P. DeVries, *Rec. Trav. Chim.*, **56**, 985 (1937).
381. E. A. Galashin, E. P. Senchenkov, and K. V. Chibisov, *Dokl. Akad. Nauk SSSR*, **181**, 124 (1969); E. A. Galashin and E. P. Senchenkov, *Zh. Nauch. Prikl. Fotogr. Kinematogr.*, **16**, 339 (1971).
382. P. J. Hillson, *J. Photogr. Sci.*, **22**, 31 (1974); **23**, 15 (1975).
383. F. Troutweiler, *Photogr. Sci. Eng.*, **17**, 138 (1968); W. Jaenicke, *J. Photogr. Sci.*, **20**, 2 (1972).
384. D. Kirstein, E. Kahrig, G. Dreyer, J. Erpenbeck, and F. Lange, *Z. Wiss. Photogr.*, **61**, 165, 171 (1967).

Chapter **IV**

REACTION RATE MEASUREMENTS IN SOLUTION ON MICROSECOND TO SUBNANOSECOND TIME SCALES

Neil Purdie, Edward M. Eyring, and Licesio Rodriguez

1 INTRODUCTION

In the microsecond to subnanosecond time range, the relaxation methods [1–3] frequently are best suited for the measurement of reaction rates in liquid solutions. Common exceptions are irreversible reactions that in some cases are studied successfully instead by flash-photolytic techniques [4]. The general principle of relaxation methods is that some external parameter (pressure, temperature, electric field) is changed so rapidly that the chemical reaction lags behind. In the ultrasonic technique [3], the relaxation method to which we devote the greatest attention in this chapter, temperature and pressure are varied periodically, as opposed to a transient disturbance (T-jump, P-jump, etc.). The effect of the periodic perturbation is felt in chemical reactions that establish equilibrium very rapidly. Perturbations of equilibrium phenomena by ultrasound cover the frequency range 10^4 to 10^9 Hz, equivalent to an approximate time scale for relaxation from 10^{-5} to 10^{-10} sec (Brillouin scattering excluded). Unlike flash photolysis [4, Chapter 10], disturbances are small (on the order of a few percent from equilibrium) and are a compromise between displacements that are large enough to be easily observed and those small enough so that the interpretive mathematics can be reduced to linear equations.

The T-jump [3] and E-jump (electric field jump) [3] relaxation methods that we discuss at the end of this chapter are suitable for measuring relaxation times as short as tens of nanoseconds (10^{-8} sec). In addition to the time factor, the solutes and solvent systems susceptible to study by these two techniques on a submicrosecond time scale are more restricted than those accessible to ultrasonic investigation.

2 ASPECTS OF RELAXATION KINETIC PROCESSES

In a relaxation kinetic experiment, the only quantity usually of interest has been the relaxation time. Information concerning rate constants and activation parameters is easily considered in terms of that quantity and the equilibrium concentrations. We are not interested here in a complete discussion of the kinetic basis of chemical relaxation processes, since excellent accounts of the subject can be found elsewhere [1–3]. We focus in the present section on the treatment, in a succinct form, of multiple equilibria, mainly in connection with another valuable but rarely invoked quantity, the relaxation amplitude. The information contained therein may be used in a more precise assignment of the kinetic pathway in moderately complex systems.

The authors are aware that a rigorous description of the handling of this problem may produce in the novice more desperation than understanding. For a scientist who is unfamiliar with the relaxation methods of measuring rates of chemical reactions in liquid solutions, it may therefore be more

Table 4.1 Thermodynamic Data for Aqueous Crown Ether Complexes at 25°C Reported by Izatt et al. [9]

Ligand	Cation	log K	ΔH (kcal/mole)	ΔS (cal/deg-mole)
15-Crown-5	Na^+	0.70 ± 0.10	-1.50 ± 0.04	-1.8
	K^+	0.74 ± 0.08	-4.1 ± 0.1	-10.4
	Rb^+	0.62 ± 0.10	-1.90 ± 0.01	-3.5
	Cs^+	0.8 ± 0.2	-1.3 ± 0.2	-0.5
	Ag^+	0.94 ± 0.08	-3.23 ± 0.03	-6.5
	Tl^+	1.23 ± 0.04	-4.01 ± 0.05	-7.8
18-Crown-6	Na^+	0.80 ± 0.10	-2.25 ± 0.10	-3.7
	K^+	2.03 ± 0.10	-6.21 ± 0.01	-11.4
	Rb^+	1.56 ± 0.02	-3.82 ± 0.11	-5.7
	Cs^+	0.99 ± 0.07	-3.79 ± 0.10	-8.1
	Ag^+	1.50 ± 0.03	-2.17 ± 0.09	-0.4
	Tl^+	2.27 ± 0.04	-4.44 ± 0.04	-4.5

instructive to discuss a particular reaction system than to deal in generalities.

The present authors and collaborators have invested much recent effort in submicrosecond time scale relaxation method kinetic studies of solutions of crown ethers [5–8] and various metal ions. Crown ether ligands are typified by 15-crown-5,

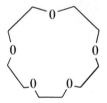

which in the nomenclature originated by Pedersen [5] means a neutral cyclic polyether with 15 atoms including five oxygen atoms in the ring. This organic ligand is water soluble and strongly complexes metal ions such as the alkali metal cations. Some equilibrium data reported by Izatt et al. [9] for these complexation reactions in aqueous solution at 25°C are shown in Table 4.1. These quantities were determined by calorimetric titration. Similar information concerning 18-crown-6 complexation reactions appears in this same table:

From these few data it is clear that 18-crown-6 shows a strong selectivity for aqueous potassium ions over three other alkali metal cations (Na^+, Rb^+, and Cs^+) that is not matched by 15-crown-5. Is the greater selectivity of 18-crown-6 achieved by the greater flexibility of the ring or possibly by its more capacious center hole? Are rates of complex ion formation essentially identical for all ions and both ligands because of a diffusion-controlled encounter between reactants? Do specific rates of complex ion dissociation play the decisive role in determining the peculiar stability constants of Table 4.1? These are a few of the many questions that occur to the kineticist on examining Table 4.1.

We can begin our search for an understanding of the experimental tools suitable to answer these questions by considering Chock's pioneering T-jump study [10] of crown ether complexation reactions. With a Joule heating T-jump apparatus Chock set out to measure relaxation times associated with the reaction

$$M^+ + CR_2 \rightleftharpoons \text{complex ion}$$

where M^+ denotes a monovalent cation such as Na^+, K^+, or NH_4^+ and CR_2 denotes dibenzo-30-crown-10,

dissolved in methanol. The subscript 2 on CR denotes a particular solution conformation of the crown ether molecule. Chock used methanol as a solvent since this crown ether is comparatively insoluble in water. The T-jump instrument had a rise time of less than 1 μsec and was used with absorbance detection at an ultraviolet wavelength of ~280 nm characteristic of this crown ether.

In the case of sodium ion complexation by methanolic dibenzo-30-crown-10, Chock found [10] an immeasurably fast absorbance change whose amplitude led, in a manner discussed below, to an enthalpy change for complex formation $\Delta H = -4 \pm 1$ kcal. Unfortunately, all that could be said on the basis of Chock's data about the rate constant for the reaction

$$Na^+ + \text{dibenzo-30-crown-10} \rightarrow \text{complex}$$

was that it is greater than $1.6 \times 10^7 M^{-1}$ sec^{-1} in methanol at 25°C. Ammonium ion complexation by the same crown ether was similarly too fast for Chock's Joule heating T-jump apparatus, whereas the fast absorbance transients associated with K^+, Rb^+, Cs^+, and Tl^+ complexation by this same crown ether were all observable on a time scale of tens of microseconds.

The reaction mechanism postulated by Chock [10] to account for his experimental data involved two distinct steps:

$$\begin{array}{l} CR_1 \\ \updownarrow \\ CR_2 + M^+ \rightleftharpoons \text{complex ion} \end{array}$$

An immeasurably rapid conformational change of methanolic dibenzo-30-crown-10 was thought to precede the less rapid formation of complex ion.

For those kineticists fortunate enough to have immediate access to a T-jump, E-jump, and/or an ultrasonic absorption spectrometer capable of measuring submicrosecond relaxation times, the impulsive, most expeditious way to discover the fast relaxations that had eluded Chock is to initiate laboratory experiments. However, the judicious first approach to the problem available to everyone involves a preliminary use of relaxation theory [2, 3, 11–13]. This theory will aid the instrumentally well endowed experimentalist in recognizing to what chemical phenomena his first observed relaxations should be attributed. To those without relaxation method instrumentation, theory will suggest the choice of relaxation method equipment to be borrowed or assembled.

In a fast reaction system consisting of two or more apparently interdependent chemical equilibria, it may be a gross oversimplification to say that one reaction is unobservable because it is too fast for the instrumentation whereas another reaction is slow enough to be detected. Years ago Eigen and DeMaeyer [2, 3] provided the mathematical tools for analyzing such systems in terms of both relaxation times and relaxation amplitudes. Unfortunately, the use of these tools did not begin to command the attention of relaxation kineticists generally until comparatively recently. Ruthild Winkler's impressive success [14] in analyzing the kinetics of complexation of alkali and alkaline earth metal cations by murexide ion contributed significantly to this renaissance of interest in relaxation amplitude calculations.

In the following arithmetic manipulations we emphasize the how and neglect the why. The hope is that the reader will be favorably impressed by the usefulness of the resulting expressions and will therefore not mind seeking detailed explanations elsewhere [2]. The particular format of the calculations presented here was suggested by Professor R. D. White.

Let us write Chock's tentative reaction mechanism in the following general notation:

$$A_1 \underset{k_{b1}}{\overset{k_{f1}}{\rightleftharpoons}} A_2 \tag{1}$$

$$A_2 + A_3 \underset{k_{b2}}{\overset{k_{f2}}{\rightleftharpoons}} A_4 \tag{2}$$

Each of these reactions may be represented by the still more general notation of Gibbs

$$\Sigma \mu_{ri} A_i = 0 \tag{3}$$

where the subscripts $r = 1, 2, 3, \ldots$ denote the reactions, the subscripts $i = 1, 2, 3, \ldots$ denote the chemical species, and the μ_{ri} are stoichiometric coefficients. Thus for reactions (1) and (2) above we have the following stoichiometric coefficients:

$$\mu_{11} = -1 \qquad \mu_{12} = 1 \qquad \mu_{13} = 0 \qquad \mu_{14} = 0$$

$$\mu_{21} = 0 \qquad \mu_{22} = -1 \qquad \mu_{23} = -1 \qquad \mu_{24} = 1$$

These stoichiometric coefficients may be expressed as a matrix

$$\mu = \begin{pmatrix} -1 & 1 & 0 & 0 \\ 0 & -1 & -1 & 1 \end{pmatrix} \tag{4}$$

It next is desirable to construct a matrix

$$g = \mu \, C^{0-1} \, \mu^T \tag{5}$$

in which μ is the stoichiometric coefficient matrix of eq. (4), the matrix μ^T is its transpose, and C^{0-1} is the reciprocal of the diagonal equilibrium concentration matrix

$$
C^0 = \begin{pmatrix} C_1^0 & 0 & 0 & 0 \\ 0 & C_2^0 & 0 & 0 \\ 0 & 0 & C_3^0 & 0 \\ 0 & 0 & 0 & C_4^0 \end{pmatrix}
\tag{6}
$$

In the special case of reactions (1) and (2) the **g** matrix is

$$
\mathbf{g} = \begin{pmatrix} -1 & 1 & 0 & 0 \\ 0 & -1 & -1 & -1 \end{pmatrix}
\begin{pmatrix} \dfrac{1}{C_1^0} & 0 & 0 & 0 \\ 0 & \dfrac{1}{C_2^0} & 0 & 0 \\ 0 & 0 & \dfrac{1}{C_3^0} & 0 \\ 0 & 0 & 0 & \dfrac{1}{C_4^0} \end{pmatrix}
\begin{pmatrix} -1 & 0 \\ 1 & -1 \\ 0 & -1 \\ 0 & 1 \end{pmatrix}
$$

$$
= \begin{pmatrix} -1 & 1 & 0 & 0 \\ 0 & -1 & -1 & -1 \end{pmatrix}
\begin{pmatrix} -\dfrac{1}{C_1^0} & 0 \\ \dfrac{1}{C_2^0} & -\dfrac{1}{C_2^0} \\ 0 & -\dfrac{1}{C_3^0} \\ 0 & \dfrac{1}{C_4^0} \end{pmatrix}
$$

$$
= \begin{pmatrix} \dfrac{1}{C_1^0} + \dfrac{1}{C_2^0} & -\dfrac{1}{C_2^0} \\ -\dfrac{1}{C_2^0} & \dfrac{1}{C_2^0} + \dfrac{1}{C_3^0} + \dfrac{1}{C_4^0} \end{pmatrix}
\tag{7}
$$

Since **g** will always be a symmetrical matrix, it is desirable to check the accuracy of the derived g by noting that $g_{12} \equiv g_{21}$.

At chemical equilibrium there is a diagonal reaction velocity matrix **S**.

that in the present particular case of two reactions can be written in any one of the four following numerically equivalent ways:

$$S^0 = \begin{pmatrix} k_{f1}C_1^0 & 0 \\ 0 & k_{f2}C_2^0 C_3^0 \end{pmatrix} = \begin{pmatrix} k_{b1}C_2 & 0 \\ 0 & k_{b2}C_4^0 \end{pmatrix}$$

$$= \begin{pmatrix} k_{b1}C_2^0 & 0 \\ 0 & k_{f2}C_2^0 C_3 \end{pmatrix} = \begin{pmatrix} k_{f1}C_1 & 0 \\ 0 & k_{b2}C_4^0 \end{pmatrix} \quad (8)$$

If one knew the numerical values of all the terms in each of these matrices, it would be immediately apparent that there is only one S^0 matrix.

The next objective is the calculation of the eigenvalues of the matrix $S^0 g$:

$$S^0 g = \begin{pmatrix} k_{f1}C_1^0 & 0 \\ 0 & k_{f2}C_2^0 C_3^0 \end{pmatrix} \begin{pmatrix} \dfrac{1}{C_1^0} + \dfrac{1}{C_2^0} & -\dfrac{1}{C_2^0} \\ -\dfrac{1}{C_2^0} & \dfrac{1}{C_2^0} + \dfrac{1}{C_3^0} + \dfrac{1}{C_4^0} \end{pmatrix}$$

$$= \begin{pmatrix} k_{f1} + k_{b1} & -k_{b1} \\ -k_{f2}C_3^0 & k_{f2}(C_3^0 + C_2^0) + k_{b2} \end{pmatrix} \quad (9)$$

whence it follows that the secular determinant is

$$\begin{vmatrix} k_{f1} + k_{b1} - \lambda & -k_{b1} \\ -k_{f2}C_3^0 & k_{f2}(C_2^0 + C_3^0) + k_{b2} - \lambda \end{vmatrix} = 0 \quad (10)$$

in which λ denotes the eigenvalues. The extent to which the two chemical reactions (1) and (2) are closely coupled will depend on the size of the off-diagonal terms $-k_{b1}$ and $-k_{f2}C_3^0$. For convenience a further simplification in notation is appropriate. Rewriting (10) as

$$\begin{vmatrix} a_{11} - \lambda & a_{12} \\ a_{21} & a_{22} - \lambda \end{vmatrix} = 0$$

and expanding, one obtains

$$\lambda = \frac{1}{2} \left[a_{11} + a_{22} \pm \left((a_{11} - a_{22})^2 + 4a_{12}a_{21} \right)^{1/2} \right] \quad (11)$$

In this special case in which $4a_{12}a_{21}$ is negligibly small, the following trivial result obtains:

$$\lambda = \frac{1}{\tau_I} = a_{11} = k_{f1} + k_{b1} \quad \text{and} \quad \lambda = \frac{1}{\tau_{II}} = k_{f2}(C_2^0 + C_3^0) + k_{b2}$$

These define the relaxation times τ^* of reactions (1) and (2) as if they were completely uncoupled. In general, if reasonable estimates of the various quantities in (10) were available, at this juncture simplifying assumptions would be advantageous. In the absence of such information, let us proceed instead to consider the case of full coupling of reactions (1) and (2) that will require no assumptions regarding magnitudes of relaxation times, rate constants, and so forth. In the general case that several consecutive and parallel equilibria are established, the coupling among them arises from the fact that the individual relaxation times do not correspond to the isolated reaction steps. The analogy of this behavior to the coupling of the vibrations of a multiple oscillator system has been pointed out [2, 15–18]. It appears then that the relaxation times of this system are related to normal modes of reaction rather than to the reactions themselves as they are treated in the mechanistic scheme.

The problem of finding suitable normal reaction modes can be looked at as a simple transformation between two coordinate systems by which the first space of concentrations is connected to the other space representing the independent relaxation modes of the chemical system.

Returning again to the particular case now being considered, the transformation matrix would be built up from the column-eigenvectors of S^0g:

$$\begin{pmatrix} a_{11} - \tau_i^{-1} & a_{12} \\ a_{21} & a_{22} - \tau_i^{-1} \end{pmatrix} \begin{pmatrix} \alpha_{1i} \\ \alpha_{2i} \end{pmatrix} = 0 \tag{12}$$

where $i = 1$ or 2.

A homogeneous system of equations follows which enables us to determine only the quotient of the eigenvector elements:

$$\frac{\alpha_{1i}}{\alpha_{2i}} = -\frac{a_{12}}{a_{11} - \tau_i^{-1}} = -\frac{a_{21}}{a_{22} - \tau_i^{-1}} \tag{13}$$

wherein both expressions are equivalent.

We may assign to the α_{2i} factors the value $\alpha_{2i} = 1$. This choice, of course, is arbitrary and will only affect the stoichiometric coefficients of the normal-mode reactions as a simple scale factor; their arbitrariness will not be propagated within the calculated relaxation amplitudes.

Taking into account the above, the resulting transformation matrix is

$$\mathbf{A} = \begin{pmatrix} -\dfrac{a_{12}}{a_{11} - \tau_I^{-1}} & -\dfrac{a_{12}}{a_{11} - \tau_{II}^{-1}} \\ 1 & 1 \end{pmatrix} \tag{14}$$

*Defined as the time it takes for the displacement to decrease to $1/e$ of its initial value.

The stoichiometric matrix for the normal reactions is made up as follows:

$$\tilde{\mu} = \mathbf{A}^T \mu = \begin{pmatrix} -\dfrac{a_{12}}{a_{11} - \tau_I^{-1}} & 1 \\[2ex] -\dfrac{a_{12}}{a_{11} - \tau_{II}^{-1}} & 1 \end{pmatrix} \begin{pmatrix} -1 & 1 & 0 & 0 \\ 0 & -1 & -1 & 1 \end{pmatrix} \tag{15}$$

resulting in

$$\tilde{\mu} = \begin{pmatrix} \dfrac{a_{12}}{a_{11} - \tau_I^{-1}} & -\dfrac{a_{12}}{a_{11} - \tau_I^{-1}} - 1 & -1 & 1 \\[3ex] \dfrac{a_{12}}{a_{11} - \tau_{II}^{-1}} & \dfrac{a_{12}}{a_{11} - \tau_{II}^{-1}} - 1 & -1 & 1 \end{pmatrix} \tag{16}$$

by which the normal mode reactions are seen to be

$$\frac{a_{12}}{\tau_I^{-1} - a_{11}} \mathrm{CR}_1 + \mathrm{M}^+ \rightleftharpoons \left(\frac{a_{12}}{\tau_I^{-1} - a_{11}} - 1 \right) \mathrm{CR}_2 + \mathrm{MCR}_2^+$$

$$\frac{a_{12}}{\tau_{II}^{-1} - a_{11}} \mathrm{CR}_1 + \mathrm{M}^+ \rightleftharpoons \left(\frac{a_{12}}{\tau_{II}^{-1} - a_{11}} - 1 \right) \mathrm{CR}_2 + \mathrm{MCR}_2^+ \tag{17}$$

In principle, thermodynamic information on the activation energies for the different processes can be obtained by measuring the relaxation times at different temperatures. However, a detailed study of the relaxation amplitudes can lead in addition to the determination of thermodynamic quantities of reaction such as enthalpy, entropy, and volume changes. The relaxation amplitude analysis implies the calculation of the "reaction capacity" [2], expressed by

$$\Gamma_r = \left[\sum_i \frac{\tilde{\mu}_{ri}^2}{C_i} + \Sigma \, \tilde{\mu}_{ri}^2 \frac{d \ln \gamma_i}{dC_i} \right]^{-1} \tag{18}$$

wherein $\tilde{\mu}_{ri}$ are the elements of the stoichiometric matrix (16), C_i are the equilibrium concentrations, and γ_i are the activity coefficients. In the case of a multiple equilibrium system, this analysis would be impossible without a previous knowledge of the reaction normal modes and their stoichiometry.

As a consequence of his amplitude considerations, Chock, in his study mentioned above, suggests that a conformational equilibrium occurs before the cation complexation by the crown ether. Assuming a mechanism based on a straightforward complexation,

$$\mathrm{CR} + \mathrm{M}^+ \rightleftharpoons \mathrm{MCR}^+ \tag{19}$$

with an equilibrium constant K, the Γ function for this single-step scheme would be

$$\Gamma = \frac{1}{2K}\left[\frac{K^{-1} + [CR]_0 + [M^+]_0}{\{([CR]_0 - [M^+]_0)^2 + 2K^{-1}([CR]_0 + [M^+]_0) + K^{-2}\}^{1/2}} - 1\right] \tag{20}$$

The relaxation amplitude in his T-jump experiments would then have taken the form

$$A = \Delta p\, \Gamma\, \frac{\Delta H}{RT^2}\, \delta T \tag{21}$$

where $\Delta p = (\epsilon_{MCR^+} - \epsilon_{CR})l$, ϵ is the extinction coefficient, and l is the length of the light path. Relaxation amplitudes were obtained with a fixed high voltage to produce a constant δT from different initial temperatures. Were the mechanism (19) correct, a plot of A/Γ versus $1/T^2$ should give a straight line. The experimental results clearly dismissed this possibility. Several alternative explanations for this deviation were given including a more consistent mechanism given by (1) and (2). Chock's conclusion was that at least three configurations were present in the polycyclic ether studied: (1) an unreactive species; (2) an open configuration, which is the predominant species in the absence of cations, and (3) a closed configuration, which is stabilized by the monovalent cation.

This speculation led the present authors and collaborators to investigate the existence of such conformational equilibria in other water-soluble crown ethers by a subnanosecond ultrasonic technique. 15-Crown-5 [8] and 18-crown-6 [5–7] solutions in water presented concentration-independent relaxation frequencies at 23 and 101 MHz, respectively [5, 8], which were attributed to conformational equilibria in those particular crown ethers. Therefore Chock's two-step mechanism seemed more justifiable, and coupling between both reactions could be expected.

A study inspired by a method of Farber and Petrucci [19], which is discussed in another section, allowed the calculation of the equilibrium constant [7]:

$$K_{conf} = \frac{k_{b1}}{k_{f1}} = \frac{[CR_1]}{[CR_2]} = 1.6 \times 10^{-2}$$

for the 18-crown-6 conformational change. This, along with the fact that only a single relaxation frequency was experimentally observed for each cation complexation and that this one was at least three times slower than that observed for the crown alone, should make the off-diagonal terms in (10) so small that the coupling would not affect the independent equilibria.

In the case of 15-crown-5, the relaxation frequencies, both for the conformational change and for the cation complexation, fell in a narrow frequency range so that coupling should be considered.

Since the rate constants for the complexation process were known through consideration of the only observed relaxation time, information about the equilibrium constant for the conformational change had to be obtained from the relaxation amplitudes.

Using the Γ function with the proper normal mode reaction stoichiometry and guessing different values for $K_{conf} = k_{b1}/k_{f1} = [CR_1]/[CR_2]$ up to 0.1, the resulting calculated amplitudes did not show any concordance with those experimentally obtained.

Assuming no coupling between both reactions, the normal modes should be

$$CR_1 \rightleftharpoons CR_2$$

$$\frac{a_{12}}{a_{22} - a_{11}}CR_1 + M^+ \rightleftharpoons \left(\frac{a_{12}}{a_{22} - a_{11}} - 1\right)CR_2 + MCR_2^+ \qquad (22)$$

since in this case

$$\tau_I^{-1} = k_{f1} + k_{b1} = a_{11} \qquad (23)$$

$$\tau_{II}^{-1} = k_{f2}([CR_2] + [M^+]) + k_{b2} = a_{22}$$

The calculations based upon this new stoichiometry and upon values of $K_{conf} < 10^{-2}$ gave acceptable agreement with the experimental relaxation amplitudes. This is therefore a very interesting case since even though the two reactions occur at similar rates, the fact that $K_{conf} < 10^{-2}$ results in the two reactions occurring independently of one another.

3   ULTRASONIC METHODS

This section is devoted entirely to the discussion of ultrasound absorption, and specifically as it is applied to the study of equilibria in the liquid state. We shall concern ourselves with the phenomenon of sound absorption, information retrieval, instrumentation, and the types of systems which are amenable to study. Mathematical expressions needed to interpret ultrasound absorption spectra are given without the rigorous derivations for which the reader can refer to more exhaustive treatises (20–24).

Sound Absorption Phenomena

Acoustic power (P) and intensity (I) undergo an exponential loss with distance of penetration d as a plane progressive sound wave traverses an absorbing medium, according to the expressions

$$P = P_0 \exp(-\alpha d), \qquad I = I_0 \exp(-2\alpha d) \qquad (24)$$

where P_0 and I_0 are the values at incidence, $d = 0$. The quantity α is referred to as the sound absorption coefficient. It has units of dB/cm or neper/cm (8.686 dB = 1 neper, abbrev. Np) and varies in magnitude with the frequency of propagation f, increasing as the frequency is increased. Appropriately, chemical information is contained in the variation of α with f as opposed to transient perturbation techniques where the information is extracted from the time-dependent exponential decay of the initial displacement [25].

Two absorption phenomena contribute toward the magnitude of α, namely, classical absorption and relaxational absorption, whose coefficients are designated by α_{Cl} and α_{xs} respectively. The different frequency dependences of α_{Cl} and α_{xs} usually allow for their separation from the experimentally measured α value in (24).

Classical absorption is composed of two terms, one due to viscous losses, the other to thermal conductance or heat losses. Theoretical expressions for viscous losses [26] (Stokes) and heat losses [27] (Kirchoff) together provide us with an equation to calculate α_{Cl}:

$$\alpha_{Cl} = \frac{2\pi^2 f^2}{v^3 \rho} \left(\frac{4}{3}\eta + \frac{\gamma - 1}{c_p}K \right) \tag{25}$$

for any absorbing fluid, where v is sound velocity, ρ is density, η is coefficient of shear viscosity, γ is C_p/C_v, and K is thermal conductivity. It is apparent from (25) that α_{Cl}/f^2 is a constant, or that α_{Cl} increases linearly with f^2. For liquids, viscous losses far exceed heat losses [24], and adiabatic propagation is assumed. Commonly α(experimental) $\gg \alpha_{Cl}$. The difference is attributed to relaxational absorptions associated with the displacement of structural or isomerization equilibria, and so on, in the pure liquids [28]. In water, for example, $\alpha \simeq 3\alpha_{Cl}$ up to several hundred MHz, in accord with a very rapid equilibrium interchange between bulk water structures (see the "flickering clusters" model [29]). In theory the high-frequency limit of α should be α_{Cl} as the frequency of propagation f is increased to a value much shorter than the relaxation time(s) τ_R for solvent equilibrations ($2\pi f_R = \tau_R^{-1}$). Over the frequency range of a relaxation, the increase in α with f^2 is less rapid than the increase outside the range, and α/f^2 is not constant but decreases toward α_{Cl}/f^2. The situation is different for solute equilibrations in *solutions* in that as α/f^2 decreases it approaches the *experimental* value for the solvent (Fig. 4.1a). Data over an appropriate and sufficiently wide frequency range should enable us to kinetically characterize the relaxation processes, although in practice it is frequently very difficult.

There are two mechanisms for dissipating acoustic power in *relaxational* absorptions. These are associated with the pressure and temperature variations that accompany the sinusoidal acoustic wave. Efficient

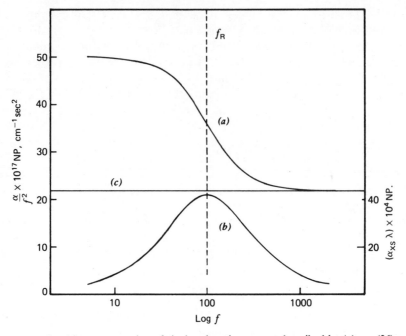

Fig. 4.1. Graphic representation of single relaxation curves described by (a) eq. (26) and (b) eq. (27); horizontal line (c) is the solvent background value. The curves were calculated using $f_R = 100$ MHz, $A = 28 \times 10^{-17}$ Np sec^{-1} cm^2 and $B = 22 \times 10^{-17}$ Np sec^{-1} cm^2.

coupling to the chemical or physical (i.e., structural) equilibrium requires that either ΔV^0 and/or ΔH^0 of the process be nonzero [3, 20]. The dominant work term in aqueous media is $P\,\Delta V^0$ since temperature fluctuations are very small [24] $(C_p \simeq C_v)$, $\Delta T \simeq 10^{-3}$°C. The reverse is true in many nonaqueous solvent media. In principle, the duration of the displacement is so short that the reaction cannot reestablish equilibrium instantly. A phase lag between pressure and volume results, and energy is absorbed. Absorption is greatest when the time it takes the reaction to return to the steady state (the relaxation time) is on the order of the inverse frequency of the acoustic wave. At much lower frequencies the pressure–volume changes are in phase with the acoustic signal, and at much higher frequencies pressure variations are so rapid that little if any displacement occurs.

From the mathematical theory of damped oscillatory impulses [12] the dependence of α on frequency is shown to be

$$\frac{\alpha}{f^2} = \frac{A}{1 + \omega^2 \tau_R^2} + B = \frac{A}{1 + (f/f_R)^2} + B \qquad (26)$$

for a *single* relaxation [see Fig. 4.1*a* for plot of eq. (26)], where ω, the angular frequency, equals $2\pi f$, τ_R is the relaxation time, A is the frequency-independent amplitude parameter, and B is the background or solvent (very high frequency, $\omega \gg \tau^{-1}$) absorption. An alternative equation is often used to analyze absorption data [30, 31]:

$$\alpha_{xs}\lambda = 2(\alpha_{xs}\lambda)_R \frac{f/f_R}{1 + (f/f_R)^2} \tag{27}$$

where $\alpha_{xs} = \alpha - Bf^2$, the subscript R refers to the characteristic frequency and amplitude of maximum absorption (Fig. 4.1*b*), and $\alpha\lambda$ is a dimensionless quantity referred to as the absorption per wavelength. Other relationships which arise from this analysis are that $\omega_R\tau_R = 1$; $\omega_R = 2\pi f_R$; $\tau_R^{-1} = 2\pi f_R$; and $A = 2(\alpha_{xs}\lambda)_R/vf_R$. A single relaxation extends over at least a decade in frequency [24, 32], and, in order to characterize it, data must be obtained over a wide frequency range. A fundamental and perhaps erroneous assumption commonly made is that the values of B and v are identical for solvent and solution. At low solute concentrations the approximation appears to be valid.

A more serious difficulty occurs when multiple chemical relaxations are present, since these are distinguishable as single relaxations only if successive f_R values differ by at least a factor of 2. Multiple relaxation spectra are treated as additive single relaxations:

$$\frac{\alpha}{f^2} = \sum_{i=1}^{i} \left[\frac{A_i}{[1 + (f/f_{R,i})^2]} \right] + B \tag{28}$$

Experimental error in α and the uncertainty in i compounds the problem of fitting experimental data to eq. (28). We return to this point in the next section.

Kinetic information about the systems is contained in the relaxation frequency (see later), and thermodynamic information is contained in the amplitude of maximum absorption. Factors that affect the magnitude of $(\alpha_{xs}\lambda)_R$ are the solvent, the temperature, the mole fractions of reactants and products, and most importantly the adiabatic compressibility [11, 20]. Dividing the adiabatic compressibility into two terms, one the compressibility at infinite frequency, and the other a frequency-dependent term, a full mathematical treatment of the latter provides an equation that relates A to ΔV^0 and ΔH^0, the standard molar volume and enthalpy changes for the system perturbed:

$$A = \frac{2\pi^2 \rho v \tau_R}{RT} \left(\Delta V^0 - \frac{\theta_0}{\rho C_p} \Delta H^0 \right)^2 K_c \frac{\partial \theta}{\partial K_c} \tag{29}$$

where ρ is the density, θ_0 is the coefficient of thermal expansion, θ is the extent of the displacement from equilibrium, and K_c is the equilibrium constant. The expansion of $K_c(\partial\theta/\partial K_c)$ depends upon the particular reaction under investigation [20]. In aqueous media, A is commonly taken to be proportional to $(\Delta V^0)^2$, but Stuehr has emphasized the caution that one must exercise in making such an assumption [33] because of contributions from ΔH^0.

In summary then, accurate ultrasound absorption data collected over a very wide frequency range can potentially provide us with important kinetic and thermodynamic information about equilibrium systems that relax on the appropriate time scale and that efficiently couple with the acoustic wave. In passing, it should be noted that a velocity dispersion also occurs over the frequency range of the relaxation. For liquid systems, however, the magnitude of the dispersion is too low ($\sim$2%) to be of value [21, 33], especially since very accurate velocity measurements are exceedingly difficult to make.

Data Analysis

For accuracy in absorption measurements one would obviously want α_{xs}/f^2 to be much larger than B. Since the total α is measured and this increases rapidly with f^2, the smaller the absolute value of B is, the better is the accuracy. Water, for example, has a comparatively low B value, equal to 22×10^{-17} Np(sec^2)/(cm) at 25°C, or a value for α of 55 Np/cm at 500 MHz. Excess absorptions, expressed as α_{xs}/f^2, are most often in the range of (10 to 100) $\times 10^{-17}$ Np(sec)^2/(cm). A is also dependent upon τ_R (i.e., inversely proportional to $2\pi f_R$) [see eq. (29)]. At high frequencies, therefore, A (and as a result α_{xs}/f^2 will be small, usually toward the lower limit of the above range, and is calculated as the difference between two very large numbers, $(\alpha/f^2) - B$. An additional difficulty at high frequencies is the lack of precision in measuring the very short distances d. At low frequencies absorptions are very small. For example, at 10 MHz α ($= Bf^2$) for water is only 0.022 Np/cm. In this case errors in the measurement of α can be as large as the absolute magnitude of α_{xs}. Distances are so long that it becomes inconvenient to measure them in the laboratory. If asked to define an arbitrary frequency range where experimental accuracy might be considered "optimum," we would, from our experiences, propose the range from 25 to 350 MHz. Fortuitously the majority of chemical and physical equilibria in liquids studied have had relaxation times that do fall within this range.

To predict whether an absorption will be observed, one can only project an approximate value for A [eq. (29)]. This would require prior knowledge of the thermodynamic parameters for the system under investigation.

Predictions can be made using known values of either ΔH^0 or ΔV^0 as well as K_c. In some instances it is possible to substitute ΔS^0 for ΔV^0 via Maxwell's equations [34, 35]. A rule of thumb, as good as any, is that K_c should be on the order of 10^3. There is no a priori way to predict even a possible frequency range for f_R.

The problems in data interpretation are many, apart from the inherent errors in measurement at low and high frequencies. There is sufficient evidence to substantiate the fact that the solvent B value is changed on the addition of solute [36, 37]. The convenient experimental range of 25 to 350 MHz does not encompass even two decades in frequency, and many times even a single relaxation spectrum is incomplete. Deconvoluting a multirelaxation spectrum into its component single relaxations can lead to more than one plausible and defensible interpretation [38–40]. Lastly, strong coupling between relaxations complicates kinetic interpretations in that there is no longer a one-to-one correspondence between a relaxation time and the rate constants for a particular step in the overall mechanism [41]. A very wide experimental frequency range is mandatory, necessitating usually the availability of at least two different ultrasonic absorption measuring instruments in a given laboratory.

The consequences of presuming prior knowledge of the magnitude of B have been well described by Stuehr [20]. A small error in B ($\pm5\%$) at high frequencies can significantly increase the f_R value or, worse, produce a "relaxation" which in reality does not exist. Fitting experimental data to (27) either numerically or using an overlay or template is discouraged in that the factor B has to be subtracted out. The most exemplary analysis is to keep B as an adjustable parameter and to fit the data to eqs. (26) or (28). One can later resort to a log–log plot of (27) which gives an excellent pictorial presentation of the relaxation step(s) (Fig. 4.1b).

Two approaches are employed to fit data to (26) for a *single* relaxation, either graphically or by a computer-fitting procedure. Proceeding graphically, (26) is rearranged into a *linear* equation such as

$$\frac{1}{\alpha/f^2 - B} = \frac{1}{A} + \frac{f^2}{f_R^2 A} \tag{30}$$

according to Piercy and Lamb [42], or into

$$\frac{\alpha}{f^2} - B = -\frac{\alpha - Bf^2}{f_R^2} + A \tag{31}$$

an equation whose original use is attributed to Andreae [43]. As a word of caution we would add that the experimental data will fit these linear equations *only* over the frequency range [43] of the relaxation and *only* if the proper B value is used [20]. Deviations from linearity which are due to improper B values have been described [20]. But there is no satisfactory way to determine if, in an effort to find the proper B and obtain linearity, a low-amplitude absorption has been sacrificed. A commonly used pragmatic procedure is to restrict the search for B by incrementing its value over a narrow range which is plus or minus a few percent of the pure solvent value [44]. Two widely separated single relaxations can be analyzed similarly with one modification. The B value will apply to the higher-frequency relaxation only. The minimum in α/f^2 for the low-frequency relaxation will be $A_2 + B$, where A_2 is the amplitude of the high-frequency absorption.

Strongly coupled multiple relaxation spectra can only be solved conveniently by computer fitting [45] data to eq. (28). The procedure can be run without assigning a single value to any of the $2i + 1$ parameters [44, 46], although it is economical to assign reasonably wide limits to each. An iterative grid-search procedure is then used to converge upon the narrowest limits which are consistent with the minimum root-mean-square deviation between the point-by-point simulated and experimental curves [44, 46]. That is, the expression

$$s = \left\{ \sum_1^n \left(\frac{[(\alpha/f^2)_{\text{exp}} - (\alpha/f^2)_{\text{calc}}]^2 W}{n - p} \right) \right\}^{1/2} \qquad (32)$$

is minimized for each chosen value of i, where n is the number of data points. Obviously the greater i is, the lower the value of s. One must again rely upon chemical intuition to decide upon the number of contributing relaxations. The weighting of the data W as a function of frequency must be considered with great care. A useful analysis of how errors in A, B, and f_R are propagated has been demonstrated for single and double relaxations [47] and how these might be distinguished in cases of close coupling between relaxations.

Data Interpretation

The primary utility of the ultrasound absorption technique is as a kinetic tool. With improved instrumentation and procedures for data analysis we can expect more and more thermodynamic interpretations from the spectra.

The time of relaxation of a perturbed equilibrium is a function of both the forward and backward rate constants which can be expressed collectively by an effective specific rate constant, k_{eff} [3, 12, 48]:

$$k_{\text{eff}} = \tau_R^{-1} = 2\pi f_R \qquad (33)$$

The analytic form of k_{eff} depends upon the molecularity of the reversible steps. For the equilibrium

$$A \underset{k_b}{\overset{k_f}{\rightleftharpoons}} B$$

$k_{eff} = k_f + k_b$, which shows no dependence of f_R on solution concentration, and therefore the two rate constants are not separable. For

$$A + B \underset{k_b}{\overset{k_f}{\rightleftharpoons}} AB$$

$k_{eff} = k_f(\bar{A} + \bar{B}) + k_b$, and it is possible to determine both constants from a concentration dependence of f_R (the bars refer to equilibrium concentrations of A and B). Expressions for k_{eff} for other stoichiometries are available [2, 3, 48]. These increase in complexity if (a) the overall reaction is multistep and (b) the reactant concentrations contribute substantially to the ionic strength of the medium, for example, in the absence of excess inert electrolyte. If one is dealing with a single-step equilibrium and a single relaxation, or with widely separated relaxation frequencies in a multistep reaction, the interpretation is relatively straightforward, especially if the overall equilibrium constant is available. Strong coupling between relaxations greatly complicates the interpretation in that the number of steps in the mechanism may exceed the number of separate relaxations observed [3, 11]. In these cases more than two rate constants are associated with each f_R. Equations for these interpretations have been worked out in detail [3, 11].

The most serious criticism of the technique is the uncertainty in the number of relaxations. Contrast this with relaxation methods which use spectrophotometric detection where changes in optical density at a chosen wavelength directly identify the process that has been perturbed. The only criterion we have to test that a proposed mechanism has validity is that the rate constants form an internally self-consistent set with the equilibrium constant for the overall reaction [48]. It is essential therefore that the ultrasound absorption measurements are made under precisely the same experimental conditions of temperature and ionic strength that were used in the determination of the equilibrium constant.

Activation parameters are obtained from a temperature-dependent study of specific rate constants or of the relaxation time. A very interesting and detailed derivation of the latter is described by Farber and Petrucci [19] for a reaction of the type $A \rightleftharpoons B$ and for which it will be remembered that τ_R is concentration independent. According to the transition state theory, they show that

$$\ln\frac{\tau^{-1}}{T} = \left[\ln\frac{k}{h} + \frac{\Delta S^{\ddagger}}{R}\right] - \frac{\Delta H^{\ddagger}}{RT} \tag{34}$$

which yield $\Delta S^{\ddagger}$ and $\Delta H^{\ddagger}$ from the slope and intercept, respectively. Equation 34 can be used only where K_c is known because τ_R is a function of *two* rate constants and we must be able to express one in terms of the other, namely, $k_f = K_c k_b$.

Reference has already been made to the care which must be exercised in deriving thermodynamic parameters from amplitudes A, even when K_c is known, eq. (29). It is possible, but inherently more difficult, to determine both ΔV^0 (or ΔH^0) and K_c simultaneously under special conditions [19]. One would like to have a system where either $\Delta V^0 >> \Delta H^0$ (or vice versa) so that essentially only one of the functions contributes to A. Making no assumption about the value of K_c in the reaction of type $A \rightleftharpoons B$, eqs. (35) and (36), which describe the temperature dependences of τ_R and $\alpha_{xs}\lambda$, can be solved simultaneously [19] to give K_c, ΔH^0, ΔS^0, $\Delta H^{\ddagger}$, and $\Delta S^{\ddagger}$:

$$\frac{d \ln (\tau^{-1}/T)}{d (1/T)} = -\frac{\Delta H^{\ddagger}}{R} - \frac{K_c}{1 + K_c}\frac{\Delta H^0}{R} \tag{35}$$

$$\frac{d \ln [(\alpha_{xs}\lambda)_R T/\rho v^2]}{d(1/T)} = \frac{K_c}{1 + K_c}\frac{\Delta H^0}{R} \tag{36}$$

The solution requires the use of (34) and the fact that $\Delta X^0 = \Delta X_f^{\ddagger} - \Delta X_b^{\ddagger}$, *where* X is either H or S.

The need for very precise absorption data for all of these interpretations cannot be overemphasized.

Experimental Methods

The experimental techniques developed for ultrasound absorption measurements have been described in the greatest detail a number of times [3, 12, 20, 49]. Rather than simply repeat these details here, we have chosen to give a general outline of the methods, which are summarized as a function of their useful frequency ranges in Fig. 4.2, and to enlarge upon the most recent developments. The rapid increase in α (Np/cm) and the concomitant rapid decrease in the distance parameter d needed for measurement of α means that neither can be measured with the necessary precision over the entire frequency range on only one instrument. For water at 10 and 500 MHz, for example, $\alpha = 0.022$ and 55 Np/cm, respec-

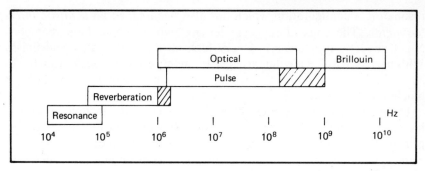

Fig. 4.2. Experimental techniques vs. frequency range of applicability. The hatched areas correspond with recent, innovative extensions.

tively, and the corresponding d intervals for a measured 10 Np change in acoustic power are 454.4 and 0.18 cm.

Complete instruments are not commercially available but are relatively easily assembled from standard components [49]. The beginner needs only a modest knowledge of electronics and optics (in some instances) and access to a first-class machine shop. Rarely are all of the methods available to one investigator. Where more than one technique has been used to characterize a system, there has usually been good spectral correspondence between the methods (this includes measurements made in different laboratories (see Fig. 4.4)).

The design characteristics of the methods employed for frequencies < 1 MHz were evolved to meet the problems that absolute absorptions are very small and propagation distances must therefore be very long. The *resonance* and *reverberation* methods are technically the same. They both depend upon multiple passes through the absorbing media by reflection from the walls of the container. In the resonance procedure unique spherical vibrational modes are excited to cover the frequency range from a few kilohertz to several hundred kilohertz. In contrast in the reverberation technique, a large number of modes are simultaneously excited. Attentuation of power is measured either by measuring the rate of decay after the sound is switched off or from linewidths of resonance peaks. The usual difficulties encountered are losses on reflection, which are accounted for either by successive measurements on solvent and solution or by calibration of the vessel using a liquid of known absorption [50] and the fact that volumes must be very large, often as much as several liters.

A recent innovative development [51] has increased the versatility of resonance methods. The assembly is a cylindrical rather than a spherical resonator, a configuration which has also been called an *acoustic inter-ferometer*. The ends of the cylinder are two matched, large-diameter, X-cut acoustic transducers of very high mechanical quality and finish. With perfect parallel alignment, measurements could be made from 0.5 to 30 MHz on 1 ml samples, making the system ideal for the study of biochemical systems.

The *pulse* method has been the workhorse of all the techniques covering an approximate 5 to 250 MHz range. It works on the send–receive principle and requires a matched pair of transmitter–receiver transducer elements. In some modifications one transducer serves both functions after reflection of the signal. The short duty-cycle eliminates excessive local heating. In this operating mode a second comparison pulse is separated by a time delay circuit. The first pulse traverses the sample while the comparison pulse is passed through a network of precision attenuators. The rest of the circuitry (pulse shaping, modulation, amplification, etc.) is common to both signals, which are ultimately displayed for comparative matching on an oscilloscope. The transducer elements are usually fastened to delay rods made from low-attenuation materials to better separate the signals and to eliminate spurious signals from "cross-talking." Although α is usually obtained as the slope of the line of attenuation versus distance at each frequency (the fundamental and odd harmonics of the transmitter crystal), a system has been developed where the path length is kept constant [51]. This approach requires precise calibration but does allow for measurements to be made on 1 ml samples.

In the *optical* method [52, 53], sound attenuation is determined by measuring the directly proportional changes in intensity of a diffracted light beam as a function of d. A plane progressive ultrasonic wave sets up a series of regions of compression and rarefaction in the medium that act as a diffraction grating to an incident light beam [54]. The amount of the change in the index of refraction caused by the acoustic wave, and therefore the efficiency of the grating, is proportional to the acoustic power [55]. Linear correspondence of sound and light intensities exists only for the zero (undiffracted) and first-order diffractions. Absorption measurements are normally made on the first-order diffracted beam, and velocities can be obtained from the spacings of higher orders where present. Using white light as source, the useful acoustic range of the instrument is 5 to 100 MHz.

A *laser acousto-optic assembly* [55–57] has now been developed (Fig. 4.3). This system operates in the low MHz ultrasonic region of ≤ 25 MHz, where the interaction of light and sound is known as the Debye–Sears effect [52], and exceeded all expectations by extending the upper limit to ~350 MHz, that is, over the Bragg diffraction [54] frequency range, > 25

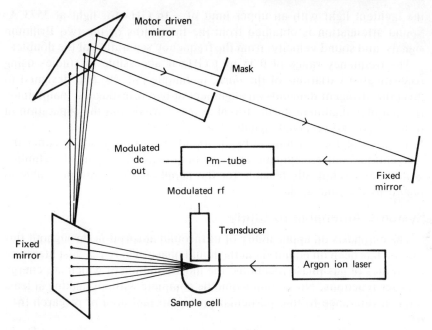

Fig. 4.3. Schematic diagram of the laser acousto-optic technique (LAOT) developed at the University of Utah.

MHz. At low frequencies many orders of diffraction are observed but only the first-order diffracted spot is evident in the Bragg region. The entire frequency range can be covered using only one X-cut quartz crystal, for example, a 5 MHz fundamental crystal can be successfully driven up to the 69th harmonic. Intensity measurements as a function of d are made either by parallel displacement of the transmitter or the laser beam. Alignment problems are not as critical as they are in the pulse technique. Differential solution–solvent spectra are technically easier to obtain [56]. Not only are the data more precise ($< \pm 2\%$) than those typically available from the pulse technique ($\pm 5\%$), but also the points can be accumulated at least ten times as rapidly through interfacing the system to a computer [44, 46, 55, 57]. Key contributing factors that make possible the increased range, speed, and precision are the stabilized argon ion laser light source and the management of alignment and transducer displacement via a system of computer-controlled stepping motors [57].

Beyond 350 MHz in frequency, the intensity of the first-order diffracted beam is so low as to be almost undetectable. From 1 GHz and upward in frequency, *Brillouin scattering* [37, 58], another optical method, is used. The practical acoustic frequency ranges depend upon the wavelength of the incident light with an upper limit of ~16 GHz for light at 2573 Å. Sound attenuation is obtained from the linewidths of the two Brillouin signals, and sound velocity, from the frequency separation of the doublet.

The frequency range of 0.35 to 1 GHz is accessible to studies using sophisticated variations of the *pulse* technique [40, 59, 60], designed to meet the stringent demands on alignment of two transducers, maintaining this parallel alignment during travel as d is varied, and the separation of transmitted and received signals.

To investigators wishing to begin sound absorption studies, either the conventional pulse technique or, even better, the laser-optical technique is the most technically feasible for the neophyte and more profitable in regard to the time scale.

Systems Amenable to Study

The emphasis on applicability of ultrasound absorption throughout this section has been on equilibria in the liquid phase. There is a vast literature on studies of gas-phase reactions, the most common of which are energy transfer reactions. No section would be complete without making at least a token reference to that particular and important area of research [61–63].

Repeating the requirements for an observable absorption in liquid media, these are: the reaction must be in dynamic equilibrium; τ_R^{-1} should be in the frequency range 10^4 to 10^9 Hz; and ΔH^0 and/or ΔV^0 should be nonzero. Obviously a wide variety of equilibria meet these requirements. The range of applicability is potentially wider than for T-jump, P-jump, and E-jump because of additional requirements for these techniques imposed by the methods used for detection, for example, spectrophotometry and conductance. It should be remembered, however, that it is sometimes difficult to associate an acoustic relaxation with a particular equilibrium process.

We have classified systems studied under *pure liquids*, *binary liquids*, and *dilute solutions*. Examples from each are reviewed in turn. Also included in each category are *critical mixtures*, which are really different enough to constitute a fourth subdivision.

Pure Liquids

Equilibria studied include *structural, rotational, conformation,* and *dimerization* equilibria and processes in *liquid crystals*.

Under *structural* equilibrations we include those processes for which the pure solvent absorption is in excess of the classical value [36], eq. (25). The working model is to consider the liquid (ionic melts are included) as having a pseudolattice structure [64] and for a very rapid equilibrium to exist between structured and "free" solvent molecules. An estimated τ_R for water [65] is $\sim 10^{-12}$ sec, which is beyond the highest accessible ultrasonic frequency obtained. The contribution from ultrasound investigations therefore is to study the dependence of B on temperature, pressure, and solute (especially $1:1$ electrolytes and urea [66]) and to interpret these in terms of changes in equilibrium distributions in the structural aggregates.

The remaining four categories would be identified by relaxations at much lower frequencies, that is, a frequency-dependent α/f^2. Kinetically *rotational* and *conformational* equilibria are described by the reaction type $A \rightleftharpoons B$. Rate constants are not separable unless K_c is known. Rotational equilibria studied include C–O, C–N, and C–C bond rotations in esters, amines, aldehydes, and substituted alkanes [67]. Spectra are usually simple, single absorptions, with strongly temperature-dependent relaxation times, from which activation and thermodynamic parameters can be obtained, eqs. (34) to (36). The classical works on conformational equilibrium relaxations are on the substituted cyclohexanes [24, 68], which were identified not as boat–chair interconversion but equilibria between axial and equatorial positions for the substituents. Heterocyclic ring structures have also been investigated. In these instances boat–chair equilibria are perturbed [69]. Although originally it was assumed that $\Delta V^0 \simeq 0$ ml mole^{-1}, there is reason to believe the assumption is in error.

Dimerization is covered under the section on dilute solutions.

Ultrasound absorption has been applied to the study of nematic phases of *liquid crystals* [70] and to phase transitions [71], for example, isotropic–nematic, isotropic–cholesteric, and cholesteric–smectic. Like other spectroscopic results, the absorption α/f^2 by the ordered nematic mesophase, at a fixed frequency, is anisotropic, varying with the angle of orientation θ_i of the acoustic signal relative to the plane of alignment of the liquid molecules. The angular dependence of attenuation is related theoretically [72] to five viscosity coefficients three of which are frequency independent; the two remaining bulk viscosities exhibit a marked relaxation. While α/f^2 at a fixed θ_i is constant with frequency for some nematics [72] (*p*-azoxyanisole), a low-frequency "relaxation" is observed for others [74] (MBBA). The singularity and origin of the "relaxation" are in question. These problems can only be solved when more low-frequency data are available.

Absorption data for phase transitions from isotropic to either nematic or cholesteric mesophases show that these are typical first-order transitions [71], being quite analogous, for example, to data for xenon at the transition from vapor to liquid at the critical density [75]. The behavior of plots of α/f^2 versus temperature over the transition range is equivalent to that for binary mixtures at and around the critical temperature (see next section). On the other hand, the cholesteric-to-smectic transition for cholesteryl caprate is totally different [71], showing no very sharp peak in the plot of $\alpha\lambda$ versus temperature.

Binary Liquids

Mutual solvation and *molecular aggregation* are the two equilibrium processes that are amenable to an acoustic study in these systems [76]. Information is obtained from four presentations of α/f^2, namely, versus molar composition X_A, frequency, temperature, and pressure. In each case the remaining three parameters are constant. The change in α/f^2 with mole fraction is not monotonic over the entire composition range, and, where molecular association is present, it frequently maximizes at a PSAC (peak sound absorption composition) [77], from which models for solvation or aggregation can be derived. The kinetics of dimerization or higher aggregation equilibria are examined in the usual way, that is, from α/f^2 versus f. Most of the systems studied were short-chain alcohols as mixtures or with water.

An unusually high α/f^2 is observed at the critical temperature T_c of a binary mixture, defined as the temperature at which a homogeneous mixture at a certain X_A separates into two homogeneous phases. In fact, theory predicts that at T_c, $\alpha/f^2 \rightarrow \infty$, which can be demonstrated experimentally in a plot of α/f^2 versus T. An early theory of aggregates scattering the sound beam [78] has been dismissed in favor of localized composition-fluctuation theories [79]. The excessive sound absorption is linked with an anomalously high heat capacity of the mixture near the critical point.

Dilute Solutions

Equilibria discussed in this section involve *electrolytes, dimerization, conformation changes, aggregation,* and processes in *metal-ammonia* solutions.

Two categories of *electrolyte* equilibria which have been of primary interest are proton transfer [80] and metal ion association reactions [11, 22, 23]. The classical examples of each are the ionization of aqueous ammonia [81] and the ion association [82, 83] of $MgSO_4$ which accounted for the excess absorption in sea water compared to "fresh" water. Both were subsequently studied as a function of applied hydrostatic pressure [84, 85].

Expressing proton transfer in general terms as one-step reactions

$$H_3O^+ + A^- \rightleftharpoons HA + H_2O \tag{37}$$

and

$$OH^- + HA \rightleftharpoons H_2O + A^- \tag{38}$$

the reciprocal relaxation time is given by

$$\tau_R^{-1} = k_f \Pi_f (\Sigma \ \bar{C}_R) + k_b \tag{39}$$

where Π_f is an activity coefficient correction introduced to account for the change in ionic strength during the perturbation and $\Sigma \ \bar{C}_R$ is the sum of the reactant equilibrium concentrations. Except in cases of reactions involving hindered molecules, k_f approaches the diffusion-controlled limiting rate constant [86] $k_D > 10^{10} \ M^{-1}$ sec $^{-1}$. Variations in equilibrium constants are therefore reflected in the values for k_b. As a general rule ΔH^0 for these reactions are small in aqueous solution, and amplitudes are principally determined by the ΔV^0 terms which are large (~ 20 ml/mole) due to solvent electrostriction by the ions. Studies have since been extended to include a number of dibasic acids [87]. Little if anything has been done in solvents other than water.

For nonassociating salts (1:1 electrolytes) the solvent B value is progressively lowered with increasing concentration [36, 37], a change which is attributed to a modification of the water structure. Salts of higher stoichiometry produce multiple relaxations over a wide frequency range [11, 12, 20, 22, 23, 48] (Fig. 4.4). Presumably they also alter the value of B. This cannot be proved experimentally, but it certainly justifies leaving B as an adjustable parameter in the analysis of spectra.

There is general agreement that the multiple relaxations are the result of a stepwise approach of the ions from infinite separation to ultimate associative contact, via an uncertain number of intermediate solvent-separated ion pairs of fair stability [88]. Initially, the approach is diffusion controlled. Subsequent steps are slower as the primary hydration sheaths of the ligand (anion) and metal ion are opened to substitution. The final substitution is rate determining. The gist of the uncertainty over the exact number of steps is the uncertainty in the number of relaxations. Much of this is due to the difficulty in obtaining sufficient accuracy at high frequency. For example, early reports [81, 83] placed a high-frequency relaxation around 200 MHz, while more recent reports [40, 89] suggest that 400 MHz is a more realistic value.

Metal complexations have been the subject of most reviews on sound absorption in solutions [3, 11, 12, 20, 22, 23, 48, 64]. The reader is referred to these for the many details of the mechanism, spectral interpretation, and data handling, as well as the controversies over the number of steps. Divalent and rare earth [48] metal sulfates and acetates are the systems

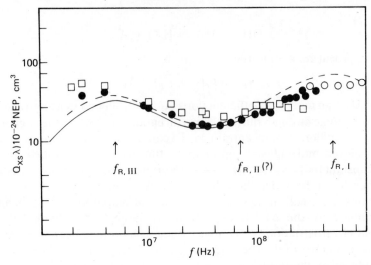

Fig. 4.4. Montage of recent experimental data for aqueous $MnSO_4$. Q_{xs} is substituted for α_{xs} to normalize all data to the same concentration and is called the absorption cross section per molecule. The three relaxation times correspond with three steps in the ion association mechanism. $f_{R,II}$ is controversial. With permission from N. Purdie and M. M. Farrow, *Coordination Chemistry Reviews*, **11**, 189 (1973).

that have been of most interest. Association of divalent metal sulfates in alcohol–water mixtures [11, 22, 90] appears to proceed by the same mechanism because of preferential primary solvation of the ions by water. Acoustic studies of association, even aggregation [11, 64], of ions of 1:1 tetraalkylammonium salts and metal perchlorates in nonaqueous solvents have contributed to our understanding of the noncontinuum macroscopic properties of these liquids. There is evidence also of cation dimerization [64]. The complexation of cations by crown ether [5–8] discussed earlier in this chapter is of course another example of metal complexation that has been fruitfully investigated by ultrasonic absorption.

Dimerization is classified as a $2M \rightleftharpoons D$ reaction. A single low-frequency relaxation is typical of these reactions for which

$$\tau_R^{-1} = 2\Pi f_R = k_{\text{eff}} = k_b + 4k_f\bar{C}_M \tag{40}$$

where $\bar{C}_M$ is the equilibrium monomer concentration, or on rearrangement

$$\tau_R^{-2} = k_b^2 + 8k_f k_b C_0 \tag{41}$$

where C_0 is the analytic concentration. A number of hydrogen-bonded dimerizations in aqueous and nonaqueous media have been examined [20, 91]. Equilibrium constants and ΔV^0 from acoustics agree favorably with the values obtained by more conventional methods. The forward rate constant k_f again approaches the limiting diffusion-controlled value.

Conformational changes that have attracted much interest are the syn–anti rotations in nucleosides [92] and helix–coil transitions in polypeptides and polyamino acids [93]. Relaxations are in the KHz to low MHz frequency range. Possible complications from proton transfer equilibria are present which would require a careful pH-dependent study of the relaxations to be made to simplify the interpretation. Binding, even of univalent counterions, is a further complicating feature to be considered.

The solvent absorption for pure *liquid ammonia* is in excess of the classical value by an amount almost equivalent to that for water [94–96]. A solvent structural equilibrium is presumably in effect. There is no evidence for a chemical relaxation in dilute metal-ammonia solutions ($\sim 10^{-4} M$) since absorption and velocity measurements on these solutions at 1 and 10 MHz using the pulse technique and by Brillouin scattering [96] at 6 GHz are identical to those measured for the pure solvent. Another study of the ultrasound absorption of lithium-ammonia solutions as a function of concentration [95] shows a peak in α/f^2 at the critical mixture composition of 4.25 mole-% metal. As would be expected data are sparse for such an experimentally difficult problem, which may account in part for the apparent failure of the data to fit the theoretical models [79] that describe the critical phenomena. A "single" relaxation is again reported in the plot of α/f^2 versus frequency, but once again insufficient data are available either to fully characterize it or to identify its origin.

Conclusion

It is clear that many systems have kept relatively few researchers enthralled with the problems of making accurate ultrasound absorption measurements and with data interpretation. It should be eminently clear that much remains to be done. We believe it is an auspicious time for more investigators to become involved. The obvious immediate projections where activity will increase are in the areas of equilibria in nonaqueous solvents and in biological and biochemical systems. We can also expect an increased emphasis in obtaining thermodynamic data from absorption amplitudes.

4 TRANSIENT METHODS

In the time scale range with which we are concerned, only certain versions of the temperature jump (T-jump) and electric field jump (E-jump) relaxation techniques can be used to measure relaxation times shorter than a microsecond. We devote this section to the discussion of some of those versions that have been recently developed.

Temperature Jump

In the few years since Chock's pioneering temperature jump study, nuclear magnetic resonance and ultrasonic absorption techniques have produced a wealth of kinetic data on monovalent cation complexation by a variety of crown ethers [5–8, 97–100]. While the trends in the resulting rate constants (see Table 4.1) are interesting, more thermodynamic information would have been made available if these kinetic studies had been carried out on the necessary submicrosecond time scale with temperature jump instrumentation. Workers in the fast reaction field have long recognized the greater intrinsic power of the temperature jump relaxation technique to elucidate rapid reactions and have therefore invested a great deal of effort in the development of new instrumentation that would extend T-jump measurements to shorter (and longer) time scales.

Working in Eigen's laboratory, Hoffman [101] found that he could effect a 10°C temperature rise in a sample solution by Joule (conductive) heating in a time as short as 50 nsec. Other techniques (dielectric or laser heating) can achieve the same result, but only at much greater cost. As we see in Fig. 4.5, Hoffman used a high-voltage (100 kV) coaxial cable impedance matched to a minute (40 microliters in volume) sample cell as the source of energy. Spectrophotometric detection is conveniently possible since the gold or platinum electrodes of the sample cell are well

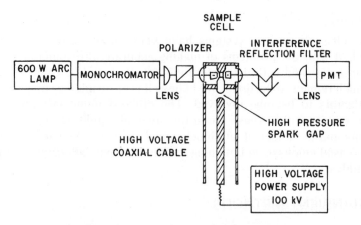

Fig. 4.5. Schematic of the Hoffman [101] cable discharge temperature jump apparatus with the optical detection system suggested by Pörschke [102].

separated by a distance of ~4 mm. A long time limit, imposed by convective cooling, of ~1 sec, is not worse than that of less sophisticated T-jump instruments. The use of a light pipe to carry the sample beam to a remotely located photomultiplier is one way of overcoming electronic noise pickup from the high-voltage discharge that would otherwise plague the detection of the small, fast signal. The only significant potential disadvantage of this coaxial cable discharge T-jump apparatus is that the sample cell must have an electrical resistance of not more than about 50 Ω. This can only be achieved in a high dielectric solvent (such as water) to which enough ions have been added to produce an ionic strength of ~0.1M. The optical detection system of Hoffman's apparatus has since been modified by Pörschke [102] to suppress artifacts such as electrodichroism and electroluminescence.

In addition to Joule (conductive) rapid heating of solutions, a number of studies have been reported of ultrarapid dielectric and laser heating of liquids. We will refer here only to recent, though otherwise representative, studies of each of these heating methods. The interested reader will find extensive bibliographies to the older literature in the papers cited.

At the time of writing this review, the most recent paper concerning a microwave (dielectric) T-jump apparatus was one written by Egozy and Weiss [103]. These authors noted that only five earlier descriptions of this type of T-jump instrumentation had been published including the first by Ertl and Gerischer [104] in 1961. Egozy and Weiss used a surplus radar power supply to drive a 5.3 GHz nominal frequency magnetron. A particular advantage of their apparatus (see Fig. 4.6) is that it can be operated in a repetitive pulse mode (~200 Hz) that facilitated optical absorbance signal averaging and therefore compensates for the modest temperature jumps that can be achieved (2°C in water, 0.5° in 1,2-dichloroethane, and only 0.3° in chloroform). The sample cell has a volume of only 70 μl and a respectably long optical light path of 10 mm, both of which are favorable features. The short time limit of the apparatus (dictated by electronic noise) is a rather long 5 μsec, whereas the long time limit of ~100 msec is set by convective cooling. If the potential subject of a microsecond time scale kinetic study is a low ionic strength solution in a polar solvent, a microwave T-jump apparatus, if available, would in principle be a promising experimental tool. A microwave temperature jump apparatus in the laboratory of Professor Ted Caldin at Canterbury has, for example, become one of the principal research tools there for studying proton transfers in aprotic solvents.

As the rather extensive recent literature [105] on the subject would suggest, the laser temperature jump techniques have been much more

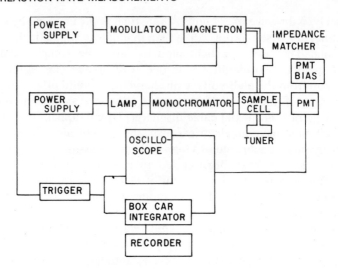

Fig. 4.6. Egozy and Weiss microwave T-jump apparatus [103].

successful than the microwave temperature jump in capturing the imagi-
nation of relaxation kineticists. Since laser physicists can now obtain
picosecond duration pulses of energy from mode-locked lasers, it is reasonable to ask how soon a laser heating T-jump apparatus will permit
routine relaxation method rate measurements on a similar subnanosecond
time scale. The problem is basically one of absorbing enough energy from
a single subnanosecond laser pulse to produce a spectroscopically detectable change in concentrations of a reactant species dissolved in the heated
solvent. One joule of energy from a laser pulse completely absorbed by
0.1 ml aqueous solution will raise the temperature of the solution by
~2.5°. There are temperature-sensitive acid–base indicator equilibria
such as those of aqueous azo dyes for which even this rather small
temperature rise produces a significant change in visible light absorbance.
The key problems are (1) finding a laser wavelength that is absorbed
neither too much nor too little by the sample solution, (2) compensating
for differences in ΔT at various points in the sample solution arising from
the fact that the laser pulse heats the layer of solution through which it
first passes more than subsequent layers (Beer–Lambert Law), and (3)
assembling a photodetector circuit sufficiently free of noise and sensitive,
yet fast enough to measure a small-amplitude nanosecond or even faster
transient signal.

The most interesting recent resolution of these difficulties was achieved
by Reiss [106]. The 25-psec pulses from a mode-locked neodymium(III)
solid-state laser were Raman shifted in liquid N_2 to a 1.41-μm wavelength

and used to heat an aqueous solution in which the 16 ± 5 nsec relaxation time of the equilibrium

$$I_2 + I^- \rightleftharpoons I_3^-$$

was measured spectrophotometrically. This particular piece of kinetic information was not new. Turner and co-workers (107) achieved the same result several years earlier using a Q-switched, Raman-shifted neodymium laser. The intriguing features of the Reiss experiment are (1) the potential for resolving much faster chemical relaxation times than those accessible to Turner et al., (2) the much higher (possibly fourfold) conversion efficiency of $\lambda = 1.06$ to 1.41 μm radiation by stimulated Raman scattering with 25-psec pulses (exceeding 80%), and (3) the absence of potential complications from Brillouin scattering in the Raman cell as well as other nonlinear effects that might have been anticipated from high-power laser pulses [108].

The laser temperature jump experiments of Turner and co-workers [107, 109–113] remain the most impressive of those done on a submicrosecond time scale because a substantial body of new chemical information has followed from the instrumentation effort. No important new chemical knowledge has so far followed reports [114] of success in T-jumping aqueous sample solutions with neodymium laser pulses Raman shifted in gaseous hydrogen (to 1.3 μm) or the use of 1.35-μm pulses from a gaseous iodine photodissociation laser [115]. This latter type of high-power laser is especially attractive since it is less costly to build than a comparably powerful neodymium laser. For under $10,000 in components, one can build a laser oscillator that emits ~ 5 J in a 2-μsec pulse and ~ 0.5 J in 100 nsec. Perhaps even more significantly, the gaseous laser medium is indestructible in contrast to the fragility of expensive glass or YAG laser rods. Neodymium(III) dissolved in an aprotic solvent such as $POCl_3$ can be made to lase [116] and clearly is less susceptible to damage than rods. However, there has only been one reported use [117] in a laser T-jump experiment, and the chemical systems studied (aqueous nickel(II) reacting with ammonia and Tiron) react slowly and have been kinetically investigated previously by less exotic methods.

If the worth of an experimental kinetic technique is measured by the number of chemical systems it has aided in elucidating, the nominal pulse ruby laser T-jump studies by Caldin and co-workers sweep the field [118]. Since these studies however have been carried out on time scales of milliseconds or longer, they lie outside the province of the present survey of methods.

Since none of the submicrosecond laser T-jump techniques mentioned above lends itself to inexpensive implementation, it is interesting to consider what would be required in a more versatile instrument of the same general high cost. The ideal apparatus would be adaptable to both flash kinetic spectroscopy of excited states and T-jump relaxation method studies of ground-state kinetics. The laser used to rapidly heat a sample solution would be a mode-locked, wavelength-tunable, near-infrared laser that could rapidly heat any of several different organic solvents as well as water by tuning to the appropriate solvent vibration wavelength. For the flash photolytic work, the same laser would produce a variety of shorter exciting wavelengths in the ultraviolet and visible. The pulse repetition rate would be high enough to make signal averaging possible, and each individual exciting pulse would have a power below the threshold for producing stimulated emission in a sample, thus avoiding spurious measured lifetimes of sample excited states [119]. For kinetic measurements on a time scale of 1 nsec or longer, the ultraviolet–visible analyzing light source could be a tunable CW laser or, more practically, an intense flash lamp–monochromator combination.

The more critical choice is the selection of the fast photosensitive detector. For instance, an RCA C31024 photomultiplier tube could be teamed with a Tektronix R7912 Transient Digitizer (0.7 nsec risetime, 10 mV/cm maximum sensitivity) to record relaxation times of the order of a nanosecond or longer. For shorter time scales, a portion of the exciting laser beam could be split out, delayed by reflections through a somewhat longer light path, and focused to create a white-light spark, analyzing continuum for photographic recording with a streak camera.

Many features of this idealized laser T-jump have been incorporated into an experiment described by Giannini [120]. He has used a Nd:YAG solid-state laser (Chromatix Model 1000) coupled with a tunable optical parametric oscillator (Chromatix Model 1020) to heat aqueous solutions with 100-nsec, ~ 3 mJ, $\lambda = 1.36$ μm pulses at a 75-Hz repetition rate. Water absorption is good at this wavelength, and a $\Delta T \sim 1°C$ is achieved by focusing the laser beam into a smaller volume within a ~ 1.5 ml flowing sample cell. Rüppel and Witt have described [121] the signal-averaging techniques on which Giannini has relied. When an inexpensive pulsed laser becomes available that tunes over a broader range than the 600 nm to 3 μm accessible with Giannini's Chromatix laser, a wider utilization of the laser T-jump technique will surely develop.

Electric Field Jump

Just as the laser T-jump technique of Turner et al. [107, 122] has been described in sufficient detail to be fairly readily imitated [123], the electric field jump (E-jump) relaxation technique has also been frequently described [3, 124–133] but somehow not in a manner—either because of language barriers or an excess of detail on the latest innovations—to inspire frequent imitation. It should be conceded, however, that except in rather specialized circumstances, the E-jump relaxation technique is a less promising method of measuring microsecond to subnanosecond reaction processes in sample solutions than the high-voltage (Hoffman), microwave, or laser T-jump techniques.

For a chemical reaction to be susceptible to study by the E-jump relaxation technique, it is necessary that the equilibrium constant be affected by the application of a strong external electric field. Such reactions fall into two different classes: (1) those which involve a separation of electric charge, such as dissociation of weak electrolytes [134], and (2) those for which the chemical transformation involves a change in the dipole moment [135]. We will only be concerned in this section with those of the first type, to which so far the most attention has been dedicated.

Interaction of Electric Fields with Ionic Equilibria

It was long ago observed by Wien that the conductance of electrolytes increases with the intensity of the applied external electric field, conferring on Ohm's law a strict limitation in its application. For solutions of strong electrolytes [136], this effect is rather small (first Wien effect), and its interpretation has been successfully based upon the destruction of the "ionic atmosphere." In the case of weak electrolytes [137], the enhancement of conductance is much more pronounced, presenting a linear increase over a considerable range of field intensity; the limit of this increase corresponds to the complete dissociation of the electrolyte present. This fact, known as the second Wien effect, or "dissociation field effect," was theoretically explained by Onsager [134] with application to very dilute solutions and on the basis of interionic interaction in an electric field.

Onsager's theory expresses the relative increases of the dissociation constant by a series expansion:

$$\frac{K(E)}{K(0)} = 1 + 2\beta q + \frac{(4\beta q)^2}{2!3!} + \frac{(4\beta q)^3}{3!4!} + \cdots \tag{42}$$

where

$$q = \frac{-z_1 z_2 e_0^2}{2DkT} > 0$$

$$2\beta = \frac{|Ee_0 (z_1 u_1 - z_2 u_2)|}{kT(u_1 + u_2)} \tag{43}$$

$K(0)$ is the equilibrium constant for the system in the absence of an external field, $z_i e$ are the ionic charges of the ions, E is the intensity of the electric field (in electrostatic units), D is the dielectric constant of the solution, and u_i the ionic mobilities. The quantities q and $(2\beta)^{-1}$ have the dimensions of distance. The quantity q is called the "Bjerrum radius" representing the distance at which two approaching ions acquire coulombic interaction energies greater than $2kT$. The quantity $(2\beta)^{-1}$ determines the distance at which two charges of opposite sign and equal magnitude will form a dipole with an energy $-kT$ when aligned in the same direction of the electric field.

According to (42), the variation of the equilibrium constant with the applied field E can be expressed in an asymptotic approximation by

$$\frac{d\ln K}{d|E|} = \frac{z_1 u_1 - z_2 u_2}{u_1 + u_2} \frac{|z_1 z_2| e_0^3}{2D(kT)^2} \tag{44}$$

In the most frequently used version of the electric field jump experiments, applied to a chemical equilibrium involving an incompletely dissociated electrolyte, a change in the electric field intensity ΔE of the order of 10^5 V/cm is effected so rapidly in the sample solution that it is essentially a step function on the time scale of the chemical reaction that is responding to the dissociation field effect. A classic example is the dissociation of aqueous acetic acid:

$$CH_3-COOH(aq) \rightleftharpoons CH_3-COO^-(aq) + H^+(aq)$$

which in an intense electric field greater than ~40 kV/cm is displaced by a few percent toward products [124].

Experimental Methods

We shall assume that the reader has at least a casual familiarity with pages 988 through 994 of Ref. 3 so that we may limit ourselves here to a practical discussion of some of the difficulties encountered with the square-pulse electric field jump method.

The rectangular-pulse electric field jump method lends itself to either conductimetric or spectrophotometric detection. In either case, a sample cell typically contains two bright platinum, gold, or stainless steel electrodes each with a 2-cm-diameter flat circular surface with the flat faces parallel to one another and separated by a distance of about 0.5 cm.

For conductimetric detection the exclusion of air and other impurities from the sample cell is critical since the sought-after change in solution conductivity will in many cases be less than 1%. Thus electrodes are usually sealed into an all-glass cell with a fairly elaborate arrangement of

Teflon stopcocks. The sample cell constitutes one arm of an asymmetrical, high-voltage Wheatstone bridge [126]. The objective is to observe a microsecond or submicrosecond exponential excursion from balance of the Wheatstone bridge. Since this can only be measured with a voltage on the bridge, the probes located at opposite corners of the bridge must withstand voltages of the order of 50 kV (to achieve an electric field strength of 100 kV/cm in a sample cell with a 0.5-cm interelectrode distance). Furthermore, the capacitance and inductance of the probes must be adjustable to compensate for distortions they would otherwise introduce into the oscilloscopically observed relaxation curve. The adjustment of a probe such as the Tektronix P6015 is a tedious art that does not lend itself to computerization, thus arguing strongly for spectrophotometric detection whenever such is possible.

The sample cell for an electric field jump apparatus utilizing spectrophotometric detection necessarily must have windows that allow a monochromatic beam of light to pass through the sample liquid between the two electrodes and on to the photocathode of a fast photomultiplier tube. The cell can be constructed of a plastic such as Kel-F, and the windows may be made of glass although quartz is clearly preferable since UV-spectroscopic measurements may be necessary. A low background of conductivity arising from impurities in the sample liquid can usually be tolerated, so rigorous exclusion of air (CO_2 is the principal offender) from the spectrophotometric sample cell is not crucial, particularly when working with at least slightly acidic aqueous solutions. On the other hand, high concentrations of supporting electrolyte cannot be tolerated, and it is therefore necessary to know equilibrium constants for sample systems in the limit of zero ionic strength.

The deduction of equilibrium constants from relaxation amplitudes does not proceed as readily in the case of electric field jump experiments as with the temperature jump and ultrasonic absorption relaxation experiments. The necessary determination of amplitude of the Second Wien Effect as a function of electric field strength has been described by Ilgenfritz [125] and Patterson [138].

Modern transient recorders that digitize the experimental decay of a Wheatstone bridge or photomultiplier voltage on time scales as short as a nanosecond greatly facilitate the accumulation of electric field jump relaxation data. However, they do not obviate the need for eliminating electronic and other noise from the experiment. In the case of spectrophotometric detection it is particularly important to use an intense quasi-CW light source [139] so that the number of monochromatic photons passing through the sample cell in the very short duration of the experiment will be sufficient to produce a decent signal-to-noise ratio. The design of lenses [125, 129] for gathering the light from the source and passing it through the sample cell is of course no less important.

Systems Amenable to Study

In the slightly more than two decades since the invention of the electric field jump relaxation method [140, 141] it has been used most frequently to investigate the rates and mechanisms of diffusion-controlled or near-diffusion-controlled protolytic reactions in aqueous solutions [3]. The reaction of the hydronium ion with aqueous acetate is a classic example [142, 143]. The somewhat slower reaction of a hydrolyzed metal cation such as $AlOH^{2+}$(aq) with hydronium ion provides another illustration [144]. Massey and Schelly [145] have unraveled the hydrolysis reactions occurring in aqueous and ethanolic phenolphthalein by an ingenious utilization of E-jump and T-jump techniques in the same set of experiments.

More recently several E-jump kinetic studies have been made of the helix–coil transition in aqueous polyamino acids [146]. This is a still unresolved basic research problem for which the enthusiasm of biochemists has waned since it is now clear that there is much less alpha-helical structure in enzymatic proteins than was formerly thought. Similar E-jump studies of helix–coil transitions in single-stranded polynucleotides [147] will evoke greater interest.

Pörschke [148] has used the E-jump technique to investigate the interaction of Mg^{2+} and Ca^{2+} cations with polynucleotides. This is a somewhat more sophisticated use of the technique than the loosely related determination of metal cation complexation kinetics by simpler ligands in water [149] and in acetonitrile [150].

One of the most elegant recent applications of the square-wave E-jump relaxation method was to the kinetic properties of cetylpyridinium iodide micelles [151]. Other variations of the basic E-jump method [152] have been found more suitable for kinetic studies of electrolytes in low-polarity solvents [153, 154]. Conceivably all E-jump techniques for studying rapid reactions of electrolytes could be superseded by Zare's ingenious utilization of the Dember effect in liquids [155] since the restriction to low ionic-strength media would no longer apply.

The E-jump methods are more likely to be laid to a fitful rest by funding agencies determined to foster science "relevant" to national needs than by any shortage of chemical systems susceptible to kinetic study by E-jump techniques. In particular, kinetic studies of fast reactions in nonaqueous media and of polyelectrolytes have scarcely gotten underway.

ACKNOWLEDGMENT

This work was sponsored by a contract from the Office of Naval Research and by Grant No. AFOSR 73-2444C from the Directorate of Chemical Sciences, Air Force Office of Scientific Research. L. J. R., on leave from the University of Salamanca, was supported by a stipend from the Commission for Cultural Exchange between the United States and Spain.

References

1. C. F. Bernasconi, *Relaxation Kinetics*, Academic Press, New York, 1976.
2. M. Eigen and L. DeMaeyer, in *Techniques of Chemistry*, Vol. VI, Part 2, G. G. Hammes, Ed., Wiley-Interscience, New York, 1974.
3. M. Eigen and L. DeMaeyer, in *Techniques of Organic Chemistry*, Vol. VIII, Part 2, A. Weissberger, Ed., Wiley-Interscience, New York, 1963.
4. G. Porter and M. A. West, in *Techniques of Chemistry*, Vol. VI, Part 2, G. G. Hammes, Ed., Wiley-Interscience, New York, 1974.
5. G. W. Liesegang, M. M. Farrow, N. Purdie, and E. M. Eyring, *J. Am. Chem. Soc.*, **98**, 6905 (1976).
6. G. W. Liesegang, M. M. Farrow, F. A. Vazquez, N. Purdie, and E. M. Eyring, *J. Am. Chem. Soc.*, **99**, 3240 (1977).
7. L. J. Rodriguez, G. W. Liesegang, M. M. Farrow, N. Purdie, and E. M. Eyring, *J. Phys. Chem.*, **82**, 647 (1978).
8. L. J. Rodriguez, G. W. Liesegang, R. D. White, M. M. Farrow, N. Purdie, and E. M. Eyring, *J. Phys. Chem.*, **81**, 2118 (1977).
9. R. M. Izatt, R. E. Terry, B. L. Haymore, L. D. Hansen, N. K. Dalley, A. G. Avondet, and J. J. Christensen, *J. Am. Chem. Soc.*, **98**, 7620 (1976).
10. P. B. Chock, *Proc. Natl. Acad. Sci. USA*, **69**, 1939 (1972).
11. S. Petrucci, in *Ionic Interactions*, Vol. II, S. Petrucci, Ed., Academic Press, New York, 1971.
12. E. F. Caldin, *Fast Reactions in Solution*, Blackwell, Oxford, 1964.
13. G. Czerlinski, *Chemical Relaxation*, Marcel Dekker, New York, 1966.
14. R. Winkler, Dissertation, Max-Planck-Institut, Goettingen, and Technical University of Vienna, 1969.
15. K. Kustin, D. Shear, and D. Kleitman, *J. Theor. Biol.*, **9**, 186 (1965).
16. G. Schwarz, *Rev. Mod. Phys.*, **40**, 206 (1968).
17. G. W. Castellan, *Ber. Bunsenges Phys. Chem.*, **67**, 898 (1963).
18. M. Eigen, in *Fifth Nobel Symposium*, S. Claesson, Ed., Interscience, New York, 1967, p. 333.
19. H. Farber and S. Petrucci, *J. Phys. Chem.*, **80**, 327 (1976).
20. J. Stuehr, in *Techniques of Chemistry*, Vol. VI, 3rd ed., G. G. Hammes, Wiley, New York, 1974, Chap. 7.

21. J. Lamb, in *Physical Acoustics*, Vol. II, Part A, W. P. Mason, Ed., Academic Press, New York, 1965, Chap. 4.

22. J. Stuehr and E. Yeager, *ibid.*, Chap. 6.

23. K. Tamm, "Akustik I," in *Handbuch der Physik*, Band XI, Springer-Verlag, Berlin, 1961.

24. R. O. Davies and J. Lamb, *Q. Rev. Chem. Soc.*, **11**, 134 (1957).

25. This chapter, Section 4, Transient Methods.

26. G. Stokes, *Trans. Cambridge Phil. Soc.*, **8**, 287 (1845).

27. G. Kirchhof, *Ann. Phys.*, **134**, 177 (1868).

28. L. Hall, *J. Acoust. Soc. Am.*, **24**, 704 (1952).

29. H. S. Frank and W. Y. Wen, *Disc. Faraday Soc.*, **24**, 133 (1957).

30. K. F. Herzfeld and F. O. Rice, *Phys. Rev.*, **31**, 691 (1928).

31. L. Hall, *ibid.*, **73**, 775 (1948).

32. It is actually greater than one decade, see Ref. 20, p. 249.

33. See Ref. 20, p. 248.

34. L. G. Hepler, *J. Phys. Chem.*, **69**, 965 (1965).

35. N. Purdie and A. J. Barlow, *J. Chem. Soc. Faraday II*, **68**, 33 (1972).

36. K. G. Breitschwerdt, H. Kistenmacher, and K. Tamm, *Phys. Letts.*, **24A**, 550 (1967).

37. A. R. Maret and E. Yeager, *J. Acoust. Soc. Am.*, **54**, 668 (1973).

38. P. Hemmes and S. Petrucci, *J. Phys. Chem.*, **74**, 467 (1970).

39. L. G. Jackopin and E. Yeager, *J. Phys. Chem.*, **70**, 313 (1966); *ibid.*, **74**, 3766 (1970).

40. K. Fritsch, G. J. Montrose, J. L. Hunter, and J. F. Dill, *J. Chem. Phys.*, **52**, 2242 (1970).

41. See Section 2 of this chapter.

42. J. E. Piercy and J. Lamb, *Trans. Faraday Soc.*, **52**, 930 (1956).

43. J. H. Andreae, P. L. Joyce, and R. J. Oliver, *Proc. Roy. Soc. (London)*, **75**, 82 (1960).

44. M. M. Farrow, N. Purdie, and E. M. Eyring, *J. Phys. Chem.*, **79**, 1995 (1975).

45. M. J. Blandamer, D. E. Clarke, N. J. Hidden, and M. C. R. Symons, *Trans. Faraday Soc.*, **63**, 66 (1967).

46. M. M. Farrow, N. Purdie, and E. M. Eyring, *Inorg. Chem.*, **14**, 1584 (1975).

47. J. Rassing and H. Lassen, *Acta Chem. Scand.*, **23**, 1007 (1969).

48. N. Purdie and M. M. Farrow, *Coord. Chem. Rev.*, **11**, 189 (1973).

49. F. Eggers and K. Kustin, in *Advances in Enzymology*, Vol. XVI, K. Kustin, Ed., Academic Press, New York, 1969.

50. J. Stuehr and E. Yeager, *J. Chem. Phys.*, **46**, 1222 (1967).

51. F. Eggers and T. Funck, *Rev. Sci. Instrum.*, **44**, 969 (1973); F. Eggers and T. Funck, *J. Acoust. Soc. Am.*, **57**, 331 (1975); F. Eggers, T. Funck, and K. H. Richman, *Rev. Sci. Instrum.*, **47**, 361, 378 (1976).

52. P. Debye and F. W. Sears, *Proc. Natl. Acad. Sci. USA*, **18**, 409 (1932).

53. R. Lucas and P. Biquard, *Compt. Rend.*, **194**, 2132 (1932).

54. T. Nowicki, *Electro-Optical Systems Design*, **23**, Feb. (1974).

55. M. M. Farrow, N. Purdie, A. L. Cummings, W. Hermann, Jr., and E. M. Eyring, in *Chemical and Biological Applications of Relaxation Spectrometry*, E. Wyn-Jones, Ed., Reidel, Dordrecht, Holland, 1975, p. 69.

56. T. Tanaka, *Rev. Sci. Instrum.*, **43**, 164 (1972).

57. M. M. Farrow, S. L. Olsen, N. Purdie, and E. M. Eyring, *Rev. Sci. Instrum.*, **47**, 657 (1976).

58. R. S. Krishnan, in *The Raman Effect*, Vol. I, A. Anderson, Ed., Marcel Dekker, New York, 1971, p. 343.

59. K. G. Plass, *Acustica*, **19**, 236 (1967/1968).

60. F. Dunn and J. E. Breyer, *J. Acoust. Soc. Am.*, **34**, 775 (1962).

61. K. F. Herzfeld and T. A. Litovitz, *Absorption and Dispersion of Ultrasonic Waves*, Academic Press, New York, 1959.

62. T. L. Cottrell and J. C. McCoubrey, *Molecular Energy Transfer in Gases*, Butterworth, London, 1961.

63. D. Secrest, *Ann. Rev. Phys. Chem.*, **24**, 379 (1973).

64. M. J. Blandamer, *Introduction to Chemical Ultrasonics*, Academic Press, New York, 1973, p. 83.

65. This estimate was obtained from inelastic neutron scattering.

66. G. G. Hammes and P. R. Schimmel, *J. Am. Chem. Soc.*, **89**, 442 (1967).

67. Ref. 64, Chap. 7.

68. J. Lamb and J. Sherwood, *Trans. Faraday Soc.*, **51**, 1674 (1955).

69. V. M. Gittins, R. F. M. White, and E. Wyn-Jones, in *Internal Rotation in Molecules*, W. J. Orville-Thomas, Ed., Wiley, New York, 1974, pp. 425–480.

70. K. A. Kemp and S. V. Letcher, in *Liquid Crystals and Ordered Fluids*, Vol. II, J. F. Johnson and R. S. Porter, Eds., Plenum, New York, 1974.

71. P. D. Edmonds and D. A. Orr, in *Liquid Crystals*, *Proc. Int. Conf. Liquid Crystals*, 1965, coordinated by G. H. Brown, G. J. Dienes, and M. M. Labes, Gordon & Breach, New York, 1967, p. 395.

72. D. Forster, F. C. Lubensky, P. C. Martin, J. Swift, and P. S. Pershan, *Phys. Rev. Lett.*, **26**, 1016 (1971).

73. K. A. Kemp and S. V. Letcher, *Phys. Rev. Lett.*, **27**, 1634 (1972).

74. G. C. Wetsel, R. S. Speer, B. A. Lowry, and M. R. Woodard, *J. Appl. Phys.*, **43**, 1495 (1972).

75. A. G. Chynoweth and W. G. Schneider, *J. Chem. Phys.*, **20**, 1777 (1952).

76. Ref. 64, Chaps. 10 and 11.

77. J. H. Andreae, P. D. Edmonds, and J. F. McKellar, *Acustica*, **15**, 74 (1965).

78. A. E. Brown, *Acustica*, **18**, 169 (1967).

79. M. Fixman, *J. Chem. Phys.*, **36**, 1965 (1967).

80. M. Eigen, W. Kruse, G. Maass, and L. DeMaeyer, *Progr. React. Kinet.*, **2**, 285 (1964).

81. K. Tamm, G. Kurtze, and R. Kaiser, *Acustica*, **4**, 380 (1954).

82. R. Leonard, P. Combs, and L. Skidmore, *J. Acoust. Chem. Soc.*, **21**, 63A (1949).

83. M. Eigen and K. Tamm, *Z. Elektrochem.*, **66**, 93, 107 (1962).

84. E. H. Carnevale and T. A. Litovitz, *J. Acoust. Soc. Am.*, **30**, 610 (1958).
85. F. H. Fisher, *J. Acoust. Soc. Am.*, **40**, 805 (1963).
86. P. Debye, *Trans. Electrochem. Soc.*, **82**, 265 (1942).
87. T. Sano and T. Yasunaga, *J. Phys. Chem.*, **77**, 2031 (1973).
88. H. Diebler and M. Eigen, *Z. Phys. Chem. (Frankfurt am Main)*, **20**, 229 (1959).
89. A. Bechtler, K. G. Breitschwerdt, and K. Tamm, *J. Chem. Phys.*, **52**, 2975 (1970).
90. P. Hemmes, F. Fittipaldi, and S. Petrucci, *Acustica*, **21**, 228 (1969).
91. G. G. Hammes and A. C. Park, *J. Am. Chem. Soc.*, **91**, 956 (1969).
92. L. M. Rhodes and P. R. Schimmel, *Biochemistry*, **10**, 4426 (1971).
93. E.g., A. D. Barksdale and J. E. Stuehr, *J. Am. Chem. Soc.*, **94**, 3334 (1972).
94. D. E. Bowen, in *Electrons in Fluids*, J. Jortner and N. R. Kestner, Eds., Springer-Verlag, Berlin, 1973.
95. D. E. Bowen, *J. Phys. Chem.*, **79**, 2895 (1975).
96. W. F. Love, C. T. Walker, and W. S. Glaunsinger, *J. Phys. Chem.*, **79**, 2948 (1975).
97. M. Shporer and S. Luz, *J. Am. Chem. Soc.*, **97**, 665 (1975).
98. E. Shchori, J. Jagur-Grodzinski, S. Luz, and M. Shporer, *J. Am. Chem. Soc.*, **93**, 7133 (1971).
99. E. Shchori, J. Jagur-Grodzinski, and M. Shporer, *J. Am. Chem. Soc.*, **95**, 3842 (1973).
100. T.-P. I and E. Grundwald, *J. Am. Chem. Soc.*, **95**, 2879 (1974).
101. G. W. Hoffman, *Rev. Sci. Instrum.*, **42**, 1643 (1971).
102. D. Pörschke, *Rev. Sci. Instrum.*, **47**, 1363 (1976).
103. Y. Egozy and S. Weiss, *J. Phys. E. Sci. Instrum.*, **9**, 366, 367 (1976).
104. G. Ertl and H. Gerischer, *Z. Elektrochem.*, **65**, 629 (1961).
105. J. T. Knudtson and E. M. Eyring, *Ann. Rev. Phys. Chem.*, **25**, 255-274 (1974).
106. C. Reiss, in *Lasers in Physical Chemistry and Biophysics*, J. Joussot-Dubien, Ed., Elsevier, Amsterdam, 1975, pp. 239–245.
107. D. H. Turner, G. W. Flynn, N. Sutin, and J. V. Beitz, *J. Am. Chem. Soc.*, **94**, 1554 (1972).
108. Z. A. Schelly, J. Lang, and E. M. Eyring, *Monatsh, Chem.*, **104**, 1672 (1973).
109. D. H. Turner, G. W. Flynn, S. K. Lundberg, L. D. Faller, and N. Sutin, *Nature*, **239**, 215 (1972).
110. J. K. Beattie, N. Sutin, D. H. Turner, and G. W. Flynn, *J. Am. Chem. Soc.*, **95**, 2052 (1973).
111. C. Creutz and N. Sutin, *J. Am. Chem. Soc.*, **95**, 7177 (1973).
112. D. H. Turner, R. Yuan, G. W. Flynn, and N. Sutin, *Biophys. Chem.*, **2**, 385 (1974).
113. R. F. Pasternack, N. Sutin, and D. H. Turner, *J. Am. Chem. Soc.*, **98**, 1908 (1976).
114. S. Ameen and L. De Maeyer, *J. Am. Chem. Soc.*, **97**, 1590 (1975); S. Ameen, *Rev. Sci. Instrum.*, **46**, 1209 (1975).

115. J. Holzwarth and H. Wolff, *Ber. Bunsenges.*, **79**, 1154 (1975).
116. C. Brecher and K. W. French, *J. Phys. Chem.*, **73**, 1785 (1969); A. Heller, *J. Am. Chem. Soc.*, **88**, 2058 (1966).
117. S. Saigo, Y. Husimi, and A. Wada, *Jap. J. Appl. Phys.*, **14**, 1209 (1975).
118. See, for example, E. F. Caldin, M. W. Grant, and B. B. Hasinoff, *J. Chem. Soc. D*, 1351 (1971); B. B. Hasinoff, *Can. J. Chem.*, **54**, 1820 (1976).
119. A. J. Campillo, V. H. Kollman, and S. L. Shapiro, *Science*, **193**, 227 (1976).
120. I. Giannini, in *Chemical and Biological Applications of Relaxation Spectrometry*, E. Wyn-Jones, Ed., D. Reidel, Dordrecht, Holland, 1975, pp. 49–53; I. Giannini, in *Lasers in Physical Chemistry and Biophysics*, J. Joussot-Dubien, Ed., Elsevier, Amsterdam, 1975, pp. 229–237.
121. H. Rüppel and H. T. Witt, "Fast Reactions," in *Methods in Enzymology*, Vol. XVI, K. Kustin, Ed., Academic Press, New York, 1969, pp. 316–379.
122. D. H. Turner, Ph.D. Thesis, Columbia University, New York, 1972, 93 pp.
123. M. M. Farrow, N. Purdie, A. L. Cummings, W. Hermann, Jr., and E. M. Eyring, in *Chemical and Biological Applications of Relaxation Spectrometry*, E. Wyn-Jones, Ed., D. Reidel, Dordrecht, Holland, 1975, pp. 69–83.
124. M. Eigen and J. Schoen, *Z. Elektrochem.*, **59**, 483 (1955).
125. G. Ilgenfritz, Ph.D. Thesis, Georg-August University, Goettingen, Germany, 1966, 81 pp.
126. D. T. Rampton, L. P. Holmes, D. L. Cole, R. P. Jensen, and E. M. Eyring, *Rev. Sci. Instrum.*, **38**, 1637 (1967).
127. L. C. M. De Maeyer, in "Fast Reactions," *Methods in Enzymology*, Vol. XVI, K. Kustin, Ed., Academic Press, New York, 1969, pp. 80–118.
128. S. L. Olsen, R. L. Silver, L. P. Holmes, J. J. Auborn, P. Warrick, Jr., and E. M. Eyring, *Rev. Sci. Instrum.*, **42**, 1247 (1971).
129. H. H. Grünhagen, *Biophysik*, **10**, 347 (1973).
130. H. H. Grünhagen, *Messtechnik*, **1**, 19 (1974).
131. L. De Maeyer and A. Persoons, in "Investigations of Rates and Mechanisms of Reactions," *Techniques of Chemistry*, Vol. VI, Part 2, 3rd ed., G. G. Hammes, Ed., Wiley-Interscience, New York, 1974, pp. 211–235.
132. S. L. Olsen, L. P. Holmes, and E. M. Eyring, *Rev. Sci. Instrum.*, **45**, 859 (1974).
133. A. P. Persoons, *J. Phys. Chem.*, **78**, 1210 (1974).
134. L. Onsager, *J. Chem. Phys.*, **2**, 599 (1934).
135. G. Schwarz, *J. Phys. Chem.*, **71**, 4021 (1967).
136. M. Wien, *Phys. Z.*, **29**, 751 (1928).
137. M. Wien, *Phys. Z.*, **32**, 545 (1931).
138. J. F. Spinnler and A. Patterson, Jr., *J. Phys. Chem.*, **69**, 500 (1965).
139. S. L. Olsen, L. P. Holmes, and E. M. Eyring, *Rev. Sci. Instrum.*, **43**, 859 (1974).
140. M. Eigen, *Z. Phys. Chem. (N.F.)*, **1**, 176 (1954).
141. M. Eigen, *Discuss. Farady Soc.*, **17**, 194 (1954).
142. M. Eigen and J. Schoen, *Z. Elektrochem.*, **59**, 483 (1955).

143. J. J. Auborn, P. Warrick, Jr., and E. M. Eyring, *J. Phys. Chem.*, **75**, 2488 (1971).

144. P. Hemmes, L. D. Rich, D. L. Cole, and E. M. Eyring, *J. Phys. Chem.*, **74**, 2859 (1970); *ibid.*, **75**, 929 (1971).

145. M. W. Massey, Jr., and Z. A. Schelly, *J. Phys. Chem.*, **78**, 2450 (1974).

146. T. Yasunaga, T. Sano, K. Takahashi, H. Takenaka, and S. Ito, *Chem. Lett. (Japan)*, 405 (1973); A. Cummings and E. M. Eyring, *Biopolymers*, **14**, 2107. (1975).

147. D. Pörschke, *Biopolymers*, **15**, 1917 (1976).

148. D. Pörschke, *Biophys. Chem.*, **4**, 383 (1976).

149. M. M. Farrow, N. Purdie, and E. M. Eyring, *Inorg. Chem.*, **13**, 2024 (1974).

150. H. Hirohara, K. J. Ivin, J. J. McGarvey, and J. Wilson, *J. Am. Chem. Soc.*, **96**, 4435 (1974); H. Hirohara, K. J. Ivin, and J. J. McGarvey, *J. Am. Chem. Soc.*, **96**, 3311 (1974).

151. H. H. Grünhagen, *J. Colloid Interfac. Sci.*, **53**, 282 (1975).

152. A. P. Persoons, *J. Phys. Chem.*, **78**, 1210 (1974).

153. R. F. W. Hopmann, *J. Phys. Chem.*, **78**, 2341 (1974).

154. F. Nauwelaers, L. Hellemans, and A. Persoons, *J. Phys. Chem.*, **80**, 767 (1976).

155. A. Bergman, C. R. Dickson, S. D. Lidofsky, and R. N. Zare, *J. Chem. Phys.*, **65**, 1186 (1976).

Chapter **V**

CHEMICAL MEASUREMENTS IN THE PICOSECOND AND SHORTER TIME RANGE

Stephen C. Pyke and Maurice W. Windsor

---

1   INTRODUCTION AND BACKGROUND

Why should we be interested in making chemical observations on a time scale as short as picoseconds (10^{-12} sec)? The first response that comes to mind is, "Because it is possible," reminiscent of the well-known replies to the questions, "Why should we climb Mt. Everest?" "Why should we go to the moon?" On reflection, we find this initial response a bit too hasty and unguarded. Many things are now possible that one might not want to do. It is possible for us to wipe out most, if not all, life on earth. It is possible to construct a building one mile tall. It is possible to recombine lengths of DNA to make genetic messages that have never before existed and that might prove a godsend to those suffering from genetic diseases such as diabetes and sickle cell anemia. Few (we would wish none) would want to implement the first possibility. The economic sense of the second is open to serious question—the first hundred floors of such a structure might be entirely consumed by elevator shafts and duct work for ventilation and other services. As to the third, a controversy is now raging as to whether the putative benefits of recombinant DNA research are worth the risk of possibly creating new viruses against which man might have no natural defenses. It is clear then that we should ask further questions before deciding to go ahead and do something just because it is possible, questions such as, "Does it make sense?" "Is it dangerous?" "Is the expected return likely to be worth the effort?" "Could we achieve the same results in some other way?" Let's examine some of these questions in relation to the matter of making chemical measurements in the picosecond and shorter time range.

First, does it make sense? To examine this question, we need to ask what chemical and physical molecular processes take place on the picosecond time scale. We note from the time scale shown in Fig. 5.1 that the average encounter time between gas molecules at 1 atm pressure at ordinary temperatures is about 100 psec; that proton and electron transfer reactions characteristically occur in times ranging from 100 psec to much less than 1 psec. There is recent evidence that the primary electron transfer

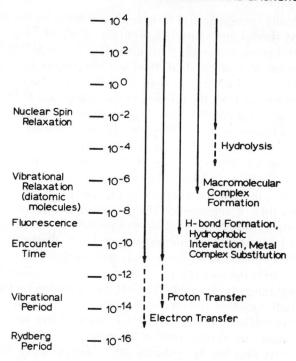

Fig. 5.1. Time scale of chemistry.

step in photosynthesis is a picosecond process. The periods of molecular vibrations lie in the range of 10^{-14} to 10^{-12} sec. Bond rupture and dissociation can occur on the same time scale, and so too can various kinds of isomerization. Internal conversion between states of the same multiplicity is thought to take place in a few picoseconds, as shown by the lifetimes of certain dyes used for laser modelocking. It appears that intersystem crossing between the lowest excited singlet state S_1 and the triplet manifold can also occur in picoseconds in favorable cases such as aromatic ketones. Intersystem crossing in inorganic systems is also fairly fast ($\sim 10^{11}$ sec^{-1}). Thus there seems to be plenty of incentive to attempt chemical (and physical) observations on the picosecond time scale. How about still shorter times? Should we try to extend observations to the femtosecond region (10^{-15} sec)? The picosecond–femtosecond time range could prove fruitful with regard to information on reactions involving proton or electron transfer. Photodissociation of hemoglobin–carbon monoxide complexes has recently been reported to occur in less than 0.5 psec using subpicosecond pulses from a passively mode-locked dye laser. Owing to the uncertainty principle, the information we gather becomes spectroscopically (energetically) less and less precise as the time of observation gets shorter and shorter. Eventually, at about 1 fsec (10^{-15} sec) chemistry must come to an end. What a strange statement that is, and yet it must be so because of the dictates of the uncertainty principle. If the time of observation is no greater than 10^{-15} sec, then the corresponding uncertainty in energy from the relation $\Delta E \cdot \Delta t \sim h$ or $\Delta \nu \cdot \Delta t \sim 1$ turns out to be 30,000 cm^{-1} or about 90 kcal/mole. This figure is of the same order as the energies of chemical bonds. Thus the chemical identity of the species we are hoping to study will become indeterminate as we approach the femtosecond barrier. This same barrier poses no problem to the nuclear physicist, because an uncertainty of 90 kcal/mole or $\sim$4 eV is still negligible compared to the binding energies of nuclear particles.

In summary then, it appears that the nanosecond-to-picosecond region is very important from the point of view of primary processes in chemical reactions, and this time range embraces many kinds of elementary chemical steps. This conclusion has been amply substantiated by many experimental studies over the past few years. We are just beginning to probe the subpicosecond range experimentally. Useful information on reactions involving photodissociation and on primary processes involving electron or proton transfer will probably be gained over the next five years as experimental techniques are further refined for subpicosecond studies.

Finally, let us deal briefly with the other questions we have posed. Could we get the same information another way? Certainly there are other ways of getting picosecond information. Studies of spectral line broadening can give kinetic information in the picosecond and subpicosecond ranges. However, the information obtained is less specific than that obtainable using picosecond absorption spectroscopy, and the systems that can be studied are more limited. Last, is it dangerous? We can see no special danger in picosecond laser studies, although eye protection is a concern when working with high-power pulsed lasers. Picosecond laser-induced thermonuclear fusion has already been demonstrated, and laser triggering of hydrogen bombs is therefore likely to be possible. But the attendant increase in danger is very small compared to the large potential return from controlled fusion as a new source of energy. It seems likely also that picosecond studies will be crucial in unraveling the complex processes that Nature uses to capture solar energy in photosynthesis. Armed with a detailed knowledge of these processes, we may be able to go one better than Nature and develop artificial photosynthetic systems for making fuels such as methanol and hydrogen or for converting sunlight directly into electricity.

Most reviews of picosecond spectroscopy and kinetics have discussed experimental technique only briefly and then only with reference to specific applications and from the point of view of the specialist. The purpose of this review is to present a discussion of techniques to chemists not particularly well versed in laser physics and who might be interested in using picosecond techniques in their research. We have tried to identify and describe instrumental features that are particularly important in the development of dependable sources of single picosecond pulses. Finally we decided to stress a verbal rather than a mathematical discussion of the more complex subjects, including theory of mode-locked pulse production, sum and difference frequency generation, and correlation methods used in pulsewidth measurement. However, to assist those who wish to pursue these subjects further, we have provided references that cover each of these topics in greater depth. In addition, these resource materials provide alternative points of view, so that the reader can opt for a discussion that suits his or her particular background.

Our survey of the literature covers papers published through mid-1977. For the sake of simplicity, the formal charge of 3+ on Nd has been omitted from the text. For Nd:glass, the host is silicate glass unless otherwise noted (phosphate glass is becoming more widely used). Finally, Nd:YAG refers to Nd in a crystalline host (yttrium–aluminum–garnet).

2 INSTRUMENTAL TECHNIQUES

This section emphasizes the techniques used in generating and characterizing picosecond pulses as well as the ways picosecond pulses can be used to study the spectroscopy and decay kinetics of short-lived transient species.

We first describe some of the aspects of modelocking, single pulse selection, pulsewidth limitations, ways to change the wavelength of the pulses, and pulsewidth measurement, all of which we have found important in developing a dependable picosecond laser system. We originally had planned to include commercial information and guidelines for those committed to buying or building a picosecond laser system. However any information that we might have included soon would have been out of date. *Laser Focus* magazine publishes a comprehensive listing of manufacturers in their annual *Buyers' Guide*, which is the best resource for current information regarding commercially available laser systems, subsystems, and components. There are one or two additional points that are worth mentioning. Commercially available mode-locked laser systems are generally more expensive than normal laser systems, whereas the modelocking often requires little more than putting a modelocking dye cell in the cavity and adjusting the dye concentration and pump energy for the shortest pulses (see the section on pulsewidth measurements). The more elaborate vibration-isolated honeycomb tables are generally not required. However some sort of rigid platform is advisable if one or more beams are to be focused onto a target after traveling more than a few feet. Otherwise frequent realignment of the system may be required. Finally, it is generally cheaper to purchase items such as Pockels cells, Nd and ruby rods, and doubling crystals directly from the manufacturer. They can be incorporated into an existing system with a little extra effort.

The section on fluorescence lifetime measurements outlines the methods used to characterize emission decay and spectra on the picosecond time scale. The section on production of tunable picosecond and subpicosecond pulses using dye lasers reviews the fairly recent development of mode-locked CW dye lasers and their use and application in measurements where subpicosecond resolution is required.

Production and Characterization of Picosecond Laser Pulses

Modelocking

The key to producing picosecond laser pulses lies in modelocking the laser cavity. Both active and passive techniques have been used, and a number of reviews are available for the reader who desires more detail [1–8]. Active modelocking involves the use of either an electro-optic or acousto-optic modulator which modulates the amplitude or the frequency of the radiation inside the laser cavity. The modulation frequency is tuned to the frequency separation between the modes of the cavity, $\omega = c/2L$, where L is the cavity length and c is the velocity of light. Many modes of the cavity are spanned by the emission profile of the gain medium ($\sim 10^4$ for a 1-m Nd:glass oscillator, $\sim 10^3$ for ruby), and consequently the energy stored in the radiation field inside the cavity will build up preferentially in these modes. As this occurs, the radiation field can be formulated as a single pulse, in time, traveling back and forth in the cavity with a width approximately equal to the inverse of the bandwidth of the gain medium, $\Delta\tau = 1/\Delta\nu$ (see section on pulsewidth limitations). At the output mirror, which is only partially reflecting, part of the pulse escapes the cavity, and the total output appears as a train of pulses, with each pulse separated by the time required for the pulse inside the cavity to make one complete round trip, $2L/c$. The pulse train will continue until the medium no longer provides gain. The action of the modelocker is to ensure that all the cavity modes oscillate in phase. In the frequency domain, the following description can be used. The mode of highest gain will appear first, and the modulation of its frequency will progressively introduce *in phase* other cavity modes that lie within the gain curve. Transformation of this assembly of inphase frequencies to the time domain (see later) gives a pulse of very short duration. The more cavity modes that are locked in phase, the shorter the resultant pulse. This is why a medium with a broad bandwidth gain curve, such as a laser dye, can give shorter pulses than say a ruby or Nd:glass laser.

While acousto-optic modulators were first used in active modelocking, electro-optic modulators were found to be much more dependable. With the application of saturable dyes and passive modelocking, the production of picosecond pulses by active modelocking has virtually disappeared for pulsed lasers, although active modelocking is still employed for CW ion and dye lasers.

Passive modelocking is accomplished by placing a saturable absorber in the cavity. Theoretical and numerical treatments for the development of picosecond pulses are available [9–11] and are beyond the scope of this review. As lasing begins, picosecond- and subpicosecond-duration spikes with a random phase relationship exist simultaneously inside the cavity. These spikes are part of the noise spectrum of the emitting medium and arise because the emission by excited states is a statistical process. If a bleachable dye solution is placed inside the cavity, only those spikes with sufficient intensity will bleach the solution and penetrate the dye cell, and only these spikes will gain intensity on passing through the gain medium. If the dye absorbance is chosen properly and the laser operated at or very near threshold, ideally only one pulse should gain intensity. Figure 5.2 shows how the dye selects a pulse and shrinks the pulsewidth on successive passes. The dye absorbs the first part of the pulse before saturation, and if the recovery time of the dye is short compared with the pulsewidth, the trailing low-intensity portion of the pulse is also absorbed. Several dyes have been used in passive modelocking. Eastman Kodak dyes No. 9860 and No. 9740 and the nickel dithiene bis(4-dimethylaminodithiobenzyl)nickel(II) (BDN) [12] have been used to modelock the Nd:glass oscillator [56, 134], whereas the polymethine dyes 1,1'-diethyl-2,2'-dicarbocyanine iodide (DDI) and 1,1'-diethyl-4,4'-carbocyanine iodide (cryptocyanine) have been successfully employed to modelock the ruby laser. The polymethine dye 3,3'-diethyloxadicarbocyanine iodide (DODCI) has been used to modelock the rhodamine 6G (R-6G) and rhodamine B flashlamp-pumped dye lasers, and DODCI has been employed to modelock CW dye lasers as well [13]. Double modelocking (using one laser dye to modelock a second in a single cavity) has also been accomplished. Because of the high gain, saturation of the gain medium also plays an important role in pulse shortening in dye lasers. Modelocking in CW ion and dye lasers is discussed in more detail later.

The size and the location of the dye cell in the cavity have been important considerations. Bradley and co-workers have maintained that the shortest pulses are obtained when the dye cell is thin ($\sim$30 to 100 μ)

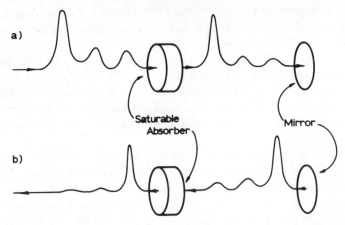

Fig. 5.2.   Effect of a saturable absorber on spikes of different intensity. Since the transmittance of the dye varies as the square of the intensity, the largest spikes will be absorbed less strongly than the weaker ones. The leading edge of each pulse will lose more intensity than the center for the same reason, and if the recovery time of the dye is short compared with the pulsewidth, a similar loss will occur for the trailing edge of the pulse. The dye concentration and the flashlamps are adjusted so that the gain-to-loss ratio is greater than unity for only the most intense spike. (a) poles propagating left to right; (b) poles propagating right to left.

and when the dye is in contact with one of the cavity mirrors [14]. Figure 5.3 outlines this type of cavity configuration. While this arrangement may reduce the pulsewidth somewhat, its primary advantage may be convenience. Since the dye volume is small, a flowing cell is advisable to avoid breakdown of the dye, and a large reservoir of dye solution can be employed. In the alternative arrangement, a dye cuvette with relatively small volume (~10 ml) and 10 mm path length is used, and the dye solution must be changed regularly. When the dye cell is removed and replaced, a slight alteration of the cavity characteristics is inevitable. The cavity retuning required is minimal but can be avoided by using the

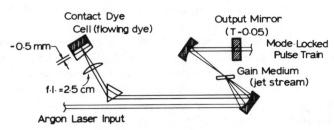

Fig. 5.3.   Schematic of mode-locked CW dye laser showing the contact dye cell [85].

optically contacted flowing dye cell. Yu and Alfano [15] maintained that satellite pulses cannot be excluded from the pulse train unless an optically contacted dye cell is used, and Garmire and Yariv [16] pointed out that a nonoptically contacted dye cuvette may sometimes lead to the formation of subcavities. These can produce secondary pulse trains separated from the primary train by $2L'/c$, where L' is the distance between the dye cuvette and one of the cavity mirrors. This problem can be avoided with an optically contacted dye cell but need not preclude the use of the cheaper dye cuvette in passive modelocking, as long as subcavities are avoided by careful placement and orientation of the cell.

Although passive modelocking has proved to be the simplest way to produce picosecond pulse trains in Nd:glass and ruby laser systems, repeatable performance is a much more sensitive function of cavity stability, flashlamp intensity, and rod temperature for ruby than for Nd:glass systems. Quality of the ruby rod is also important. Cavity configuration seems to have little effect in producing mode-locked pulse trains in that modelocking in ruby has been achieved with concentric cavities and plano–plano cavities. Confocal cavities are not advised, however, because mirror damage is much more likely.

Single Pulse Selection

While the entire pulse train from a modelocked laser has often been used as a source of excitation, complications often ensue. For Nd:glass lasers, the pulsewidth increases from the beginning to the end of the train by 20 to 30%. For flashlamp-pumped dye lasers, on the other hand, the opposite is true, because of gain saturation. Figure 5.4 illustrates the effect of a greater number of cavity round trips in the flashlamp-pumped dye laser [17]. Consequently, when the entire pulse train is used for measuring transient phenomena, true lifetimes will be convoluted with an average pulsewidth, and power levels will be averaged over the whole train. Thus both lifetimes and pulse intensity will be difficult to determine

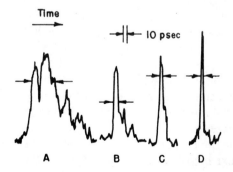

Time →

10 psec

A   B   C   D

Fig. 5.4. Microdensitometer traces of streak camera records showing progressive shortening of the output pulse from a rhodamine 6G dye laser as a function of the number of cavity round trips: (A) seven cavity round trips; (B) 20 round trips; (C) 28 round trips; and (D) 45 round trips into the pulse train. Full widths at half-maximum are indicated by arrows [17].

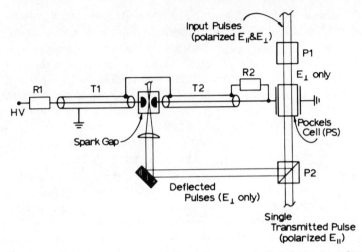

Fig. 5.5. Schematic of a single-pulse selector using an optically triggered spark gap and Pockels cell.

precisely. Other complications that can occur are saturation of the ground-state transition; absorption of a second photon by long-lived transients ($\tau > 2L/c$) produced by the first part of the pulse train; stimulated emission; and excited-state annihilation occurring only at high powers where excited-state concentration is high. For these reasons, it is generally more convenient to select out a single pulse to excite the system under study. Intracavity pulse selectors were used initially, but extracavity methods were developed to avoid component damage from the high power densities inside the cavity. Figure 5.5 is a diagram of a popular pulse selector utilizing a Pockels cell and optically triggered spark gap. The laser rod is cut so that the ends are at Brewster's angle, and the dye cuvette is also situated so that its windows are at Brewster's angle with respect to the cavity mirrors. Thus the output radiation is polarized predominantly normal ($E_\perp$) to the plane of the diagram. The pulse train is then passed through a Glan–Foucault prism (a Glan polarizer with an air gap between the prism components) which eliminates the remaining light polarized in the plane ($E_\parallel$) of the diagram, and through a Pockels cell which is oriented so that, when relaxed, it passes only $E_\perp$ polarized light. A second Glan-Foucault prism passes only light with $E_\parallel$ polarization and reflects the first part of the pulse train (with $E_\perp$ polarization) into a pressurized spark gap with quartz windows. As the height of successive pulses grows, one pulse finally triggers the spark gap, and a high-voltage pulse is applied to the Pockels cell for a short time. The high voltage induces a birefringence, and the next picosecond optical pulse to reach the Pockels cell has its polarization rotated by 90° and is then able to pass through the second Glan–Foucault prism. A more detailed description of the electro-optic effect can be found in Ref. 18.

The timing of this optical trigger, which is critical to the successful selection of one picosecond pulse, is controlled by simple RC time constants of a circuit as outlined in Fig. 5.5. The threshold pulse causes the spark gap to break down, and the high voltage begins to affect the Pockels cell in a time determined by the length of cable T2 between the spark gap and the Pockels cell and the speed of a signal in RG-8/u cable (~0.6 the speed of light). The cable length is adjusted so that the half-wave voltage applied to the Pockels cell is synchronized with the arrival of the next optical pulse. The total time that the Pockels cell feels the high voltage is determined by the length of cable between the resistance box R1 and the spark gap. The cable length T1 is usually adjusted to provide a window ~5 nsec in duration. The duty cycle of the Pockels cell (time required before another pulse can be selected) is determined by the RC time constant of the resistance $R1$ and the cable capacitance from R1 to the spark gap, and since $R1 \cong 10^9 \, \Omega$, $RC \sim 80$ msec. The pulse trains are typically less than a few milliseconds in duration. Consequently only one pulse in each train is capable of activating the Pockels cell. A detailed description of a similar pulse selection method using a Kerr cell was presented by von der Linde, Bernecker, and Laubereau [19]. High-voltage pulse power supplies with very low jitter are now available and are begining to replace spark gap-activated Pockels cells. These power supplies can be triggered by photodiodes and will deliver a triangular HV pulse with a width, peak voltage, and delay relative to the trigger pulse which can be changed to permit pulse selection from different points in the train. In the case of spark gaps, this can be accomplished by adjusting the electrode separation, the nature of the filling gas (usually N_2), or (most easily) the gas pressure. With a spark gap, the pulsewidth is adjusted by using different lengths of cable T1.

Pulsewidth Limitations

The theoretical limit of the minimum pulsewidth is related to the bandwidth of the emission profile by a Fourier transform:

$$F(\omega) = \frac{1}{\sqrt{2\pi}} \int_{-\infty}^{\infty} f(t) \, e^{i\omega t} \, dt$$

The symmetry inherent in the Fourier transform yields

$$f(t) = \frac{1}{\sqrt{2\pi}} \int_{-\infty}^{\infty} F(\omega) \, e^{-i\omega t} \, d\omega$$

and results in the following constraint: $\Delta\nu \cdot \Delta t \sim 1$, where $\Delta\omega = 2\eta \, \Delta\nu$. Since the emission bandwidth of the lasing medium $\Delta\nu \sim 100 \text{ cm}^{-1}$ for Nd:glass, the theoretical minimum pulsewidth is > 0.3 psec. For ruby the bandwidth is smaller, $\Delta\nu \sim 10 \text{ cm}^{-1}$, because of the crystalline host, and $\Delta t \gtrsim 3$ psec. For Nd:YAG, $\Delta\nu \sim 1 \text{ cm}^{-1}$, and $\Delta t \sim 30$ psec. Subpicosecond pulses have been obtained from dye lasers whose bandwidth is 10 to 100 cm^{-1}; and for ruby, dye, and Nd:YAG lasers nearly transform-limited pulsewidths have been achieved. However modelocked Nd:glass lasers, which have been used most frequently in the study of picosecond physical and chemical processes, have rarely produced transform-limited pulses.

For all passively mode-locked solid-state laser systems, the relaxation time of the modelocking dye, τ_r, must be fast compared with the reciprocal of the gain width of the lasing medium in order to achieve transform-limited pulsewidths ($\tau_r < 1/\Delta\nu$). The situation in which $\tau_r > 1/\Delta\nu$ leads to longer than transform-limited pulsewidths [20, 21]. This is clearly the case for Nd:glass lasers, where $1/\Delta\nu \sim 0.3$ psec. A short dye relaxation time is necessary in order that the weaker photon noise be eliminated by the saturation behavior of the dye both before and after the intensity maximum as shown in Fig. 5.2. When $\tau_r > 1/\Delta\nu$, the result is a skewed pulse shape due to the inability of the absorber to recover fast enough to attenuate photon noise following the peak of the pulse. Self-focusing and resulting self-phase modulation (SPM) at high power densities leads to nonlinear dispersion and contributes to temporal broadening of the pulse. In addition, spectral narrowing of the gain width of the Nd:glass laser occurs in the initial linear growth region of amplification. The reason for this is that the modes nearest the emission maximum will store energy at a faster rate than those modes on the wings of the emission band. This reduces the effective bandwidth of the medium near the lasing threshold and results in an increase in the effective pulsewidth. Von der Linde and Rodgers [22] have used an intracavity Fabry–Perot etalon to reduce the rate of spectral narrowing in the linear amplification regime of an Nd:glass laser by tuning the etalon so that the transmission minimum coincides with the gain maximum. This gives those modes in the wings of the gain width more of a chance to grow before threshold is reached. Dye lasers, on the other hand, produce nearly transform-limited pulsewidths because of gain saturation. With each cavity round trip the nonlinear absorption of the modelocking dye attenuates the leading edge of the pulse, but it is the gain saturation of the lasing dye that leads to reduced gain for the trailing edge of the pulse [23], the bulk of the excited state population being exhausted by the earlier portions of the pulse. In addition, pulsewidths are shortened somewhat by nonlinear optical elements such as doubling crystals and Kerr cells which discriminate against the lower power levels at the beginning and end of each picosecond pulse. Finally, the dye cell thickness and position described above in the section on modelocking has been found to be important in the reduction of pulsewidths in passively mode-locked lasers [14].

Harmonics and Wavelength Tuning in Solid-State Lasers

The high powers of typical picosecond pulses from mode-locked solid-state lasers ($\sim 10^9$ W) permit the utilization of nonlinear optical and spectroscopic properties, notably sum and difference frequency generation and stimulated Raman scattering (SRS) to alter the fundamental lasing frequency.

Sum and difference frequency generation (SFG and DFG) are processes that occur in materials with high nonlinear susceptibility or that respond to the applied optical electric field of high-power laser pulses with a change in the polarization, which is dependent on the square (or higher powers) of the applied field. Polarization corresponds to the induced distortion of the equilibrium distribution of electrons around the nuclei. Formally known as three- and four-wave mixing, the theoretical description of SFG and DFG is beyond the scope of this review and is well described in a monograph by Zernike and Midwinter [24] and a review by Yariv and Pearson [25]. Second harmonic generation (SHG) is the most common form of sum frequency generation (SFG), and nonlinear materials and their use in SHG are listed in Zernike and Midwinter [24]. The most frequently used materials are those of the crystallographic point group $\overline{4}2m$: KH_2PO_4 (KDP), $(NH_4)H_2PO_4$ (ADP), and their deuterated analogs abbreviated KD*P and AD*P, respectively. The main advantages are their high efficiency (up to 70 to 80% is attainable, but 10 to 20% is more usual), high damage threshold, relatively low cost, and the availability of large optical-quality crystals. Their disadvantage is that they are all hygroscopic. The most general application of SFG involves the input of two different frequencies, and the materials used in SHG are also used in this application. KDP and ADP have thresholds for UV absorbance below 200 nm [26] and therefore can be used where the sum frequency occurs at a wavelength longer than about 200 nm. However phase matching in the near UV is more difficult, since differences in group velocity limit the use of KDP or ADP very near the absorption threshold. Group velocity dispersion refers to the different speeds of the input and output frequencies within the crystal. This can lead to a breakdown of the precise phase-matching conditions required for efficient SFG and generally leads to a broadened output pulsewidth. Where the peak power is sufficiently high, group velocity dispersion can be minimized, at the cost of some efficiency, by using crystals with a short path length.

Difference frequency generation (DFG) also requires two photons to generate a third or in some circumstances requires three photons to produce a fourth, otherwise called three- and four-wave mixing. The theory is the same as that used to describe SHG and SFG, except the output photon frequency is the difference rather than the sum of the energies in two or three other photons. The same nonlinear media are generally also employed. One percent conversion efficiencies have been observed in two-photon generation of tunable IR pulses from a mode-locked Nd:glass laser and spontaneous parametric fluorescence in $LiNbO_3$ [27]. Picosecond pulses from Nd:glass lasers can also be mixed with pulses generated by stimulated Raman scattering (SRS) and self-phase modulation (SPM) to yield difference frequencies in the IR. Tunable IR output from 1.4 to 3.8 μm has been generated directly, in a cavity, by synchronously pumping (see section on production of tunable picosecond and subpicosecond pulses using dye lasers) a crystal of $LiIO_3$ with a mode-locked Nd:glass laser [28] with about 2% efficiency. Moore and Goldberg [29] mixed the output of a synchronously pumped dye laser with that of the mode-locked Nd:YAG laser in various nonlinear materials to generate SFG from 270 to 432 nm and DFG from 1.13 to 5.6 μm. Their conversion efficiency was about 2% for pulses from 1.1 to 2.3 μm, but conversion efficiency for longer wavelengths is not quoted.

Stimulated Raman scattering (SRS) in liquids has also been an efficient way of changing the frequency of a high-power picosecond laser pulse. By capitalizing on the coherent nature of the SRS process [30], one can shift the pulse frequency by an amount corresponding to the energy of a Raman-active vibrational mode of the medium while maintaining low divergence and narrow bandwidth in the converted beam. Table 5.1 outlines some of the frequency shifts that can be obtained using SRS.

Production of picosecond pulses in the vacuum UV (197 nm) has been demonstrated in strontium vapor near 600°C [31] following earlier work of Kung et al. [32] and Hodgson et al. [33]. Very recently XUV frequencies down to as short as 38 nm have been obtained by high-order harmonic generation in rare gases, using up to the seventh harmonic of a Nd:YAG laser output at 266 nm as the pump frequency [34]. Streak cameras necessary to characterize picosecond pulses in the VUV and X-ray regions are also under development [35].

Pulsewidth Measurements

Along with the production of picosecond pulses came the problem of finding a way to measure the pulsewidths accurately. Photoelectric measurement is limited by the bandwidths of commercially available oscilloscopes. At the present time the fastest oscilloscope and amplifier combination available through Tektronix is the 7904 mainframe and the 7A19

Table 5.1   Table of SRS Shifted Wavelengths for Various Pump Frequencies [30]

| | SRS Wavelength (nm) | | |
Solvent	Nd (SH) at 530 nm	Ruby at 694 nm	Ruby (SH) at 347 nm
Benzene	559	745	359
CS_2	550	729	355
Nitrobenzene	570	765	364
Chlorobenzene	560	745	359
Ethanol	627	870	386
Methanol	624	864	385
	628	872	386
Isopropanol	626	867	386
Acetone	627	871	386
1,1,1-Trichloroethane	628	872	386
Calcite	562	750	360
Water	648	912	394

amplifier, which together provide an 800-psec risetime with a sensitivity of 10 mV/cm. With direct access to the CRT, using a 7A21N plug-in, a bandwidth of 1 GHz and a risetime of 350 psec can be attained, but sensitivity is reduced to 5 V/cm. Oscilloscopes with bandwidths up to 5 GHz have been manufactured but are not readily available, and at these frequencies the writing speed becomes a limiting factor. Sampling scopes have better time resolution (~25 psec for a Tektronix S−4 sampling head); however, high repetition rates and reproducible signals are required. Sampling scopes can be applied to measure fluorescence decay but are still too slow to resolve pulsewidths generated in most mode-locked lasers. Even the fastest oscilloscope and photodiode combination lacks the resolution to measure picosecond pulses. Consequently the so-called correlation methods were developed to measure indirectly the widths of picosecond pulses.

Correlation methods have been extensively reviewed by Ippen and Shank [36] and Bradley and New [37] and will be discussed only briefly. The popular nonlinear correlation methods involve the superposition of one pulse with itself (autocorrelation) or another pulse in a nonlinear optical medium where a property such as second harmonic generation (SHG) or two-photon fluorescence (TPF), which is dependent on the square of the input power, can be observed. Figure 5.6 represents a zero-background technique developed by Weber [38] to measure the amount of second harmonic produced when pulses were split by a par-

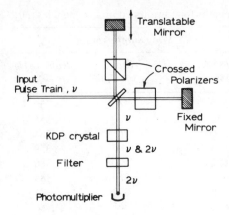

Fig. 5.6. Schematic of correlation technique developed by Weber [38] to measure picosecond pulsewidths. Orthogonal pulse polarization permits type II phase matching in KDP and the elimination of background signal due to SHG from only one pulse (orthogonal polarization is required in type II phase matching).

tially reflecting mirror, passed through polarizers to produce orthogonal polarizations in the two beams, and later recombined and passed through a crystal of KDP. The intensity of the second harmonic is recorded, the relative position of the two pulses is altered by moving a mirror, and the laser is fired again. Clearly SHG will be at a maximum when the two pulses overlap in the crystal and there is zero phase shift. The pulse profile is then plotted against mirror position X, and the width ΔX is related to the pulse duration by the speed of light ($\Delta t \cong \Delta X/c$). A similar technique using the second harmonic reflection produced at the surface of a nonlinear crystal (GaAs) was introduced by Armstrong [39] and used more recently by Tomov [40]. Weber's method has been successfully applied in measuring pulsewidths from mode-locked CW dye lasers, because the pulse trains are continuous and the signal-to-noise ratio is improved owing to signal averaging. Figure 5.7 illustrates the approach taken by Ippen and Shank. When pulsed lasers are used, the primary weakness of SHG, in both KDP and reflections from GaAs, was that many laser shots were required to construct a picture of the pulse profile, and the profiles represented the time average of all the pulses in the train. These limitations were eliminated by the development of the two-photon fluorescence (TPF) technique [41] and the measurement of the widths of single pulses. Figure 5.8 illustrates this experiment. A single pulse is extracted from the train and split by a partially reflecting mirror and recombined in a medium transparent to the lasing wavelength, but it will simultaneously absorb two photons and exhibit a fluorescence with an intensity dependent on the square of the input power of the pulse. The fluorescence is photographed, and the pulsewidths are determined by measuring the size of the fluorescence image directly with a ruler or from microdensitometer traces. No information is available about the sym-

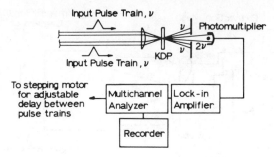

Fig. 5.7. Ippen and Shank's adaptation of Weber's technique for pulsewidth measurements of pulses from a mode-locked CW dye laser (*Laser Focus*, July 1977). Second harmonic output is at an angle to the fundamental frequency and eliminates the need for filters to separate ν from 2ν. One of the input pulse trains is chopped, and the lock-in amplifier provides synchronous detection of 2ν at the chopping frequency. This increases signal-to-noise ratio dramatically, and weak (second harmonic) signals are more easily measured.

metry of the pulse, and data from the microdensitometer must be fit to an arbitrary pulse shape (usually Gaussian) to yield a value for the pulsewidth. In TPF a peak-to-background contrast ratio of 3:1 is a general feature of a good single-pulse measurement. The 4:1 ratio, calculated from the I^2 dependence of two-photon fluorescence, is not expected because the two pulses propagate in opposite directions and a correlation must involve an average of all possible relative phase angles. In fact, a 3:1 ratio is often not observed, and background may even approach the peak intensity for pulses from a poorly mode-locked laser. The pulse intensity

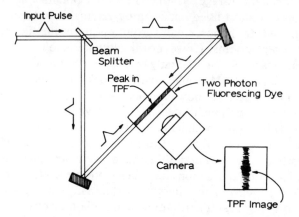

Fig. 5.8. Two-photon fluorescence (TPF) method. Pulsewidths are generally determined from a microdensitometer trace of the photographed fluorescence image.

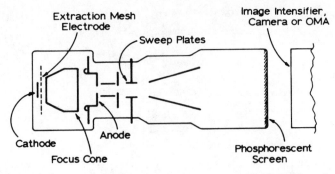

Fig. 5.9. Streak tube from an Imacon II streak camera (Hadland Photonics, Ltd.).

must be just intense enough to produce a readable track, otherwise TPF will occur with just half of the pulse with an efficiency approaching that when the pulses overlap. Rentzepis and Duguay [42] were able to reduce the background by producing a two-photon fluorescence from the super-position of the residual fundamental pulse at 1060 nm from an Nd:glass laser and the weaker second harmonic at 530 nm. Rentzepis et al. [43] measured the third-order correlation function by three-photon fluores-cence (3PF), which gives a much higher value of contrast (10:1) and the additional advantage that the pulsewidths are independent of the assumed pulse shape.

The most direct method for measuring picosecond pulsewidths is electron-optical chronography. The instrumentation used is commonly known as a streak camera. It consists of an image tube, high-voltage pulse circuitry, and a data storage system. The streak camera was proposed for kinetic studies of short-lived emission by Zavoiskii and Fanchenko [44] but was limited to nanosecond resolution until the development of picosecond laser systems. However the ultimate resolution of the streak camera was shown to be limited only by the distribution of electron transit times in the image tube [44], and several streak cameras with resolutions of between 1 to 10 psec are now commercially available. Very simply, the streak camera can be thought of as the first stage in a common photomul-tiplier tube with minor modifications. Figure 5.9 shows the Imacon II streak tube (Hadland Photonics, Ltd.). The photoelectrons emitted from the photocathode are accelerated in an electric field generated by a very high potential on an extraction mesh electrode very close to the photo-cathode (~20 kV/cm). This effectively eliminates the spread in transit times due to the energy distribution of the photoelectrons. The photo-electrons are then focused to a beam, and the beam is deflected by a fast voltage ramp which streaks the photoelectrons across a phosphorescent screen or an image intensifier. The final image is either photographed or recorded by an optical multichannel analyzer (OMA). The pulsewidths can then be related to the intensity profile of the streaked image on the phosphor and can be determined exactly by calibration. Calibration can be performed easily by reflecting a picosecond pulse from a glass slide or optically flat piece of glass or quartz of known thickness and then streaking the reflected pulses. The separation of the pulse image reflected off the first surface and that reflected off the second surface is then related to the distance between the surfaces and the speed of light in the medium.

Principles of Picosecond Measurement

In this section we classify picosecond measurement techniques according to the physical principles on which they are based. Then we give a detailed account of each type of experimental approach. New kinds of measurement techniques become possible in the picosecond-to-nanosecond time regime that are not practicable on a longer time scale. This is so because we are dealing with time intervals so short that even light can travel only fairly modest distances in the times involved. For example, light travels 1 ft, or 30 cm, in 1 nsec and only 0.03 cm in 1 psec. (These are in vacuo distances; in material media the velocity is reduced still further according to the refractive index of the medium.) Thus a new class of measurement techniques is opened up. These are based on using a short-duration monitoring pulse to take a snapshot of the fading trail of excited molecules that each picosecond excitation pulse leaves behind as it sails through the medium. Think of the wake of a ship and then imagine taking a snapshot. In front of the ship (picosecond pulse) the medium is as yet undisturbed. Immediately behind, the medium is maximally perturbed, and further down the wake, the disturbance progressively relaxes. In essence we have substituted distance for time; as we proceed back along the wake, we encounter regions that were excited progressively longer ago in time. We shall call such techniques optical wake techniques and discuss them in more detail later on. A number of variations exist, and we give several examples. If the excited molecules emit, time-resolved fluorescence studies can be made in the same manner.

Some measurement techniques used in longer time frames can be adapted for use in the picosecond regime provided certain instrumental advances can be made. The very successful technique of flash photolysis and spectroscopy can be extended to the picosecond time range, given the availability of broad-band continuum monitoring pulses of picosecond duration. Absorption spectra of transient species can be recorded photographically for exploratory work. Single-shot spectrophotometry can then be used to measure the optical density changes at a given wavelength and time delay, using a Vidicon detector (similar to a TV camera) coupled to an optical multichannel analyzer (OMA). Repeating such measurements at many wavelengths allows the absorption spectrum of an intermediate to be built up at a given time delay. Alternatively a kinetic study can be made at a chosen wavelength by taking measurements at many different time intervals after excitation. It is worth noting that the detector in picosecond flash-photolysis studies is not required to provide time resolution but simply a measurement of intensity. Time resolution is provided by the known preset time delay between the excitation pulse and the monitoring pulse. A more detailed discussion of picosecond flash photolysis is given later.

If a detector that *can* provide picosecond time resolution is available, such as a fast streak camera, the decay of excited species can be monitored temporally after a single laser shot. Fluorescence decay studies have been made and lifetimes measured in this way. Absorption decay studies are also possible, given a suitable background-monitoring pulse, but none has been done to date using a streak camera.

We devoted a separate section below to fluorescence lifetime measurements. Measurements that use a continuous train of picosecond laser pulses and a signal-averaging detection technique have been most useful in extending time resolution into the subpicosecond time domain. A later section therefore discusses dye laser techniques and subpicosecond measurements.

Additional techniques have been used to measure subnanosecond lifetimes. For instance, in measurements of emission lifetimes, intensity-modulated excitation sources and phase-sensitive detection systems (lock-in amplifier) have been used [45]. Recently an inexpensive variation was proposed [46]. For species that do not emit, other methods that yield excited-state lifetimes include saturation spectroscopy [47] and other nonlinear techniques for studying coherent transients [48]. We do not discuss these methods further. Those desiring additional details should consult the references just given.

Picosecond Absorption Techniques

Picosecond absorption measurements have several advantages over emission measurements. First, excited states that do not emit can be studied. Since there are many more states that do not emit than states that do, this is an important benefit. For example, there are many states, such as $n\pi^*$ singlet states and triplet states and higher excited electronic levels, for which the quantum yield of radiative emission is extremely low ($<$ 10^{-4}), either because the radiative rate is small or because the competing quenching processes are very fast ($\geq 10^{12}$ sec^{-1}). For such states excited-state absorption (ESA) measurements are the only way to follow the time evolution and decay of their populations. Second, using absorption techniques one can study the kinetics of the bleaching of the ground state (GSB) and its subsequent repopulation (GSR). There is of course no alternative way of studying the time history of the ground state. Third, absorption studies are more amenable to quantitative measurements than are emission studies. The reason for this is that, in absorption studies, one can measure the attenuation of a monitoring beam of known intensity and cross section on passage through a known path length of sample. By contrast, quantitative emission studies require the collection of all photons emitted within a well-defined and known solid angle. (Concern is also necessary for possible polarization effects and resultant anisotropy in the spatial distribution of the emitted radiation.) To obtain quantum yields, intensity comparisons must be made with standard systems of known yield. In making such comparisons, allowance must be made for the lifetime of the emitting state. Either the emission must be integrated over a time several times longer than the exponential decay time or the lifetime must be measured and photons collected over a well-defined time interval.

Both for absorption and for emission studies, laser pulses greatly facilitate the making of quantitative measurements compared to the use of incoherent light sources. The energy of a laser pulse can be accurately measured either by actinometry or by a well-calibrated thermopile bolometer. Because of the spatial coherence of the laser output, it is then relatively easy to focus the collimated beam onto a well-defined cross section of the sample. If a sample of high optical density is used, essentially all of the energy of the laser pulse is absorbed in a short path length (1 or 2 mm or even 0.1 mm in special cases). If absorption is not complete, the fraction absorbed can be found by measuring the energy of the pulse after leaving the sample. Thus the number of photons absorbed and therefore the number of initially excited states produced in a given sample volume can be calculated. Knowing the extinction coefficient of the ground state, one can then readily calculate the extinction coefficient for the excited-state absorption from the observed change in optical density at a particular wavelength.

We now proceed to a detailed discussion of several types of picosecond absorption measurements. A useful recent review of picosecond techniques is given by Ippen and Shank [36].

Optical Wake Techniques

The earliest picosecond studies fall in this category [49]. On longer time scales, say a nanosecond or greater, the spatial extent of the light pulse greatly exceeds the length of the sample. This is illustrated in Fig. 5.10*a*.

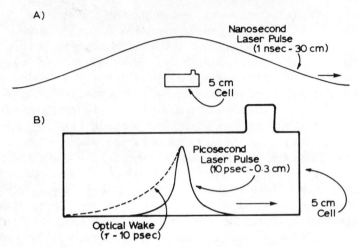

Fig. 5.10. Spatial extent of laser pulses relative to same size: (*A*) pulsewidth = 1 nsec; (*B*) pulsewidth = 10 psec.

For all intents and purposes, the medium is homogeneously illuminated and excited by the light pulse. To make a kinetic measurement, we must wait until the pulse has gone by and then send a weaker probe pulse through the sample at a known time later and measure its attenuation. This is what is done in flash photolysis. Alternatively we can use a CW probe beam and monitor its transmitted intensity continually using a detection system that provides time resolution, such as a photomultiplier and an oscilloscope. Usually we make such measurements at a chosen wavelength, and the technique is called kinetic spectrophotometry. The time resolution is limited by the falltime of the trailing edge of the pump pulse. Only events with relaxation times comparable to or longer than this time are amenable to measurement.

Now let us examine the situation when a picosecond excitation pulse is used. Picosecond pulses are so limited in spatial extent that new monitoring techniques become possible. A 5-psec pulse is a little bullet of light less than 2 mm long. Thus we have the interesting situation that we can quite conveniently make the sample longer than the pumping pulse. As it passes through the sample medium, the pump pulse leaves behind a progressively decaying trail of excited molecules—an optical wake. (The situation is analogous to that present behind a shock wave in a shock tube experiment, but here the time scale is much faster because the perturbation is optic rather than acoustic.) The picosecond light pulse and its optical wake are shown in Fig. 5.10*b*. All we have to do is stand off to one side of the beam path and take a snapshot that freezes the pump pulse and its attendant trail of excited molecules. For this to work the probe pulse must be as short as or shorter than the pump pulse, weak enough that it does not itself perturb the excited-state population and spatially homogeneous in intensity over the length of sample being monitored. We also need a detector. Several ingenious experimental arrangements based on the above principles have been developed.

In the earliest experiments by Rentzepis and co-workers [50], a 530-nm picosecond pulse was used to excite azulene in solution to an upper vibronic level of its first excited singlet state, S_1. The optical wake of excited S_1 molecules was then made visible by sending an interrogating 1060-nm picosecond pulse through the sample in the opposite direction. The energy of this pulse (9434 cm^{-1}) was sufficient to raise the S_1 excited molecules still higher, to S_2. Since azulene fluoresces from S_2 to S_0, the decaying population of S_1 molecules could be monitored by photographing the fluorescence so produced. The spatial decay of the fluorescence along the wake gave a time of about 7.5 psec for the vibrational relaxation of azulene in the third or fourth vibrational level of S_1 [50].

The above experimental technique is not readily extended to other molecules because it depends on a fortuitous matching between energy gaps in the azulene molecule and the laser frequencies and also upon the unusual circumstance that azulene fluoresces from its S_2 level. Another cause for concern is that the interrogating pulse itself must lose intensity as it proceeds down the wake, and allowance should somehow be made for this factor in extracting the relaxation time from the trace that records the decay of fluorescence intensity from S_2. This factor was not considered in these early experiments. Azulene is discussed further in the section on electronic relaxation.

A more versatile variation of the above technique, using a crossed-beam geometry, was described by Malley and Rentzepis [51]. The principle is shown in Fig. 5.11. A picosecond laser pulse is divided by a beam splitter. The stronger pulse (90%) traverses a rectangular (2 mm × 10 mm) sample cell. To eliminate inhomogeneity throughout the cross section of the weaker interrogating pulse, it is first spatially broadened by passage through a coarse (500 lines/in.) diffraction grating, called a Ronchi ruling, and is then caused to pass through the sample cell at right angles to the bleaching pulse. If the timing is right, the interrogating pulse catches the bleaching pulse within the cell and takes a snapshot, in the manner previously described, that includes a region of unexcited sample ahead of the bleaching pulse, the excited region occupied by the pulse, and the optical wake behind the pulse. The detector can be either a camera or a diode array (Vidicon or Reticon) coupled to a data storage device such as an OMA, minicomputer, or storage scope. In the reference cited, 1060-nm

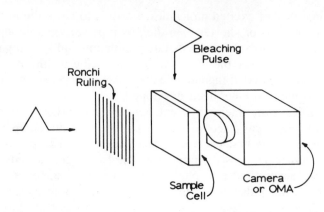

Fig. 5.11. Malley–Rentzepis crossed-beam technique [51]. The interrogating pulse is passed through a Ronchi ruling, about 1 m from the ruling, to smooth its cross-sectional intensity profile. In this way homogeneous illumination of the sample cell is accomplished. This assures reasonably accurate determination of decay kinetics from optical wake profiles.

pulses were used and a camera with Polaroid Type 413 film was the detector. The sample studied was the Kodak Q-switch/modelocking dye 9860. A value of 9 psec was found for the ground-state recovery time, in good agreement with earlier measurements. The result obtained on the photographic film is actually the convolution of the bleaching pulse, the monitoring pulse, and the recovery time of the dye; but, provided that the laser pulses are well characterized, the dye recovery time can be deduced. Time resolution is limited not only by the pulse durations but also by the thickness of the sample cell, since the bleaching pulse moves slightly during the time it takes for the interrogating pulse to pass through the cell. Thinner cells can be used, but then the sample concentration must be increased to ensure enough optical density to provide a detectable contrast ratio.

In the so-called echelon technique, if an interrogating laser pulse is passed through a stepped stack of glass plates, or *echelon*, as shown in Figs. 5.12 and 5.22 the pulse is dissected into a series of pulses equally spaced in time and distance by an interval determined by the delay, relative to air, introduced by the plate thickness. For a 1-mm-thick glass plate this delay is about 1.7 psec. The echelon technique, as originally developed by Topp, Rentzepis, and Jones [52], was first used in conjunction with an optically triggered CS_2 Kerr cell, or "light gate," to take time-resolved photographs of the stimulated emission from several laser dyes (see section on fluorescence lifetime measurements). For absorption studies, as illustrated in Fig. 5.12, the bleaching pulse and the series of

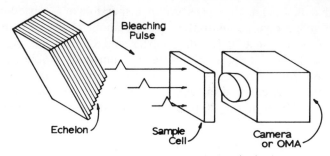

Fig. 5.12. Echelon technique developed by Topp, Rentzepis, and Jones [52]. If the bleaching pulse illuminates the sample uniformly, a microdensitometer trace of the photograph of the interrogating pulse will resemble a histogram of intensity bands reflecting decay of excited-state absorbance or return of ground-state absorbance. If a spectrograph is used, a photographic record will also include spectral information and allow the calculation of rate constants at several wavelengths.

monitoring pulses travel almost colinearly through the cell. After passing through the cell, the latter pulses are detected either on film or with the aid of a Vidicon/OMA combination as before. Because of the colinear relation of the excitation and measuring beams, the cell thickness limitation of the crossed beam technique is avoided. However, there is the disadvantage that each interrogating pulse monitors a different spatial region of the sample. Spatial uniformity of the bleaching pulse over the entire sample cross section that is being monitored is therefore required for quantitative work. Group dispersion effects must also be taken into account when several frequencies are used [53]. Such effects can be avoided by using a reflecting echelon, so that the interstep delay is provided simply by the longer optical path *in air* rather than by passage through an extra thickness of dispersive medium such as glass.

The echelon technique has been extensively used, especially by Rentzepis and co-workers. It has several advantages. It provides a kinetic record in a single laser shot; and, if a continuum monitoring pulse is used, a set of time-resolved spectra can be obtained in a single shot. This is very valuable, when we bear in mind the shot-to-shot lack of reproducibility of mode-locked lasers. It is also useful for extracting as much information as possible from a single experiment on a photolabile system, for example, rhodopsin. There is also a price to pay. It is important that both pump and probe pulse be spatially uniform in intensity over the cross section of the beam, since measurements at different time delays do not correspond to the same region of the sample. The jumps in time from segment to segment are discrete and several echelons are needed to cover adequately a time span of say 1 nsec. (However, Topp [54] has recently described a technique that uses group dispersion in a series of three prisms to give a continuous time scan over about 85 psec.) In practice, echelons are most useful for the shorter end of the time range. For example, a stack of 20 glass plates each 2 mm thick would, if used in transmission, cover from 0 to 60 psec in 3-psec steps. If 3-mm plates were used and the echelon used in reflection, the interstep delay would be a 6-mm path in air, or about 20 psec, and a 20-step echelon would cover about 400 psec.

Picosecond Flash Photolysis

This is a direct extension to the picosecond region of the well-known microsecond technique of flash photolysis using rare-gas-filled flashlamps. A mode-locked laser provides the picosecond excitation pulse. The spectroscopic flash for monitoring purposes is a picosecond continuum pulse and can be obtained by focusing a portion of the laser pulse into a variety of optical media. A variable time delay between the two pulses is obtained by changing the difference in optical path length between the two beams. The apparatus developed by Magde and Windsor [55] is shown in Fig. 5.13. A mode-locked Nd:glass laser, with single pulse selection (SPS) and one stage of amplification, produces a 5- to 8-psec pulse of 15- to 20-mJ energy at 1060 nm. Single pulse selection is accomplished using an optically triggered spark gap and Pockels cell as described in detail in the section on single pulse selection. A pulse early in the pulse train is

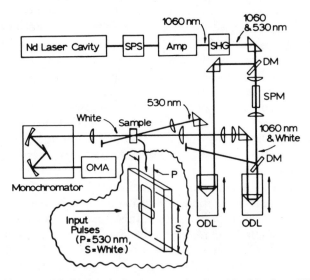

Fig. 5.13. Picosecond flash photolysis apparatus developed by Magde and Windsor (55). See text for detailed description.

selected for amplification because it has been shown that early pulses have a shorter duration than later pulses. Passage of the 1060-nm pulse through a suitably oriented crystal of potassium deuterium phosphate (KD*P, where the asterisk denotes deuteration) gives second harmonic generation (SHG) at 530 nm with a conversion efficiency of about 15%. The 530-nm and 1060-nm pulses are separated by a dichroic mirror that selectively reflects the 530-nm radiation while transmitting the 1060-nm pulse. After passage through an optical delay line (ODL), consisting of a rooftop prism mounted on an optical rail, the 530-nm pulse P (for photolytic) impinges on the sample, usually a 1- or 2-mm-thick cuvette. Any 530-nm light transmitted by the sample is intercepted by an optical stop. As a further precaution against scattered P light entering the spectrograph, it is arranged that the P beam make a slight angle (10°) with the monitoring beam S (for spectroscopic), as shown in Fig. 5.13.

The key to the technique of picosecond flash photolysis is the picosecond continuum pulse. In our experiments this is obtained by focusing the residual 1060-nm radiation into a 10-cm-long cell containing carbon tetrachloride (CCl_4). Self-focusing and filamentation occur, and a small fraction (~0.1%) of the incident beam is converted into a broad continuum. The generation of this continuum may involve several nonlinear effects, but for brevity we shall term the process SPM, for self-phase modulation. The continuum retains both the short pulse duration and much of the spatial collimation of the pumping pulse. Thus, to all intents and purposes, we have a white light laser. (Later in this section we provide a more detailed discussion of old and new work on continuum generation, specifically the SPM continuum.) Another dichroic beam splitter separates the essentially undiminished 1060-nm pulse from the continuum. The 1060-nm pulse can be used for sample excitation or can be blocked, as in Fig. 5.13. The continuum S passes through a variable optical delay line and arrives at the sample at a predetermined time relative to the P pulse. By adjusting the two ODLs we can vary the time interval between S and P between zero and about 4 nsec. Kinks in the optical rails make registration of the two beams at the sample plane unreliable for delays in excess of 4 nsec, though with better rails there is no reason why longer delays should not be used. Probably about 10 nsec is a practical limit. Negative delays (i.e., S arrives before P) are also readily obtainable. These are useful in providing a record of the sample prior to excitation for control purposes and correspond to "infinite delay" shots in conventional flash photolysis.

Let us now discuss the manner in which optical density changes induced in the sample by the P pulse are measured. The inset at bottom center of Fig. 5.13 shows the cross-section geometry of the two beams at

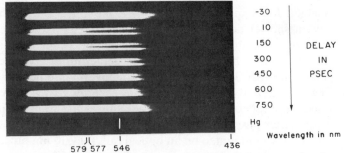

Fig. 5.14. Photographic record of excited-state decay of BDN in iodoethane using the Magde–Windsor technique [56]. The film plate is moved and the delay time between excitation pulse and monitoring pulse is changed between exposures. The dark central portion of each image is due to absorbance of the continuum by the excited state.

the sample. The detail of the geometry in this region is very important. The key point is that the probe beam is elongated along its vertical axis by means of a cylindrical lens so that it samples simultaneously the excited volume in the center and unexcited *reference* volumes both above and below the excited region. The cross section of the excited volume can be adjusted in the range 1 mm to about 5 mm by means of a lens. The entire area is then imaged at about 1:3 demagnification onto the slit of a spectrograph and, ultimately, onto either photographic film or a vidicon detector. By having a reference spectrum both above and below a purported transient spectrum, we greatly reduce the risk of misinterpreting random fluctuations in the intensity of the continuum as genuine transient effects.

The photographic method is all but essential for exploratory survey work on a previously uncharacterized system. It provides wide spectral coverage in a single shot and the time history of whatever transient changes are present can be seen at a glance on a single film or plate by taking a short sequence of shots at different time delays. This is demonstrated in Fig. 5.14 through 5.16. Figure 5.14 shows the decay of transient excited-state absorption (ESA) in the region 430-600 nm for the laser modelocking dye bis(4-dimethylaminodithiobenzyl)nickel(II) (BDN) over the time range of 0 to 750 psec [56]. Figure 5.15 shows similar picosecond flash-photolysis results for octaethylporphinatotin(IV) dichloride [(OEP) SnCl$_2$] [57]. The value of the wide spectral coverage provided by the photographic record is particularly evident here. The photograph clearly shows two regions of ESA: a short-lived transient with a main absorption peak at ~ 450 nm and absorption extensity to longer wavelengths, and a much longer-lived absorption in the blue with a peak at about 430 nm. Additional studies using the Vidicon enable these two transient absorptions to be spectrally characterized and attributed, respectively, to the

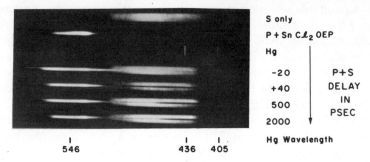

Fig. 5.15. Photographic record of excited state decay of [(OEP) SnCl₂] in 1,2-dichloroethane showing both S_1 and T_1 absorption as studied by Magde et al. [57]. The persistent excited-state absorbance due to the T_1 state is particularly evident in the 500- and 2000-psec exposures.

excited singlet S_1 state and the triplet T_1 state of the porphyrin. The photograph is very helpful in choosing suitable wavelengths for subsequent quantitative kinetic measurements.

Figure 5.16 shows a much more complicated photographic record for the molecule 3,3′-diethyloxadicarbocyanine iodide (DODCI) [55]. In the region from about 440 to 510 nm, excited-state absorption (ESA) with a lifetime of about 1 nsec can be seen. In this experiment the 530-nm excitation beam was focused to a cross section of about 1 mm, so the ESA appears as a narrow, dark band. Bleaching of the ground-state absorption in the 560- to 590-nm region is also readily seen and is especially apparent on the 20-psec exposure as a narrow, white central band that corresponds to enhanced transmission of the probe beam caused by the laser excitation

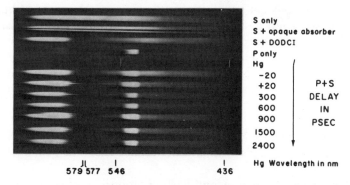

Fig. 5.16. Photographic record of excited-state decay of DODCI in ethanol as studied by Magde and Windsor [55]. Bleaching of ground-state absorbance can be seen between 560 and 590 nm as a light central band in the shorter-time exposures.

pulse. By suitable selection of wavelength, kinetic studies may be made of either ESA or ground-state repopulation (GSR). When interpreting picosecond photographic spectra, allowance must be made for "chirping" of the probe pulse—a slight dispersion in the arrival time of the various colors at the sample due to the relatively long path through dispersive media. This amounts to 16 psec between 430 and 650 nm, the blue end of the spectrum being delayed with respect to the red, and is easily accounted for in interpreting results. Of course, when kinetic measurements are subsequently made using the Vidicon/OMA, only a narrow band of wavelengths is monitored, and chirping is negligible.

It will be clear from the above discussion that, although the photographic method is of great value for survey work, it is not the most convenient method for determining precise kinetic data. Therefore once the photographic spectra reveal the best wavelengths to monitor, we replace the camera with a slit and use photoelectric detection, normally a Princeton Applied Research (PAR) 1205B Vidicon tube coupled to a 500-channel optical multichannel analyzer (OMA). We then monitor only a single narrow band of wavelengths, using the OMA to obtain an intensity profile along the length of the slit as shown in Fig. 5.13. This is exactly equivalent to a densitometer trace taken vertically across the photographic spectrogram. The natural logarithm of the ratio of the measured intensity in the reference regions (averaged over the above and below portions) to that in the central excited region gives the optical density change (ΔOD) due to the transient absorption (or bleaching). Usually four or five shots are taken at a given wavelength and time delay, in order to obtain an average value plus the standard deviation. Repeated measurements at the same wavelength and different time delays provide a plot such as that shown in Fig. 5.17, from which the decay kinetics can be determined. By keeping the time delay constant and changing the wavelength, we can obtain the difference spectrum at a given time delay. In most cases, because of spectral overlap of the ground-state and excited-state absorption regions and because, at the laser powers used, significant depletion of the ground state is common, the spectrum obtained must be considered a "difference spectrum" that represents the sum of the changes caused by ground-state depletion and production of the excited state. A typical difference spectrum for a photosynthetic system is shown in Fig. 5.18. By detailed studies it is usually possible to find wavelengths at which kinetic studies of one species can be made without interference by optical changes due to another species [58].

The advantages of picosecond flash photolysis over the echelon technique are (1) less severe constraints on the spatial uniformity of the continuum and (2) the provision of continuously variable delay times over

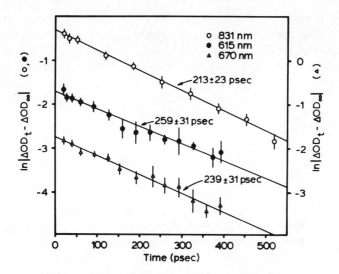

Fig. 5.17. Kinetics of one of the electron transfer steps in bacterial photosynthesis measured at three different wavelengths in reaction centers isolated from the photosynthetic bacterium *viridis*. Figure 5.18 shows the spectra of the states before and after this step. From Holten et al. [196].

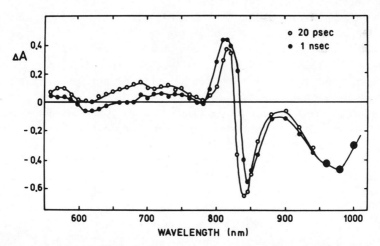

Fig. 5.18. Absorbance changes at room temperature for reaction centers of *Rhodopseudomonas viridis*, measured after time delays of 20 psec (open circles) and 1 nsec (closed circles). The change in the difference spectrum between the two time delays reflects the transfer of an electron from a molecule of bacteriopheophytin to ubiquinone. The kinetics of this process are shown in Fig. 5.17. From Holten et al. [196].

the range 3 psec to 4 nsec instead of the discontinuous jumps in time delay over a limited temporal range given by the echelon. The main disadvantage of picosecond flash photolysis is that each shot gives a measurement at only one wavelength and one time delay, necessitating a laborious series of shots to obtain either a complete spectral or complete kinetic record. This deficiency could be removed by the use of a minicomputer to retrieve data in two-dimensional fashion (optical density versus wavelength) from the vidicon surface in a single shot, an objective that we are actively pursuing at the present time.

Picosecond continua can be generated by focusing high-power picosecond laser pulses into a variety of optical media. Alfano and Shapiro [59] first described the use of picosecond continua for spectroscopic studies, although earlier papers from several groups had discussed the mechanisms by which the continuum is generated. The details of the mechanism are beyond the scope of this review and, in any case, are still not thoroughly understood. Some groups have called the process *self-phase modulation* (SPM), and we have used this as a convenient label in our own work. Others have described the mechanism as a four-photon parametric process. It seems likely that several coherent nonlinear effects combine to produce the continuum. We are concerned here with its use as a spectroscopic tool, so we provide some practical information on the generation and use of the SPM continuum without delving deeply into the mechanisms by which it is formed.

In their early work, Alfano and Shapiro [60] used samples of quartz and BK-7 glass up to 20 cm long and samples of NaCl and calcite. Using 530-nm excitation they produced continua that stretched to 300 nm in some cases. More recently Yu et al. [61] have described spectral broadening due to SPM of a 1060-nm pulse in KBr; and Ippen, Shank, and Gustafson [62] have observed SPM in CS_2-filled optical fibers 1 and 2 m long. To generate the continuum the laser pulse is focused with a short focal-length lens (5 to 10 cm) into the front portion of the sample. Self-focusing occurs and optical filaments are formed. The continuum is generated within these filaments, which remain tightly focused throughout the length of the rod and produce well-defined bright spots at the exit end. Because of this filamentation, it is wise to use a diffusing element, such as a piece of ground glass, between the SPM cell and the sample cell in order to average the beam spatially. The energy threshold for the SPM continuum is quite sparp. If a particular laser shot is a little below par, then no continuum will be observed. Also the efficiency depends on the interaction length in the generating medium. Although continua can be produced in 1- or 2-cm path lengths, it is much better to use longer optical paths up to about 10 cm. In our own experiments we were able to generate useful SPM continua in many liquids (e.g., water, ethanol, hexane, benzene) using either 530-nm or 1060-nm excitation. We found empirically that carbon tetrachloride (CCl_4), gave especially good results. For CCl_4 in a 10-cm cell, 1060-nm-generated continua stretched all the way down to about 430 nm in the visible. Using 530-nm pumping allows the continuum to be extended to about 300 nm. The big advantage of using liquids as the generating medium is that optical damage, if it occurs, is self-healing. With quartz and glass rods, on the other hand, it is not uncommon to find optical damage, in the form of small fractures a few millimeters inside the front surface, after a few laser shots. Quite recently Mataga and Nakashima [63] have found that polyphosphoric acid is a particularly favorable medium for generating picosecond continua.

Transient Grating Method

Recently Phillon, Kuizenga, and Siegman [64] presented an alternative to the basic pump and probe method discussed above. The basic principles that underlie their technique are not new, and the reader is referred to Ref. 64 for a list of previous work. The fundamentals were developed along with the development of holography and have been applied by Phillon et al. to measure transient absorbance and bleaching in the picosecond time domain. Simply stated, two pulses of the same frequency are crossed in an absorbing medium. The interference pattern at the intersection of the two beams produces a spatial modulation of intensity within the sample. This fringe pattern resembles the rulings on a plane grating, because of the coherence of the two pulses, and it produces an image (spatial modulation of the excited-state population) which persists with a lifetime equal to that of the ESA or GSB. Figure 5.19 illustrates their experimental arrangement. A probe pulse is passed through the transient grating, and its intensity is detected at the diffraction angle θ determined by the relation

$$\lambda = D \sin \theta$$

where λ is both the pump and probe wavelength and D is the fringe spacing of the interference pattern. As time progresses, the grating will relax by decay of the excited-state population or the orientational anisotropy produced in the solution. Thus the kinetics of the underlying process can be followed by delaying the probe pulse with respect to the pump pulse. The principal advantage of this technique over the normal pump and probe techniques is in measuring small absorbance values. In the normal methods, for small absorbance changes, the difference between

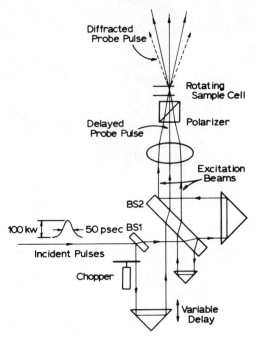

Fig. 5.19.   Schematic of the method used to measure transient kinetics by the decay of a grating (interference pattern) produced when two coherent pulses cross in a sample cell. The interference pattern is a spatial modulation in the ratio of ground state to excited state and persists with a lifetime dependent on excited-state relaxation or decay of induced orientatational anisotropy. Phillon et al. [64].

two nearly equal probe signals must be determined, whereas in the grating method a small signal is detected against a zero background. The uncertainty in absorbance values is therefore greatly reduced. In addition, the use of a Nd:YAG laser or mode-locked CW dye laser to generate a continuous train of picosecond pulses permits the use of signal-averaging techniques. In a subsequent report, Siegman [65] has proposed the use of a moving grating produced by two CW beams while one of the beams is tuned smoothly through a frequency range close to the other beam. A sample calculation showed that lifetimes of ~100 psec could be measured with input powers of ~1 W. He further suggested that a pair of mode-locked CW dye lasers with peak pulse powers of ~10^3 W could provide a means of resolving subpicosecond lifetimes. The double-mode-locked dye laser systems described in a subsequent section are excellent candidates for an experimental test of this method.

Fluorescence Lifetime Measurements

Ultrafast Optical Shutters

Fluorescence lifetime measurements on the picosecond time scale were made possible by the suggestion, in 1956 [66], and the observation, in 1964 [67, 68], that a birefringence could be induced in liquids by the intense electric fields associated with high-power laser pulses. Duguay and co-workers were the first to incorporate an optical Kerr cell in the measurement of picosecond fluorescence lifetimes [69, 70]. Typically a 1060-nm pulse is used to "gate" the Kerr cell, which contains carbon disulfide (CS_2) or some other appropriate liquid placed between crossed polarizers. Figure 5.20 illustrates the use of their design in picosecond fluorescence lifetime measurements. The high electric field of the gating pulse induces a birefringence in the liquid, which can be expressed as a convolution of the intensity profile of the gating pulse and the exponential decay of the induced orientational anisotropy of the solvent. Thus birefringence is proportional to $\Delta n(t)$, the difference in the refractive index of the solvent molecules in directions parallel and perpendicular to the electric field polarization of the gating pulse. If the lifetime for rotational relaxation is much shorter than the gating pulsewidth, the following expression for $\Delta n(t)$ results:

$$\Delta n(t) = \frac{1}{2} n_{2B} E^2(t)$$

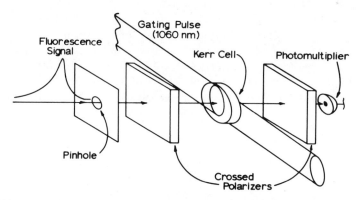

Fig. 5.20. Kerr shutter used in measuring fluorescence lifetimes. The photomultiplier detects the signal that passes through the Kerr cell coincident with the gating pulse. Delaying the arrival of the gating pulse allows the intensity of the fluorescence signal to be measured and plotted vs. the delay time. The resolution of the Kerr shutter is a function of the width of the gating pulse (~5 psec). Consequently this method works best when lifetimes are ≥ 50 psec. From Duguay et al. [69, 70].

where n_{2B} is defined as the optical Kerr constant and $E^2(t)$ is the square of the optical electric field amplitude associated with the gating pulse. Consequently the optical Kerr cell, when placed between crossed polarizers, acts as a shutter similar to the Pockels cell, rotating the polarization of the fluorescence pulse only while the medium is influenced by the optical electric field of the gating pulse. Using the formalism attributed to Fresnel, linearly polarized light can be considered as composed of two phase-matched, right- and left-hand circularly polarized beams. The activated optical Kerr cell will affect the speed of the two circularly polarized components differently and cause a differential phase lag

$$\Delta\phi(t) = \frac{\pi L n_{2B} E^2(t)}{\lambda}$$

where L is the path length of the cell and λ is the wavelength of the transmitted light. The differential phase lag determines the degree of rotation of the linearly polarized composite and the transmission of the Kerr shutter according to the expression

$$T(t) \propto \sin \frac{\Delta\phi(t)}{2}$$

Transmission is clearly a maximum when $\Delta\phi(t) = \pi$.

In principle, the aperture time of the Kerr cell can be shorter than the pump pulsewidth, since the optical Kerr effect is a nonlinear process that discriminates against the lower power in the wings of the pulse. Carbon disulfide is the preferred medium for Kerr cells because of its large optical Kerr coefficient n_{2B} and short orientational relaxation time, $\tau \sim 2$ psec [71]. The ratio of gated to ungated transmission, which is a measure of efficiency, is typically $\sim 10^3$. If the Kerr cell is gated at different times during the decay of fluorescence, only a small increment (~ 10 psec) of the emission intensity is imaged onto a photodiode or photomultiplier. The response of a photodiode is generally much slower than the speed of the shutter, but the peak in the output is directly proportional to the intensity of the incident signal. By plotting the peak voltage V_p versus the delay between the gating and fluorescence pulses, we can determine the fluorescence decay lifetime easily. Later designs permitted kinetic and spectral information to be recorded in one laser shot. Malley and Rentzepis have used a cross-beam geometry to make time-resolved measurements of stimulated emission and stimulated Raman scattering [72]. The 1060-nm gating pulse travels across a 1-mm-thick cell, and the intensity profile transmitted through the cell is recorded photographically. The intensity profile is related to the time dependence of the signal and the speed of the gating pulse as it travels across the optical Kerr cell. If a spectrometer is

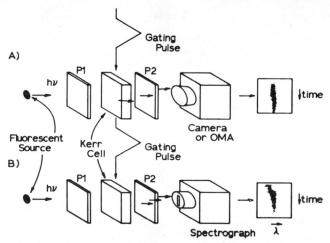

Fig. 5.21. Kerr shutter used to measure short fluorescence lifetimes ($\lesssim 50$ psec) in a single pulse: (*A*) kinetic data are obtained from OMA output from a microdensitometer trace of the photographic image; (*B*) kinetic and spectral data are both available in this design; P_1 and P_2 are crossed polarizers. From Malley and Rentzepis [72].

placed in front of the photographic film, a time-resolved emission spectrum is easily recorded (see Fig. 5.21). Topp et al. [52] used a colinear gating pulse in combination with an echelon to increase the transmission of the Kerr cell and the sensitivity of the technique (see Fig. 5.22).

A shutter that uses the high power of a picosecond pulse to saturate the absorbance of a dye was investigated by Mourou et al. [73]. They found that the extinction ratio and aperture time of this type of shutter were comparable to those of the Kerr shutter. The primary difference between the two systems was the wavelength discrimination. The Kerr shutter works at all wavelengths where the medium does not absorb light (CS_2 transmits all visible wavelengths), while the saturable absorber has a narrow bandpass (typically 20 nm or the width of the bleached absorption band). Filters must be used to prevent light from being transmitted where the dye absorbance is low.

Busch et al. [74] developed a shutter that provided an aperture time comparable to that of the Kerr shutter and the shutter that uses saturable absorbance. They called this shutter a picosecond-gated optical amplifier (PGOA). They were able to amplify the image intensity as well as gate the transmission by using the saturable absorber 3,3'-diethyloxadicarbo-cyanine iodide (DODCI). The amplification was a result of superradiance at the emission wavelength of DODCI caused by the light scattered from a dilute milk solution in a sample cell. The ungated transmission

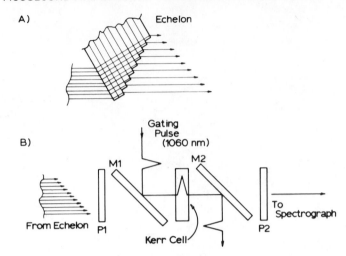

Fig. 5.22. (*A*) Detail of transmitting echelon. (*B*) Kerr shutter incorporating a colinear gating pulse and fluorescence signal. From Magde and Windsor [55].

was only $\sim 10^{-2}$, but the gated transmission was $\sim 10^2$, giving an overall efficiency tenfold better than either the Kerr shutter or the saturable absorber system proposed by Mourou. The limitation is that the bandpass is narrow, in this case being limited by the emission profile of the dye. Filters are again required to eliminate transmission where the dye does not absorb.

The Kerr shutter is useful in studies of emission lifetime and linear dispersion of the refractive index (chirp), as well as in studies of nonlinear optical properties such as continuum formation described in the section on picosecond flash photolysis. The Kerr shutter is also useful at the high repetition rates characteristic of picosecond pulses from mode-locked CW dye lasers (71).

Streak Cameras

The direct measurement of fast fluorescence lifetimes followed the development of streak cameras with picosecond resolution and obviated the use of fast shutters for this purpose. Gating is not necessary, resolution is improved, and the data for one lifetime can be obtained in a single pulse. A typical instrumental setup for picosecond emission studies using a streak camera is shown schematically in Fig. 5.23. A 530-nm pulse is split by BS1 into a reference beam and a sample beam. The sample beam is optically delayed using a rooftop prism arrangement and subsequently excites the sample. The sample fluorescence is collected and detected by a streak camera. The reference pulse also goes to the streak camera but,

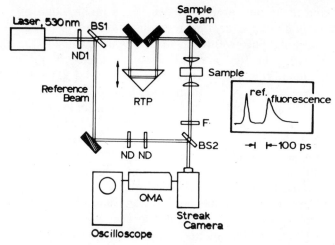

Fig. 5.23. Streak camera technique for measurement of fluorescence lifetimes. BS refers to beam splitters while ND refers to neutral density filters.

because of the shorter path traversed, arrives before the fluorescence. Thus the OMA output is displayed on a storage-type oscilloscope, as shown in Fig. 5.23. First there is a reference pulse that monitors the shape and duration of the laser excitation pulse, and then there is a pulse that shows the decay of fluorescence of the sample. Excitation intensity can be varied by changing ND1. Relative quantum yields can be obtained by integrating the fluorescence decay using the OMA and measuring the energy of the excitation pulse.

Production of Tunable Picosecond and Subpicosecond Pulses Using Dye Lasers

The development of mode-locked dye lasers has occurred along two separate and distinct paths. One approach uses pulsed excitation in which flashlamps or mode-locked pulse trains from neodymium and ruby lasers are used as the pumping source. The other involves the use of a CW ion laser as the pump and has two major advantages. First, a continuous train of pulses with a repetition rate of $\leq 10^8$ allows the use of signal averaging detection systems; and second, the lower peak pulse power ($\leq 10^3$ W) avoids the nonlinear effects described in the section on single pulse selection.

Flashlamp-pumped dye lasers have been used to produce high-power picosecond pulses in pulse trains lasting only as long as the flashlamp provides gain, which is typically only a few microseconds. The passive modelocking of flashlamp-pumped dye lasers mimics that of the solid-state lasers, with the exception that the pulsewidths are close to the transform limit and pulsewidths decrease toward the end of the train [17]. This is primarily due to gain saturation mentioned in the section on pulsewidth limitations and illustrated in Figure 5.4. The possibility of tunable subpicosecond pulses from mode-locked dye lasers has been the driving force behind their recent and rapid development. However the development of mode-locked, flashlamp-pumped dye lasers has been reviewed [75] and is not described further in this review.

Pulse trains from mode-locked ruby and neodymium laser systems also have produced picosecond pulses from dye lasers. Superradiant picosecond pulses from dye solutions excited by mode-locked pulse trains from both ruby and neodymium laser systems have been reported [72, 76, 77]. When the cavity length of the dye laser is made equal to the cavity length of the mode-locked pump laser, the dye laser is mode locked by periodic gain modulation. This has been described as synchronous pumping and has been reported for dyes pumped by the output of a mode-locked Nd:glass laser [78] and argon ion laser [79a, b].

Perhaps the most exciting recent development in the production of mode-locked pulse trains from dye lasers has been the successful production of subpicosecond pulses from CW dye lasers. Dienes et al. [80] reported the first example of a CW dye laser mode locked by an acousto-optic modulator and a 55-psec pulsewidth. Kuizenga [81] subsequently modelocked a CW dye laser using a phase modulation scheme and was able to generate ~200-psec pulses. Passive modelocking of a CW dye laser and a 1.5-psec pulsewidth was first reported by Shank et al. [82]. O'Neill [83a] was later able to reproduce these results using a contacted dye cell for the modelocking dye DODCI. In addition, Letouzey and Sari [83b] reported a passively mode-locked R-6G laser.

Typically 2 W from one or several lines of the CW argon ion laser was used to produce mode-locked pulse trains at repetition rates of 10^5 to 10^8 Hz with peak powers from 10^2 to 10^3 W. Figure 5.24 describes a multiple-mirror, folded dye laser cavity where the gain medium (R-6G) and saturable absorber (DODCI) are exposed to the pump beam in two free jet streams. The use of free jets avoids possible damage to the windows of optical cuvettes in the focal region and removes several optical interfaces from the laser cavity. Figure 5.3 shows a similar CW system, except that a dye cell is used for the modelocking dye. In both, flow configuration is required since the high average power of the CW laser compared with most pulsed systems can decompose both dyes. The pulsewidths were measured by second harmonic generation, using a modification of the zero-background method described by Weber and

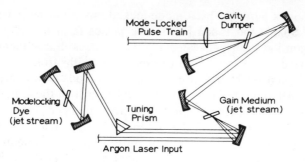

Fig. 5.24. Folded cavity configuration characteristic of the mode-locked CW dye laser systems developed by Ippen and Shank (*Laser Focus*, July 1977).

shown in Fig. 5.7. The high repetition rate and reproducible peak pulse powers made this method of pulsewidth measurement generally more convenient than when applied in measurements of mode-locked pulsed laser systems with lower repetition rates. The entire measurement takes only a few minutes, using the high repetition rates of mode-locked CW dye lasers. In early measurements the pulsewidths were obtained by fitting the experimental autocorrelation functions to either Gaussian or Lorentzian profiles, but recent theoretical treatments of modelocking under steady-state conditions using saturable absorbers predicted that the intensity profiles should show a sech^2 dependence on mirror displacement or time [11]. Later experimental results did indeed confirm that the main portion of the pulse profiles fit a sech^2 dependence.

Shank and Ippen [84] were also the first to report subpicosecond pulses. Using R-6G and DODCI in the same jet stream and dumping the cavity with an acousto-optic modulator, they were able to produce pulses of 4 kW peak power at a repetition rate of 10^5 Hz. The autocorrelation functions, however, were fit to Gaussian and Lorentzian profiles and yielded 0.7-psec and 0.5-psec pulsewidths, respectively. The pulsewidths would have been slightly smaller if the sech^2 profile were assumed. Using a mixture of two mode-locking dyes, DODCI and malachite green, to mode-lock R-6G and compressing the pulse using the method described by Treacy [85a], Ippen and Shank [85b] measured a 0.3-psec pulsewidth. Later, using an optically contacted flow dye cell for DODCI, Ruddock and Bradley [86] were able to produce 300-W transform limit pulses from R-6G ($\sim$0.3 psec) which followed the sech^2 intensity dependence.

Ippen and Shank [87] were also able to obtain subpicosecond pulses by SHG in LiIO$_3$. The second-order power dependence of SHG discriminated against the lower powers in the wings of the pulses, and UV pulses at 307.5 nm were produced with 15% efficiency.

Synchronous pumping by a mode-locked argon ion laser was first reported by Shank and Ippen [79a]. The pulses from the actively mode-locked argon ion laser generated 125-psec pulses. Subsequently several groups have demonstrated 5- to 10-psec pulsewidths from dye lasers synchronously pumped by 100-psec pulses from mode-locked argon ion lasers [79b]. Although the pulsewidths are larger than those from a mode-locked CW dye laser, several dye lasers can be pumped simultaneously by the mode-locked output of the argon ion laser giving as many synchronized and independently tunable pulse trains. A sequential arrangement of mode-locked argon ion laser–dye laser–dye laser has been found to produce subpicosecond pulsewidths. Heritage and Jain [88] reported that with a 1.5-W mode-locked argon pump power, an average mode-locked output of about 250 mW at 570 nm was produced from the first dye laser (R-6G). The 55-psec pulses from the first dye laser were then used to synchronously pump a rhodamine B (R-B) dye laser to produce 0.8-psec pulses with average output of 20 mW between 609 and 620 nm. Further reduction of the R-6G pump pulsewidth failed to shorten the R-B pulsewidth any further. Double modelocking, where one laser dye is used to modelock a second, was recently predicted and confirmed experimentally [89]. The pulses exhibited a 3:1 intensity-to-background ratio, and the intensity profile fit the expected $sech^2$ time dependence. For R-6G and a CW argon pump power of 1.8 W, a stable pulsewidth of 0.68 psec was observed at 592 nm [89]. The R-6G output was modelocked by cresyl violet, which also produced mode-locked pulses of 1.1 psec at 635 nm. When the pump power was increased to 2.5 W, the R-6G pulses were shorter (~0.55 psec), and peak power was ~150 W at a repetition rate of ~10^8 Hz.

The advantages of mode-locked CW dye lasers are (1) the demonstrated long-term stability of the pulse trains and (2) data with good signal-to-noise ratio can be obtained by time averaging the observations over a large number of relatively low power pulses. The low power is especially useful in studies where ground-state populations would otherwise be severely depleted using higher-power pulses. For weakly absorbing systems, the lower-power pulses provide no advantage, and the high-power single pulses from mode-locked Nd:glass or ruby systems are preferred. A further restriction is that the continuum (see section on picosecond flash photolysis) is not produced at these low peak powers. A subpicosecond continuum has recently been generated by pulses from a mode-locked CW dye laser amplified to 400 mW peak power [90], but until these amplified systems are put to use, the probe wavelength is constrained to be equal to or a harmonic of the pump wavelength. Synchronous pumping of two or more dye lasers and double modelocking with the attendant low jitter between pulse trains will relax this restriction and will likely approach the reliability and pulsewidths characteristic of the mode-locked CW dye laser systems.

3 APPLICATIONS

The emphasis on applications is secondary to that of the experimental techniques. However a few examples have been chosen to illustrate the diverse problems to which picosecond laser systems have been applied. Examples of the evolution of technique in the study of a single problem are also outlined.

Studies of Rotational Motion in Condensed Phase

Studies of rotational relaxation of small molecules have focused on the optical Kerr effect and have dealt with the relaxation of optically induced birefringence in CS_2, nitrobenzene, diiodomethane, and dichloroethane [69, 91]. Rotational relaxation of large dye molecules [92] has been studied by observing the time evolution of the anisotropic orientational distribution of ground-state molecules, produced by the selective absorption of a linearly polarized picosecond pump pulse by those molecules with transition moments oriented parallel to the polarization of the pump pulse (see Fig. 5.25). The decay of the induced dichroism was characterized by

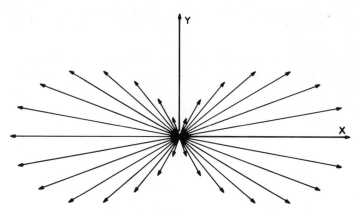

Fig. 5.25. Envelope of vectors reflecting the $\cos^2 \theta$ distribution of ground-state moments remaining after absorption of a pulse linearly polarized in the y-direction (θ is the angle off the x-axis). The excitation pulse is traveling into the page along the z-axis. Molecules with transition moments perpendicular to y remain unexcited, whereas the ensemble of molecules with moments nearly parallel with y is maximally disturbed. The induced anisotropic distribution of moments will influence the relative transmission of the polarized light and will persist with a lifetime characteristic of excited-state decay or orientational redistribution.

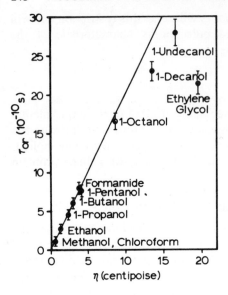

Fig. 5.26. Orientation relaxation times vs. solution viscosity for rhodamine 6G in different solvents. From Chuang and Eisenthal [93].

Eisenthal as the decay of the natural log of the ratio of intensities of transmitted probe beams parallel and perpendicular to the polarization of the pump pulse. The effect of solvent on rotation of rhodamine 6G (R-6G) was systematically studied by Chuang and Eisenthal [93] in an attempt to assess the effect of solute–solvent hydrogen bonding on rotational rates (see Fig. 5.26). They found that for the majority of solvents used, orientational relaxation times were linearly dependent on viscosity η, consistent with the Stokes–Einstein–Debye hydrodynamic model [94], in which the volume of the rotating molecule, V, is constant and the orientational relaxation time is $\tau = \eta V/kT$. This result was unexpected since it was thought that hydrogen bonding and solvent drag, in some solvents, would increase the effective volume of the solute molecule. The linear dependence was explained by assuming that the rupture and reforming of hydrogen bonds were both faster than orientational relaxation, thus reducing the effective drag of solvent hydrogen bonded to R-6G. The deviation from linearity for large alcohols was thought to be due to the breakdown of the solvent continuum approximation, since the solvent molecules are comparable in size to R-6G. The discrepancy in ethylene glycol was thought to exist because self-association is so strong in ethylene glycol that competition of hydrogen bonding to solute is negligible in comparison. By fluorescence depolarization measurements using a streak camera, Fleming et al. [95] were able to measure orientational relaxation times of eosin and rose bengal in methanol, ethanol, isopropanol, and water. They compared the observed orientational relaxation time with values calculated assuming spherical and ellipsoidal geometry. The values derived assuming ellipsoidal geometry were 0.3 to 0.6 times the measured value, whereas the calculated lifetimes based on a spherical geometry were 0.6 to 1.2 times the measured value. Based on the agreement with calculations assuming a spherical model, they concluded that solvent molecules fill out the planar shape of the solute molecule to a sphere (7 Å radius) and that the rotating species consists of both dye and hydrogen-bonded solvent. The apparent differences within the fluorescein group of dyes remain unresolved, but the authors suggest two reasons for the discrepancy between their results on eosin and rose bengal and previous studies on R-6G [93]. First, there is a charge difference between R-6G, a cation, and eosin and rose bengal, which are both dianions. Second, eosin and rose bengal presumably have more rigid structures than R-6G and can presumably sweep out a larger volume in rotation. In addition, Lessing and von Jena [96] reported that fluorescein 27, structurally similar to R-6G, eosin, and rose bengal, exhibits a lifetime for orientational relaxation of 8.5 ± 1 nsec in 1-decanol and 200 ± 50 psec in methanol, but decanol increases the apparent molecular volume $V = \pi kT/\eta$ twofold over that expected from the molecular structure, again suggesting strong solute–solvent interaction. Finally, using their transient grating technique (see Fig. 5.19), Phillon et al. [64] reproduced some of the relaxation times of R-6G measured by Chuang and Eisenthal in low-viscosity solvents, supporting those authors' earlier contention that there is no rotational hindrance due to solvent hydrogen bonding to R-6G.

Both Fleming et al. [95] and Lessing and von Jena [96] have emphasized the necessity of separating decay kinetics of electronic relaxation from that for rotational relaxation, especially where both processes have comparable lifetimes. For instance, when rotational relaxation is slow compared with electronic relaxation, $\tau_{or} \gg \tau_{el}$, the return to ground state will relax the orientational anisotropy. If $\tau_{or} \ll \tau_{el}$, orientational equilibrium will be established quickly, and decay of emission or transient absorbance will again reflect the electronic process. However, if $\tau_{or} \sim \tau_{el}$, orientational relaxation could be a large contribution to transient decay. In this case an independent experiment must be performed to separate the two processes. A simple way to perform lifetime measurements, independent of orientational effects, is to orient the pump beam polarization at an angle of 54.7° to the polarization of either the probe beam in transient absorbance measurements or emitted light in measurements of fluorescence lifetimes. This and other techniques for separating electronic and orientational contributions in transient decay are described in detail in Refs. 95 and 96.

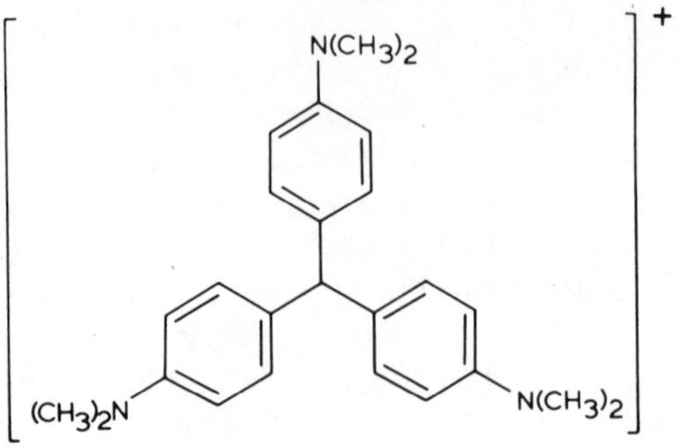

Fig. 5.27. Structure of the triphenylmethane dye crystal violet. Symmetry is D_3, with the phenyl groups at an angle to the molecular plane giving the molecule a propellerlike conformation. The structures of other triphenylmethane dyes are similar.

In addition to rotational motion of the entire molecule, internal motions of a molecule can be studied using picosecond pulses. For example, the triphenylmethane class of dyes in fluid solutions exhibits intense absorption [$\epsilon \sim 10^5$ 1./(mole) (cm)] but only a slight degree of luminescence. Radiative emission is effectively quenched by fast nonradiative decay, which occurs in the picosecond regime [97]. However the fluorescence quantum yields of several triphenylmethane dyes have a marked dependence on solvent viscosity. In fluid media the yield is low (10^{-4}), whereas in rigid media it approaches a maximum of 0.35 [98]. This dependence has been interpreted in terms of the freedom of rotation of the phenyl rings about the central carbon–carbon bond axis shown in Fig. 5.27 [97–100]. Ring rotation in the excited molecule leads to a reduction in the energy gap between the ground- and excited-state potential energy surfaces with subsequent enhancement of rates of internal conversion. Increased solvent viscosity hinders ring rotation, allowing fluorescence to compete with the slower rates of internal conversion.

Early kinetic models viewed ring rotation as either a rotational diffusion process [100] or a driven rotation due to interactions between adjacent ring ortho hydrogens [98]. In order for the rotation to be driven, the difference in potential surface minima for the two states must be large as shown in Fig. 5.28a. However, molecular orbital calculations have indicated the difference in minima to be only a few degrees [101]. With the picosecond flash photolysis apparatus of Magde and Windsor [55], it has been possible to directly monitor the kinetics of ground- and excited-state populations of the triphenylmethane dye crystal violet shown in Fig. 5.27, which was excited at 530 nm by a 5- to 10-psec pulse. Kinetics of the

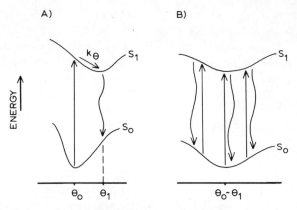

Fig. 5.28. (*A*) Driven rotation kinetic model requires the difference in ground- and excited-state potential energy minima to be large. The viscosity dependence of the ring rotation rate $k\theta$ accounts for the variation of the fluorescence quantum yield with viscosity. (*B*) Results of ground-state repopulation experiments can be explained by considering the potential surfaces to be shallow, with the excited-state surface somewhat shallower than that of the ground state. This results in excited molecules having distinct decay rates over a wide range of values. Observed viscosity dependence of kinetics results from ring rotation in low-viscosity solvents.

ground-state repopulation monitored at several wavelengths revealed nonexponential behavior in solvents having a wide range of viscosities (0.8 to 300 poises). The kinetics of ground-state repopulation also showed some wavelength dependence. As solvent viscosity is decreased, the return of ground-state absorption occurs more rapidly but the nonexponential character is retained. The existence of a purely driven rotation mechanism seems unlikely in view of the shape and temperature dependence of the kinetic curves.

The results of the ground-state repopulation measurements can be explained [102] by postulating that the ground- and excited-state potential surfaces are shallow so that thermal Boltzmann populations exist at angles "significantly different" from the minimum angle of each state. Excitation at 530 nm leads to a replica of the broad ground-state thermal distribution in the excited state. This produces an ensemble of excited molecules, each characterized by a distinct decay rate as shown in Fig. 5.28*b*. Molecules in the wings of the excited thermal population decay faster than those near the equilibrium angle because of the smaller energy gap at the larger angles. In very viscous media, ring rotation can be considered negligible on the time scale of ground-state repopulation. The decay curve then represents the sum of a large number of simple exponential decays, which agrees with our measurements. A decrease in solvent viscosity allows phenyl ring rotation to angles where internal conversion occurs more quickly than at the excited-state minimum. However, in addition to ring rotation, other conformational changes, such as an increase in the central carbon–carbon bond lengths, may play key roles in describing radiationless transitions in these molecules.

Electron Localization

Ionization, γ-radiolysis, and pulse radiolysis of glasses and crystals at low temperature have been used successfully to produce delocalized (quasi-free) electrons. Only with the developments of high-power mode-locked lasers and picosecond-pulse radiolysis techniques has the study of electron localization in polar liquids at room temperature been possible. Electron localization is the process by which an ionized or injected electron reaches equilibrium with the medium. With some nonpolar liquids (e.g., argon, krypton, and xenon), the quasi-free state is stable [103]. However in polar liquids, the quasi-free state is short lived, as evidenced by the rapid two-step evolution of the solvated electron spectrum, with an absorbance peak between 500 and 700 nm, from the initial absorbance spectrum with a peak in the infrared characteristic of the quasi-free electron [104]. The early attempts to measure the solvation rate in polar solvents using picosecond-pulse radiolysis [105] lacked the resolution, but a lower limit was determined, $k_s \geqslant 10^{11}$ s^{-1}. Somewhat later, Rentzepis et al. [106] photoionized an aqueous solution of $K_4Fe(CN)_6$ using the fourth harmonic (265 nm) of a single picosecond pulse from an Nd:glass laser system. They were able to observe the appearance of absorbance at 1060 nm, and reported a lifetime of 2 psec for the equilibration process. The same authors later used a continuum to probe for electron absorbance and were able to observe the blue shift in the spectrum as the solvent shell relaxed about the electron [107]. They observed a fast initial reaction, $\tau < 2$ psec, and revised their earlier estimate of ~2 psec to ~4 psec for complete electron localization. In this and previous papers they discussed this initial fast reaction in terms of an "intramolecular" nonradiative decay of a delocalized state of the electron, followed by solvent relaxation.

Using a stroboscopic pulse radiolysis system [108], Kenney-Wallace and Jonah (109) were able to observe the electron solvation process in six different linear alcohols and found that the relaxation times correlated well with the dielectric relaxation times of the solvent monomers. They observed an absorbance change, which occurred too quickly to measure, followed by a slower (~10 psec) change and suggested an alternative to the nonradiative process proposed by Rentzepis and co-workers. They argued that the quasi-free electrons became weakly trapped in preexisting solvent clusters. Then these electrons equilibrated more slowly depending on the orientational relaxation time of solvent monomers. Their conclusions were supported by similar studies reported by Chase and Hunt [110]. Kenney-Wallace and Jonah then extended their studies to dilute alcohol–alkane solvent systems [111]. The relaxation time did not change significantly from 0.1 to 0.6 mole fraction of butanol in hexane, but the ratio of initial to final absorbance was lower than in neat butanol, consistent with the presence of fewer clusters to act as traps. The absorption spectra of solvated electrons were the same as in neat alcohol, which suggested that the electron states are the same in the mixtures and the neat liquid. No general agreement as to the precise nature of the process preceding solvent relaxation has been reached, and the dynamics of e^- localization remain a vibrant area of chemical research.

Solvent orientation certainly contributes to electron localization, but this could be a secondary process. There appears to be some relaxation which occurs much too fast to study with picosecond techniques currently available, with the possible exception of mode-locked CW dye lasers. Further mechanistic insight will require experiments using subpicosecond techniques.

Electronic Relaxation

For many years studies of electronic relaxation were limited to linewidth data and data that could be derived from measurements of emission yields and lifetimes at low temperature. More recently kinetic studies have been possible using nanosecond flash photolysis. However in many molecules, room temperature emission is very weak, implying that nonradiative relaxation occurs very rapidly. Mode-locked lasers and picosecond flash photolysis have played an important role in providing a means of observing nonradiative processes directly and assessing the degree to which they conform with theoretical models. There are many examples of the use of both pulse laser or subpicosecond CW techniques in the study of nonradiative relaxation. A few examples are discussed below in detail in order to illustrate the variety of ways picosecond techniques can be used in the study of electronic relaxation and the depth to which they have been applied in the study of some systems. Many other systems have been studied, and a summary is given in Table 5.2.

Azulene

Azulene was one of the earliest systems studied using mode-locked lasers. Fluorescence occurs from the second excited singlet state $\Phi_f(S_2)$ ~0.2, with a lifetime $\tau_f(S_2) = 1.4$ nsec [112]. This yields a value of ~7 nsec for the natural radiative lifetime and of ~1.75 nsec for the nonradiative

Table 5.2 Picosecond Studies of Electronic Relaxation

Molecule	Excitation Source[a]	Purpose	Ref.[b]
Fluorescein	Nd:glass (SH)	Vibrational relaxation in S_1, $\tau(S_1)$	1
Eosin	Nd:glass (SH)	Vibrational relaxation in S_1, $\tau(S_1)$	1
Erythrosin	Nd:glass (SH)	Vibrational relaxation in S_1, $\tau(S_1)$	1
Naphthalene	Nd:glass (FH)	Relaxation from S_2 in gas phase	2
Anthracene	Ruby (SH)	S_1 and T_1 absorption spectra	3
Tetracene	Nd:glass (SH)	Energy relaxation in crystals	4
Tetracene	Nd:glass (SH)	Energy relaxation in crystals	5
Tetracene	Nd:YAG (SH)	Energy relaxation in crystals	6, 7
3,4-Benzpyrene	Ruby (SH)	Relaxation from S_2 in gas phase	8
Fluorenone	Ruby (SH)	Solvent dependence of $\tau(S_1)$	9
Anthrone	Ruby (SH)	Solvent dependence of $\tau(S_1)$	9
Xanthone	Nd:glass (TH)	$\tau(S_1)$ in benzene	10
Xanthione	Nd:glass (TH)	$\tau(S_1)$ and S_1 and T_1 absorption spectra	11
Benzophenone	Nd:glass (TH)	Energy transfer	12
Benzophenone	Nd:glass (TH)	Solvent dependence of $\tau(S_1)$	10
4-Phenylbenzophenone	Nd:glass (TH)	Solvent dependence of $\tau(S_1)$	10
Thionine	Nd:glass (SH)	Fluorescence quenching by Fe(II)	13
p-Dimethylaminobenzonitrile	Nd:glass (FH)	Solvent dependence of $\tau(S_1$ and $S_2)$	14

| Quinoxaline | Nd:glass (TH) | S_1 and T_1 absorption spectra | 10 |
| s-Tetrazine | Nd:glass (SH) | psec photochemistry and $\tau(S_1)$ | 15 |

[a]SH = Second harmonic; TH = third harmonic; FH = fourth harmonic.

[b]References:

1. G. Porter, E. S. Reid, and C. J. Tredwell, *Chem. Phys. Lett.*, **29**, 469 (1974).
2. P. Wannier, P. M. Rentzepis, and J. Jortner, *Chem. Phys. Lett.*, **10**, 193 (1971).
3. N. Nakashima and N. Mataga, *Chem. Phys. Lett.*, **35**, 487 (1975).
4. T. L. Netzel and P. M. Rentzepis, *Chem. Phys. Lett.*, **29**, 337 (1974).
5. G. R. Fleming, D. P. Millar, G. C. Morris, J. M. Morris, and G. W. Robinson, *Aust. J. Chem.*, **30**, 2353 (1977).
6. A. J. Campillo, R. C. Hyer, S. L. Shapiro, and C. E. Swenberg, *Chem. Phys. Lett.*, **48**, 495 (1977).
7. F. Heisl, J. A. Miehre, B. Lipp, and M. Schott, *Chem. Phys. Lett.*, **43**, 534 (1976).
8. P. Wannier, P. M. Rentzepis, and J. Jortner, *Chem. Phys. Lett.*, **10**, 102 (1971).
9. T. Kobayashi and S. Nagakura, *Chem. Phys. Lett.*, **43**, 429 (1976).
10. R. H. Hochstrasser and A. C. Nelson, in *Lasers in Physical Chemistry and Biophysics*, J. Joussot-Dubien, Ed., Elsevier, Amsterdam, 1975, p. 305.
11. R. W. Anderson, R. H. Hochstrasser, and H. J. Pownall, *Chem. Phys. Lett.*, **43**, 224 (1976).
12. R. W. Anderson, R. M. Hochstrasser, H. Lutz, and G. W. Scott, *J. Chem. Phys.*, **61**, 2500 (1974).
13. M. D. Archer, M. I. C. Ferreira, G. Porter, and C. J. Tredwell, *Nouv. J. Chim.*, **1**, 9 (1977).
14. W. S. Struve and P. M. Rentzepis, *J. Chem. Phys.*, **60**, 1533 (1974).
15. R. M. Hochstrasser, D. S. King, and A. C. Nelson, *Chem. Phys. Lett.*, **42**, 8 (1976).

lifetime of S_2. The experimental nonradiative lifetime agreed reasonably well with the estimated (based on energy gaps and Franck–Condon factors) lifetime for $S_2–S_1$ internal conversion of 4 nsec and suggested that the large $S_2–S_1$ energy gap is responsible for the relatively long nonradiative lifetime and unusually high S_2 fluorescence quantum yield [112]. Of special interest has been the lifetime of S_1 and its unusually short nonradiative lifetime. The upper limit for the fluorescence yield, $\Phi_f(S_1) \leq 10^{-4}$, and the calculated radiative lifetime suggests a lifetime of ~ 1 psec for nonradiative decay from S_1. But an estimate based on energy gap and Franck–Condon factor leads to the same value for $S_1–S_0$ and $S_2–S_1$ nonradiative decay, ~ 4 nsec. Linewidth measurements by Hochstrasser and Li [113] suggested a lifetime of 2.6 psec. Linewidth narrowing on perdeuteration led the authors to conclude that the predominant relaxation pathway is internal conversion at low temperature.

Attempts to measure directly the lifetimes for relaxation of S_1 began in 1968 with the work of Rentzepis [50], who reported a 7-psec lifetime for the relaxation of vibrationally excited states in the conversion of S_1 to S_0. Subsequently Rentzepis and co-workers have measured relaxation from various excited vibrational levels within S_1 to S_0 [50, 114–117] using a mode-locked Nd: glass laser and obtained several values all of which were greater than 6 psec. Drent et al. [118] reported a value of 4 psec for $\tau(S_1)$ and an appreciable heavy-atom effect that suggested that intersystem crossing was an important decay pathway. Using a flashlamp-pumped dye laser, Wirth et al. [119] could not resolve a decay from S_1 and postulated a lifetime of $\tau < 1$ psec. Heritage and Penzkofer [120] recently measured the decay from vibrational levels 1600 and 3510 cm^{-1} above the $S_1–S_0$ origin. They found lifetimes of 1 and 2 psec respectively. Finally, using a mode-locked CW dye laser with subpicosecond resolution (see section on production of tunable picosecond and subpicosecond pulses using dye lasers), Ippen, Shank, and Woerner [121] were able to measure a lifetime of 1.9 ± 0.2 psec for the decay of S_1 from about 2000 cm^{-1} above the $S_1–S_0$ origin. They were unable to detect measurable differences in the S_1 lifetime in cyclohexane, benzene, and chloro- and bromobenzene. This led them to concur with the interpretation that S_1 relaxation occurs predominantly via internal conversion and not by intersystem crossing, as suggested by Drent et al. [118]. Eber et al. [122] reached the same conclusion from steady-state experiments on carbonyl compounds of azulene.

The lifetime of S_1 for azulene is almost certainly very short (~ 1 to 2 psec) and the reason why, as well as the discrepancy between the measured values, has yet to be resolved. However, where intensities of pump pulses vary from one laser system to another, nonlinear effects may appear in experiments at high power (e.g., Nd:glass systems) that are absent in low-power systems (e.g., mode-locked CW dye lasers), and lifetimes may therefore be power dependent (see section on harmonics and wavelength tuning in solid-state lasers). For instance, Heritage and Penzkofer [120] were unable to eliminate S_2 fluorescence arising from absorption of two 530-nm photons for peak pulse powers from 2 to 200 mW/cm^2, thus casting doubt on the interpretations of the earlier experiments. A more critical evaluation of the possible influence of nonlinear processes on molecular properties is therefore desirable. In addition, the ability to resolve a 1- to 2-psec decay is clearly less when a 5- to 8-psec pulse from a Nd:glass laser is used than it would be for the 0.3-psec pulse from a mode-locked CW dye laser. This difference may be partially responsible for the variation in the reported S_1 lifetimes.

Acridine and Phenazine

Nitrogen heterocyclic compounds have been the subject of a number of photophysical studies because of their similarity to the analogous carbocyclic aromatic compounds [123, 124]. However, unlike the normal carbocyclic aromatics, the presence of nonbonding electrons in the nitrogen heterocyclic compounds can cause the transition energy and decay of $n\pi^*$ states to be quite solvent sensitive. For instance, while the polycyclic monoazines (quinoline, isoquinoline, and acridine) fluoresce in protic solvents, they are only weakly fluorescent in hydrocarbon solvents [123]. This dependence has been interpreted as a change in the lowest excited singlet state from $\pi\pi^*$ in protic solvents to $n\pi^*$ in hydrocarbons [125, 126]. To determine the nature of the lowest excited singlet and to characterize its relaxation pathways in hydrocarbon solvents, there have been two recent studies of the risetime of triplet–triplet (T–T) absorbance in acridine using picosecond flash photolysis [127, 128]. Hirata and Tanaka [127] used the doubled pulse (347 nm) of a mode-locked ruby laser to excite acridine in isooctane. The residual pulse at 694 nm was used to generate a picosecond continuum in CCl$_4$, and an echelon reflector was used to probe the growth of T–T absorbance at 430 nm as a function of time. Sundstrom et al. [128] used the third harmonic (353 nm) from a mode-locked Nd:glass laser, a continuum generated by residual 530- and 1060-nm pulses, and an echelon reflector in combination with an optical multichannel analyzer. They observed the growth of T–T absorbance in n-hexane at 430 nm. Hirata and Tanaka found triplet formation with a lifetime $\tau \sim 13$ psec, and Sundstrom et al. found $\tau \sim 17$ psec.

The $n\pi^*$ character of the lowest singlet was strongly suggested by a discrepancy between the experimental radiative lifetime τ_f and the value calculated for the lowest $\pi\pi^*$ based on the oscillator strength of the observed transition. The calculated value was $\tau_f \sim 10^{-8}$ sec, but the experimental value ($\tau_f = \tau/\Phi_f$) was much too large, $\tau_f > 10^{-6}$ sec, using as an upper limit $\Phi_f \lesssim 10^{-5}$. The proposal of level inversion to explain the solvent effect [125, 126] appears justified. There is however disagreement as to the mechanism for relaxation from S_1. Hirata and Tanaka assume the singlet lifetime is limited solely by the rate of intersystem crossing, whereas Sundstrom et al. have presented evidence to suggest that the rate of internal conversion and intersystem crossing may be equally important.

In a subsequent paper Hirata and Tanaka [129] have reported details of the growth of $T–T$ absorbance of 9,10-diazaanthracene (phenazine) in isooctane and methanol. The main difference is that the solvent effect on the S_1 lifetime is not significant. Using the mode-locked ruby laser and the same techniques, they found $\tau \sim 20$ psec in methanol and ~ 14 psec in isooctane. This was explained in terms of the much larger ${}^1\pi\pi^* - {}^1n\pi^*$ splitting in phenazine than in acridine and consequently no inversion of states. They conclude that intersystem crossing is the predominant relaxation pathway because phenazine and other diazine molecules phosphoresce, and they explain the fast rate, in spite of a large $S_1–T_1$ energy gap, as due to effective spin-orbit coupling resulting from one-center integrals on nitrogen. The observation of the recovery of ground-state absorbance (GSR) could eliminate conjecture if the extent of GSR is measured during the S_1 lifetime.

Benzophenone and Nitronaphthalenes

Intersystem crossing in benzophenone was first observed by Rentzepis in a study of the growth of triplet–triplet ($T–T$) absorbance [130]. The purpose was to characterize relaxation in the small molecule limit [131] where intersystem crossing, from S_1, occurs to a relatively sparse density of vibrational levels within the triplet manifold. Nitzan et al. [131] found that intersystem crossing was consistent with the small molecule limit in that $T–T$ growth was faster when excitation occurred near the $S_1–S_0$ origin, $\tau \sim 5$ psec, than when excited to a level (~ 2000 cm^{-1}) higher in energy, $\tau \sim 25$ psec. Hochstrasser et al. [132] found a solvent dependence on the growth of $T–T$ absorbance in benzophenone and concluded that more data were necessary to generate a framework for comparative solvent studies of ultrafast processes in condensed phase, but they interpreted the difference in S_1 lifetime between ethanol and benzene as involving a change in relative spacing of $n\pi^*$ and $\pi\pi^*$ singlet states.

Anderson et al. [133] proposed that intersystem crossing may be competing with vibrational relaxation in the S_1 manifold and that solvents may preferentially assist vibrational relaxation in benzophenone. In addition, solvent effects on $T–T$ growth in 1- and 2-nitronaphthalenes are not easily

Table 5.3 Mean Lifetimes for $T–T$ Growth Following 354-nm Excitation

Compound	Solvent	τ (psec)
1-Nitronaphthalene	Ethanol	8 ± 2
1-Nitronaphthalene	Benzene	12 ± 2
2-Nitronaphthalene	Ethanol	22 ± 2
2-Nitronaphthalene	Benzene	10 ± 2

accounted for by a simple change in $S_1–S_2$ energy spacing [133]. For instance, the authors observed a subtle, yet apparently significant, difference in the effect of ethanol and benzene on the rate of $T–T$ growth for the compounds listed in Table 5.3. They suggested various causes for the solvent dependence including changes in the internal conversion and intersystem crossing rate constants as well as solvent-assisted vibrational relaxation.

Inorganic Systems

One of the earliest studies of electronic relaxation in inorganic systems was that of Magde, Bushaw, and Windsor [56] on bis(4-dimethylami-nodithiobenzil)nickel(II) (BDN). This compound is useful as a Q-switch and modelocking dye for the Nd:glass laser [12, 134]. Electron-donating and heavy-atom solvents were found to increase both the decay rate of excited-state absorbance (ESA) and the rate of recovery of ground-state absorbance (GSR) by as much as fortyfold relative to benzene. The heavy-atom effect on ESA and GSR lifetimes is summarized in Table 5.4, and the single observed decay is consistent with an intersystem crossing process. However extensive ligand contribution to the nature of the excited states involved made unambiguous interpretation of decay pathways difficult.

Picosecond flash photolysis and spectroscopic techniques have also been applied in our laboratory to the study of a series of metalloporphy-

Table 5.4 Heavy-Atom Effect on ESA and GSR Lifetimes

Solvent	τ (ESA) nsec	τ (GSR) nsec
Benzene	9.0 ± 3.0	9.0 ± 1.0
1,2-Dichloroethane	3.6 ± 0.6	3.6 ± 0.5
1,2-Dibromoethane	2.6 ± 0.5	3.4 ± 0.4
Iodoethane	0.22 ± 0.03	0.26 ± 0.02

rins [57]. The photophysics of these systems has been well studied and is dominated by the π-electron system of the porphyrin macrocycle [135]. Magde et al. [57] were able to observe time-resolved spectra of S_1 and T_1 and measure the S_1 lifetime and the quantum yield of T_1 in octaethylporphinatotin(IV) dichloride [(OEP)SnCl$_2$] (see Fig. 5-15). From these data and a three-level kinetic scheme involving only S_0, S_1, and T_1, they calculated a triplet quantum yield of 0.8. Together with a fluorescence quantum yield of 0.007 [136], this indicated a quantum yield for internal conversion S_1–S_0 of about 0.19, a value intermediate between those obtained from flash photolysis [137] and flash calorimetry [138].

Electronic relaxation in coordination compounds, particularly those of Cr(III), has been studied in our laboratory by Kirk et al. [139]. The excited states are predominantly metal centered and well characterized by ligand field theory [140], and nonradiative relaxation has been of interest to those photochemists dealing with coordination compounds. In Cr(III), the lowest excited states are of quartet and doublet spin multiplicity. Photochemists studying Cr(III) have been especially interested in the lifetime of the lowest excited quartet, which is presumed to be the photoactive state [141]. Initial studies suggested that the lowest excited quartet decayed with a rate too fast to measure, $\tau \leq 10$ psec [139]. However recent work in our laboratory has established that the rise of doublet absorbance by intersystem crossing from the excited quartet in *trans*-Cr(NH$_3$)$_2$(NCS)$_4^-$ occurs in 24 psec in water [142]. In addition, for a series of thiocyanate complexes of Cr(III), there is a solvent dependence of the excited quartet lifetime, which appears to reflect the overall charge or symmetry of these ions in a manner that has yet to be determined.

Vibrational Relaxation

Experimental studies of picosecond vibrational relaxation of molecules, in the liquid state or in solution, have dealt with the relaxation of simple molecules within the ground electronic state and more complex dye molecules in both the ground and first excited singlet states. The earliest studies employed a mode-locked Nd:glass laser and a variation of the single-pulse and probe technique described in the section on picosecond absorption techniques. More recently, mode-locked CW dye laser systems have been used.

Laubereau et al. [143] were the first to apply picosecond laser pulses and detection to the study of vibrational relaxation in the ground state. They were able to measure two distinct vibrational lifetimes associated with the totally symmetrical CH$_3$ vibration in CH$_3$CCl$_3$ and ethanol, v_1. Molecules in the v_1 state were prepared via the Stokes-shifted stimulated Raman scattering (SRS) of the picosecond pump pulse. Because SRS is an efficient nonlinear process, a large fraction of the molecules in the beam are promoted to the v_1 state. In addition, the phase-matching conditions of this three-wave mixing process requires that the excited oscillators all be in phase (coherent) [144]. Such a system can be characterized by two measurable lifetimes; the decay of the excited population corresponds to one measured lifetime, whereas the decay of the *phase* relationship corresponds to a second lifetime. The dephasing time was measured by the decay of a coherent anti-Stokes Raman signal generated by a second delayed pulse at the same wavelength as the pump pulse. A precise phase relation must exist between all three waves in the SRS process, and the intensity of the *coherent* anti-Stokes Raman signal generated by the probe beam will decrease with delay as the molecules in v_1 no longer oscillate exactly in phase. Typically, dephasing times are as short as or shorter than the pump and probe pulsewidths and are difficult to determine precisely. Fischer and Laubereau [145] were able to describe dephasing times in terms of semiclassical collision theory and the interaction of vibrationally excited molecules with other nearby molecules.

The population of v_1 was monitored by observing the *incoherent* anti-Stokes Raman signal generated by the probe pulse, and a log plot of scattered intensity versus delay between the pump and probe pulses gave the lifetime of v_1. In their first paper, Laubereau et al. measured a population lifetime of 5 ± 1 psec for neat CH_3CCl_3 and 20 ± 5 psec for neat ethanol [143]. In later work Laubereau, Kirschner, and Kaiser [146] reported that the v_1 lifetime in CH_3CCl_3 was dependent on the square of the mole fraction of CH_3CCl_3 in CCl_4 solution, suggesting an intermolecular path contributing to vibrational relaxation. The decay appears to occur by way of fast energy transfer to the second harmonic of CH bending modes, since the rise and decay of an *incoherent* anti-Stokes Raman signal was observed at the CH bending frequency [146, 147]. The presence of CH bonds as well as the number of methyl groups were found to be influential in the vibrational relaxation in hydrocarbons. When a similar technique was used, the measured lifetimes varied from 11 to 60 psec [148]. Later work [147] on vibrational decay in ethanol, in which the v_1 state was pumped directly by an IR pulse derived from the pulse of a mode-locked Nd:glass laser and difference frequency generation in two $LiNbO_3$ crystals, showed an initial very fast decay of the scattering signal followed by a slower decay. The authors attributed the fast decay, $\tau \sim 2$ psec, to energy transfer to neighboring states with smaller scattering cross sections. These studies of relaxation in ethanol were performed in dilute solution at an ethanol concentration of 0.04 M in CCl_4, and the discrepancy between the slower decay, $\tau = 40 \pm 15$ psec, and the earlier reported lifetime of 20 psec was attributed to an intermolecular energy-transfer pathway similar to that observed in CH_3CCl_3. Measurements of the decay of v_1 in CH_3I in CCl_4 solution revealed a lifetime $\tau \sim 2$ psec which showed no concentration dependence, as observed in ethanol and CH_3CCl_3, and suggested therefore that intramolecular vibrational relaxation was much faster in this molecule [147].

Vibrational relaxation of larger dye molecules has been studied in a number of laboratories and using a variety of experimental techniques. Ricard et al. [149] measured the rate of thermal equilibration in the first excited singlet states of R-6G and R-B in 10^{-4} M ethanol solutions by an analysis of the wavelength dependence of the risetime of stimulated emission. A single 6-psec pulse from a mode-locked Nd:glass laser was passed through a beam splitter, and one portion of the pulse was used to excite the dye while the remaining but equally intense probe portion was used to stimulate the emission of the excited dye as a function of the delay between the two pulses. Different pump and probe wavelengths were generated by stimulated Raman scattering followed by either phase-matched sum frequency generation or second harmonic generation in KDP. The rise of stimulated emission versus the time delay between pump and probe pulses revealed vibrational lifetimes within S_1 of 8 ± 3 psec for R-6G and < 2 psec for R-B. Mourou and Malley [150] measured lifetimes of R-6G, R-B, and erythrosin B, using an optical Kerr cell, to detect the delayed appearance of spontaneous fluorescence. Subsequently Ippen and Shank [151] used the frequency-doubled output (307 nm) of their subpicosecond mode-locked CW dye laser to repeat measurements of the gain risetime in R-6G and R-B and found a delay of less than 0.2 psec in methanol and glycerol solutions. Using two different methods, Penzkofer et al. [152] found viscosity-independent lifetimes for vibrational relaxation, following absorption only several hundred wavenumbers above the excited singlet minimum, for both R-6G and R-B of $0.5 \leqslant \tau \leqslant 1.0$ psec. The apparent sensitivity of these lifetimes to the experimental method has yet to be resolved and again may be a consequence of a nonlinear power dependence, especially since the yield of stimulated emission is proportional to the square of the pump pulse intensity.

Kaiser and Laubereau [153] reported a method for measuring the rate of vibrational relaxation in the ground state of coumarin 6. By exciting a solution of the dye in CCl_4 by a picosecond pulse at about 3000 cm^{-1} and probing at 530 nm, the emission intensity can be used to measure the amount of vibrational decay that has occurred as the pump and probe pulses are separated in time. The fluorescence threshold is very near the sum of the IR and visible pulses, and any decay of the vibrational states produced in the IR pulse will prevent the second photon from exciting the molecule to the S_1 state. They measured a fluorescence intensity which decayed with the lifetime of 1.3 ± 0.4 psec at room temperature.

Photochemistry

I_2

Recently there have been a number of picosecond studies of photolabile systems in which chemical reaction is an important excited-state relaxation pathway. The most notable has been the dissociation of I_2 in solution [154]. The disappearance (bleaching) of ground-state absorbance continued for ~ 10 psec following the decay of the 5-psec pump pulse at 530 nm. The longer risetime was interpreted as the collision-induced predissociation of some excited states of I_2 to yield two nonabsorbing I atoms. In CCl_4 and hexadecane, the orientational relaxation time of I_2 is faster than the rise in transmission and argues strongly against orientational relaxation as the cause for the ~ 10-psec risetime of probe transmission. The residual bleaching at 800 psec in hexadecane and CCl_4 was interpreted as a consequence of the escape of I atoms from the solvent cage; and because the fusion of I atoms from different solvent cages probably occurs in ~ 10 nsec [154], the decay kinetics of short-time bleaching and magnitude of the residual bleaching yields the rate and extent of geminate recombination. Yields of solvent-separated I atoms estimated from the bleaching of I_2 absorbance agree reasonably with the Noyes treatment of dissociation based on the continuum hydrodynamic properties of the solvents [155]. The lifetimes for geminate recombination are 70 and 140 psec in hexadecane and in CCl_4, respectively, and comparison with values for rates of geminate recombination, estimated using Noyes random walk model, shows the experimental values to be only slightly larger than expected [156]. Comparisons of yields for different excitation wavelengths should provide an additional test of the Noyes model, since shorter excitation wavelengths ($\lambda < 530$ nm) should result in a greater yield of solvent-separated I atoms.

Tetraphenylhydrazine

Anderson and Hochstrasser [157] studied the kinetics of photodissociation of tetraphenylhydrazine (TPH) in solution at room temperature. A transient absorbance was produced with a risetime less than the 353-nm 8-psec pump pulsewidth. The spectra at 6 psec and 1 nsec agreed favorably with that observed for radical pairs in the rigid glass [158] as well as that of the longer-lived (~ 1 μsec) free radicals [159]. The similarity of radical pair and free-radical spectra did not permit observation of the escape of the diphenylamine (DPA) radical products from the solvent cage, and the very fast absorbance risetime suggested only an upper limit on the lifetime for the precursor excited singlet state, < 2 psec. The peculiar feature in the kinetics of these radicals is the apparent lack of any geminate recombination. The authors suggest that one of the radicals may have to be in an excited state for efficient recombination, a proposal supported by the formation of TPH from radical pairs under irradiation by red light [160]. An experiment using the second harmonic from a mode-locked ruby laser (347 nm) to pump the sample and the red fundamental (694 nm) to initiate recombination could be used to test this hypothesis. Delaying the fundamental relative to the 347 -nm pump pulse allowed the rate of escape from the radical cage to be determined by the amount of photoinduced recombination as a function of the time delay between the pulses.

Tetramethyl-1,2-dioxetane

A mode-locked Nd:phosphate glass laser with 10-psec pulse was used to study the photolysis of tetramethyl-1,2-dioxetane. In this study, Smith et al. [161] used a streak camera to detect the fluorescence of the acetone photoproduct in the S_1 excited state [162, 163] produced after excitation with the fourth harmonic at 264 nm. Deconvolution revealed that the fluorescence risetime was < 10 psec. This, coupled with the S_1 fluorescence decay ($\tau_f \sim 2$ nsec), suggested that all acetone was produced only in the S_1 state during the pump pulse. Wavelength-dependent yields of S_1 acetone product [164] suggested to the authors that the process leading to S_1 acetone may be competing with vibrational equilibration and that equilibration followed by intersystem crossing leads to the same biradical intermediate and T_1 acetone product observed in the thermal decomposition of tetramethyl-1,2-dioxetane [164]. Here, as in the fusion of two DPA radicals, separation of radicals into acetone products is more efficient then radical recombination, given the near-unit quantum yield for acetone formation.

Photobiology

Almost all of the application of picosecond techniques to biology and biophysics has taken place in the past two or three years. Valuable new insights have been gained, and in some areas, notably photosynthesis, quite striking advances have been made. Picosecond spectroscopy, for the first time, enables us to take a look at the elementary steps that underlie the overall complexity of biological reactions. These primary processes, involving energy migration, bond rupture, molecular rearrangements, and charge transfer, all take place on a time scale from about 10^{-14} to 10^{-8} sec. Besides new data on photosynthesis, useful results have been obtained on visual pigments, hemoglobin complexes, and energy transfer in DNA. Recent reviews by Campillo and Shapiro [165] and

Holten and Windsor [166] should be consulted for more detailed accounts of the application of picosecond techniques to biological systems. Here we attempt to do no more than summarize the new results and comment briefly on their significance.

The area that has received greatest emphasis to date is photosynthesis. This emphasis is undoubtedly appropriate since we are all dependent on photosynthesis both for our origin and for our continued survival. Picosecond fluorescence studies, in both green plants and photosynthetic bacteria, show that excitation energy migrates from the antenna or light-harvesting system (a structured pool or array of several hundred chlorophyll molecules that gather incident solar photons) to a specialized trap or reaction center (RC) in times that range from 10 psec to about 1 nsec, dependent upon species. Within the RC, the initial charge separation occurs in a time of about 200 psec, converting the electronic excitation and storing it as chemical potential [165, 166].

Picosecond techniques have been widely used to measure fluorescence lifetimes in photosynthetic systems [167]. Techniques employed include phase fluorometry [167], mode-locked CW laser methods with signal averaging [168], optical Kerr gate studies [169], and picosecond streak cameras [170–172]. Concern has been expressed about possible multiple-excitation effects when high-intensity ($> 10^{13}$ photons/cm^2) single picosecond pulses are used, leading to intensity dependence of quantum yield and lifetimes [173, 174] and in some cases to stimulated emission [175]. This has necessitated reinterpretation of some early measurements. However picosecond techniques can be valuable in measuring fluorescence lifetimes provided that care is taken in choosing the mode and intensity of excitation [172]. Techniques that avoid these difficulties are now available [36, 46, 176].

Green plants contain two photosystems, PSI and PSII, in close proximity; both absorb a photon, convert its energy to chemical potential, and cooperate in the later stages of photosynthesis. Photosynthetic bacteria contain only a single photosystem. Emission measurements on plant systems often show a biphasic decay with a fast component that decays in a few tens of picoseconds and a slower one that decays in a few hundred picoseconds. For example, Siebert and Alfano [169] studied chlorophyll a fluorescence in escarole chloroplasts and found maxima in the decay curves at 15 and 90 psec, with a dip at 50 psec. They interpreted these results as indicating fluorescence from both PSI and PSII, with respective decay times of 10 and 210 psec. Yu et al. [177] studied PSI- and PSII-enriched particles of spinach and found 60 ± 10 and 200 ± 20 psec components. Attempts to measure rates of energy migration from the antenna system to the reaction center have been made by studying the induction time for the appearance of fluorescence after excitation by a short pulse. However interpretation of the results is clouded by the possibility of excited-state annihilation processes (singlet–singlet and singlet–triplet), especially at high light intensities [165, 172]. It is not yet known exactly how fast migration of excitation occurs within the antenna system or from the antenna to the reaction center, except that the time probably lies in the range of 10 psec to 1 nsec and depends upon the species. We do not even know whether such migration occurs via an exciton mechanism that involves many short hops (a few angstroms) between closely adjacent pigment molecules [178] or over somewhat large distances (~ 10 Å) via a Förster mechanism [179]. Some very recent work [171] on photosynthetic bacteria supports a "lake" model, in which a single antenna system contains many different reaction centers any of which can receive the excitation. This is in contrast to the "puddle" model, in which independent photosynthetic units exist, each with its own reaction center and small antenna system. Given a "lake" model, it is easy to see why one would expect to observe a spectrum of trapping times. One would also expect an intensity dependence of quantum efficiency, since it takes several microseconds for an RC that has received excitation to convert it and become ready to receive again. Indeed it was by a study of such intensity dependence that the above conclusion was drawn [171].

Let us turn now to the energy conversion process that takes place within the RC. Much has been learned about this process recently from picosecond absorption studies on photosynthetic bacteria [166]. The majority of studies have been done on reaction centers of the bacterium *Rhodopseudomonas sphaeroides*, which has been well characterized as to its composition. These reactive centers contain four molecules of bacteriochlorophyll (BChl); two molecules of the magnesium-deficient analog bacteriopheophytin (BPh); one ubiquinone molecule (UQ); one nonheme iron atom, probably complexed to the UQ; and three different proteins in a 1:1:1 ratio and of molecular weight 28, 22, and 20 kdalton [180, 181]. Two of the BChl molecules form a special pair that absorbs maximally at 870 nm and is often called P870, or simply P. Receipt of excitation from the antenna system produces the excited singlet state BChl–*BChl. (In picosecond experiments on isolated reactive centers this state is excited directly or via very rapid (< 10 psec) energy transfer from a neighboring BPh that has absorbed a 530-nm laser pulse.) Picosecond spectral studies of transient intermediates [182–184], both in our own laboratory and in other laboratories, show that a special state, P^F, is formed in less than 10 psec and decays in about 200 psec with the concomitant appearance of absorption spectra characteristic of the oxidized donor BChl$\overset{+}{-}$BChl and the reduced "primary acceptor," X^-, where X is used to symbolize the

Fe–UQ complex. The net effect is that an electron is transferred from BChl–BChl to X in two stages, an initial very fast step (< 10 psec) and a second somewhat less fast step (~ 200 psec). Much interest has focused on the identity of P^F, and it is now known to have the structure (BChl$^+$– BChl)BPh$^-$ [185, 186]. It appears that Nature has good reasons for making the first step so fast, namely, to prevent the excitation from "hopping" back out of the reactive center into the antenna pool [187]. It also appears, from very recent model system studies in our laboratory by Holten et al [187], that some quite subtle effects may be involved in slowing down dramatically the reverse electron transfer, thus protecting this first crucial step in capturing and using solar energy. There is also the possibility that the additional pigment molecules in the RC have a role in coupling a portion of the energy "losses" that accompany the electron transfer to structural changes that involve the membrane proteins [186]. In summary, we note that much insight has been gained into the details of the very early events in photosynthesis. However we still need to establish the role of the "extra" pigment molecules (two additional BChl and one additional BPh) and understand better the relationships between structure and function within the reactive center.

Turning now to studies of the visual pigment rhodopsin, we find that very few picosecond studies have been made, no doubt because its photolability necessitates a fresh sample for each experiment. The few studies that have been done [188, 189] have been supplemented by experiments on a closely related pigment, bacteriorhodopsin (bR), which occurs in the purple membrane of *Halobacterium halobium*, a bacterium that occurs in salt flats. Although bacteriorhodopsin does not function as a visual transducer but rather as a crude photosynthetic system that makes ATP, its early-time photochemistry parallels that of animal rhodopsin. Furthermore, it goes through a photochemical cycle in which the original form is regenerated within a few milliseconds—a considerable experimental advantage compared to animal rhodopsin [190]. The picosecond studies of bovine rhodopsin [188] and squid rhodopsin [189] show that formation of the first intermediate, called hypsorhodopsin, occurs in a time that cannot be distinguished from the rise of the excitation pulse (< 6 psec for Nd:glass in Ref. 188 and < 20 psec for ruby in Ref. 189). Hypsorhodopsin can also be trapped at liquid helium temperatures [191]. Its formation probably involves a minor photochemically induced conformational change in the chromophore (11-cis retinal), probably around the 11th carbon atom, followed by a series of further changes that occur thermally and that can be resolved in low-temperature experiments. At physiological temperatures, conversion to the next intermediate, bathorhodopsin, occurs in about 55 psec [189] and conversion to the subsequent one, lumirhodopsin, in about 30 nsec [188]. Similar changes occur in bacteriorhodopsin, but the initially excited singlet state can be resolved, and it appears to have a lifetime of about 15 psec [176, 192, 193].

Shank, Ippen, and Bersohn [194] have used subpicosecond pulses from a CW mode-locked dye laser to study the dissociation of oxygen and carbon monoxy complexes of hemoglobin, HbO_2 and HbCO. Hemoglobin has a much stronger absorption at 615 nm than its complexes have. Thus their dissociation can be studied by following the increases in absorbance at this wavelength. For HbCO a risetime of less than 0.5 psec is observed, and the absorption decays very little in the first microsecond after excitation [165]. It is concluded that the high quantum yield for photodissociation of HbCO is due to a fast rate of dissociation combined with a slow rate of recombination [194]. In the case of HbO_2 there is also a fast rise in absorbance at 615 nm, but it decays again in about 2.5 psec. Similar results are obtained when an experiment is done on deoxygenated hemoglobin samples. This result suggests that the increased absorbance for HbO_2, unlike HbCO, is due to a short-lived excited state that absorbs more strongly at 615 nm than does the ground state. The authors conclude that the low quantum yield for photodissociation of HbO_2 is due not to rapid recombination but rather to the relatively slow rate of dissociation from the excited state, which does not compete effectively with the very fast 2.5 psec nonradiative decay of the excited state.

Shapiro et al. [195] studied energy transfer in DNA–acridine orange complexes using picosecond techniques. They used a train of 265-nm 10-psec pulses to excite the sample and a streak camera to observe the fluorescence between 530 and 600 nm. The risetime was estimated to be < 20 psec. Since the polymer-to-dye ratio was 400 to 1, most of the excitation was absorbed by the DNA. All of the fluorescence, on the other hand, comes from acridine molecules excited by energy transfer from the DNA. A transfer distance of five to ten base pairs has been estimated from quantum yield measurements. If one takes the observed risetime of < 20 psec to be the time for excitation to transfer over this distance, the value obtained for the exciton diffusion coefficient is at least 1 (base pair)^2 per picosecond. For further details of the experiments on hemoglobin and on DNA, the review of Campillo and Shapiro [165] should be consulted.

ACKNOWLEDGMENTS

The authors gratefully acknowledge the comments of Dr. C. V. Shank on the section on subpicosecond pulses, Dr. Dewey Holten for reviewing the manuscript, and Dr. Stephen Rice for comments on electron localization. Special thanks are due to David Cremers for the description of his unpublished work given in the section on rotational motion in the condensed phase. The preparation of this review was supported in part by the U.S. Army Research Office under Grant DAAG 29-76-9-0275 and by the Office of Naval Research.

References

1. A. J. DeMaria, D. A. Stetser, and W. H. Glenn, Jr., *Science,* **156,** 1557 (1967).
2. R. C. Greenhow and A. J. Schmidt, in *Advances in Quantum Electronics,* Vol. 2, D. N. Goodwin, Ed., Academic Press, New York, 1974, p. 157.
3. D. J. Bradley, *Opt. Electron.,* **6,** 25 (1974).
4. M. M. Malley, in *Creation and Detection of the Excited State,* Vol. 2, W. R. Ware, Ed., Marcel Dekker, New York, 1974, p. 99.
5. P. W. Smith, M. A. Duguay, and E. P. Ippen, in *Progress in Quantum Electronics,* Vol. II, J. H. Sanders and S. Stenholm, Eds., Pergamon, Oxford, 1974, p. 107.
6. A. Laubereau and W. Kaiser, *Opt. Electron.,* **6,** 1 (1974).
7. G. E. Busch and P. M. Rentzepis, *Science,* **194,** 276 (1976).
8. D. J. Bradley, in *Ultrashort Light Pulses,* S. L. Shapiro, Ed., Springer-Verlag, New York, 1977, p. 18.
9. C. P. Ausschnitt, *IEEE J. Quantum Electron.,* **QE-13,** 321 (1977), and references therein.
10. D. J. Kuizenga and A. E. Siegman, *IEEE J. Quantum Electron.,* **QE-6,** 694, 709 (1970).
11. H. A. Haus, *J. Appl. Phys.,* **46,** 3049 (1975); *IEEE J. Quantum Electron.,* **QE-11,** 736 (1975); *Opt. Commun.,* **15,** 29 (1975).
12. K. H. Drexhage and U. T. Müller-Westerhoff, *IEEE J. Quantum Electron.,* **QE-8,** 759 (1972); K. H. Drexhage and G. A. Reynolds, *Opt. Commun.,* **10,** 18 (1974).
13. For a good review of the structure and properties of dyes used in lasers, see K. H. Drexhage, in *Dye Lasers,* F. P. Schäfer, Ed., Springer-Verlag, New York, 1973, p. 144.
14. D. J. Bradley, G. H. C. New, and S. J. Caughy, *Phys. Lett.,* **A30,** 78 (1969); *Opt. Commun.,* **2,** 41 (1970); D. J. Bradley and W. Sibbett, *Opt. Commun.,* **9,** 17 (1973).
15. W. Yu and R. R. Alfano, *Opt. Electron.,* **6,** 243 (1974).
16. E. M. Garmire and A. Yariv, *IEEE J. Quantum Electron.,* **QE-3,** 222 (1967).
17. E. G. Arthurs, D. J. Bradley, and A. G. Roddie, *Appl. Phys. Lett.,* **23,** 88 (1973).
18. A. Yariv, *Quantum Electronics,* 2nd ed., Wiley, New York, 1975, p. 327.
19. D. von der Linde, O. Bernecker, and A. Laubereau, *Opt. Commun.,* **2,** 215 (1970).

20. E. G. Arthurs, D. J. Bradley, and T. J. Glynn, *Opt. Commun.*, **12**, 136 (1974).
21. H. Al-Obaidi, R. J. Dewhurst, D. Jacoby, G. A. Oldershaw, and S. A. Ramsden, *Opt. Commun.*, **14**, 219 (1975).
22. D. von der Linde and K. F. Rodgers, *Opt. Commun.*, **8**, 91 (1973).
23. E. G. Arthurs, D. J. Bradley, P. N. Puntambekar, I. S. Ruddock, and T. J. Glynn, *Opt. Commun.*, **12**, 360 (1974); G. H. C. New, *Opt. Commun.*, **6**, 188 (1972); G. H. C. New, *IEEE J. Quantum Electron.*, **QE-10**, 115 (1974); R. V. Ambartsumyare, N. G. Basov, V. S. Zuev, P. G. Kryukov, and V. S. Letokhov, *JETP Lett.*, **4**, 12 (1966).
24. F. Zernike and J. E. Midwinter, *Applied Nonlinear Optics*, Wiley, New York, 1973.
25. A. Yariv and J. E. Pearson, in *Progress in Quantum Electronics*, Vol. 1. J. H. Sanders and K. W. H. Stevens, Eds., Pergamon, Oxford, 1971, p. 1.
26. W. L. Smith, *Appl. Opt.*, **16**, 1798 (1977).
27. A. Laubereau, L. Greiter, and W. Kaiser, *Appl. Phys. Lett.*, **25**, 87 (1974).
28. R. B. Weisman and S. A. Rice, *Opt. Commun.*, **19**, 28 (1976).
29. C. A. Moore and L. S. Goldberg, *Opt. Commun.*, **16**, 21 (1976).
30. M. J. Colles, *Opt. Commun.*, **1**, 169 (1969).
31. T. R. Royt, C. H. Lee, and W. L. Faust, *Opt. Commun.*, **18**, 108 (1976).
32. A. H. Kung, J. F. Young, G. C. Bjorklund, and S. E. Harris, *Phys. Rev. Lett.*, **29**, 985 (1972).
33. R. T. Hodgson, P. P. Sorokin, and J. J. Wayne, *Phys. Rev. Lett.*, **32**, 343 (1974).
34. J. Reintjes, C. Y. She, R. C. Eckardt, N. E. Karangelen, R. A. Andrews, and R. C. Elton, *Appl. Phys. Lett.*, **30**, 480 (1977); C. Y. She and J. Reintjes, *ibid.*, **31**, 95 (1977), and references therein.
35. D. J. Bradley, A. G. Roddie, W. Sibbett, M. H. Key, M. L. Lamb, C. L. S. Lewis, and P. Sachsenmaier, *Opt. Commun.*, **15**, 231 (1975).
36. E. P. Ippen and C. V. Shank, in *Ultrashort Light Pulses*, S. L. Shapiro, Ed., Springer-Verlag, New York, 1977, p. 83.
37. D. J. Bradley and G. H. C. New, *Proc. IEEE*, **62**, 313 (1974).
38. H. P. Weber, *J. Appl. Phys.*, **38**, 2231 (1967); **39**, 6041 (1968).
39. J. A. Armstrong, *Appl. Phys. Lett.*, **10**, 16 (1967).
40. I. V. Tomov, *Opt. Commun.*, **10**, 154 (1974).
41. J. A. Giordmaine, P. M. Rentzepis, S. L. Shapiro, and K. W. Wecht, *Appl. Phys. Lett.*, **11**, 216 (1967).
42. P. M. Rentzepis and M. A. Duguay, *Appl. Phys. Lett.*, **11**, 218 (1967).
43. P. M. Rentzepis, C. J. Mitschele, and A. C. Saxman, *Appl. Phys. Lett.*, **17**, 122 (1970).
44. E. K. Zavoiskii and S. D. Fanchenko, *Sov. Phys. -Dokl.*, **1**, 285 (1956).
45. W. R. Ware, in *Creation and Detection of the Excited State*, Vol. 1, Part A, A. A. Lamola, Ed., Marcel Dekker, New York, 1971, p. 213.
46. Z. D. Popovic and E. R. Menzel, *Chem. Phys. Lett.*, **45**, 537 (1977).
47. A. Javan, V. S. Letokhov, M. S. Feld, and A. L. Schawlow, in *Fundamental and Applied Laser Physics*, M. S. Feld, A. Javan, and N. A. Kurnit, Eds., Wiley, New York, 1973.

48. E. Courtens, in *Laser Handbook*, Vol. 2, F. T. Arecchi and E. O. Schulz-DuBois, Eds., Elsevier, New York, 1972, p. 1259.
49. P. M. Rentzepis and W. S. Struve, in *International Reviews of Science, Physical Chemistry*, Series Two, Vol. 3, Spectroscopy, D. A. Ramsay, Ed., Butterworth, London/Boston, 1976, p. 263.
50. P. M. Rentzepis, *Chem. Phys. Lett.*, **2**, 117 (1968); *ibid.*, **3**, 717 (1969).
51. M. M. Malley and P. M. Rentzepis, *Chem. Phys. Lett.*, **3**, 534 (1969).
52. M. R. Topp, P. M. Rentzepis, and R. P. Jones, *Chem. Phys. Lett.*, **9**, 1 (1971).
53. M. R. Topp and G. C. Orner, *Chem. Phys. Lett.*, **32**, 407 (1975).
54. M. R. Topp. *Opt. Commun.*, **14**, 126 (1975).
55. D. Magde and M. W. Windsor, *Chem. Phys. Lett.*, **27**, 31 (1974).
56. D. Magde, B. A. Bushaw, and M. W. Windsor, *Chem. Phys. Lett.*, **28**, 263 (1974).
57. D. Magde, M. W. Windsor, D. Holten, and M. Gouterman, *Chem. Phys. Lett.*, **29**, 183 (1974).
58. M. G. Rockley, M. W. Windsor, R. J. Cogdell, and W. W. Parson, *Proc. Natl. Acad. Sci.*, **72**, 2251 (1975).
59. R. R. Alfano and S. L. Shapiro, *Chem. Phys. Lett.*, **8**, 631 (1971).
60. R. R. Alfano and S. L. Shapiro, *Phys. Rev. Lett.*, **24**, 584, 592, 1217 (1970).
61. W. Yu, R. R. Alfano, C. L. Sam, and R. J. Seymour, *Opt. Commun.*, **14**, 344 (1975).
62. E. P. Ippen, C. V. Shank, and T. K. Gustafson, *Appl. Phys. Lett.*, **24**, 190 (1974).
63. N. Mataga and N. Nakashima, *Spectrosc. Lett.*, **8**, 275 (1975).
64. D. W. Phillon, D. J. Kuizenga, and A. E. Siegman, *Appl. Phys. Lett.*, **27**, 85 (1975).
65. A. E. Siegman, *Appl. Phys. Lett.*, **30**, 21 (1977).
66. A. D. Backingham, *Proc. Phys. Soc.*, **B69**, 344 (1956).
67. G. Mayer and F. Gires, *C.R. Acad. Sci.* (Paris), **258**, 2039 (1964).
68. P. D. Maker, R. W. Terhune, and C. M. Savage, *Phys. Rev. Lett.*, **12**, 507 (1964).
69. M. A. Duguay and J. W. Hansen, *Appl. Phys. Lett.*, **15**, 192 (1969); *Opt. Commun.*, **1**, 254 (1969).
70. M. A. Duguay and A. T. Mattick, *Appl. Opt.*, **10**, 2162 (1971).
71. E. P. Ippen and C. V. Shank, *Appl. Phys. Lett.*, **26**, 92 (1975), and references therein.
72. M. M. Malley and P. M. Rentzepis, *Chem. Phys. Lett.*, **7**, 57 (1970).
73. G. Mourou, D. Drouin, and M. M. Denariez-Roberge, *Appl. Phys. Lett.*, **20**, 453 (1972).
74. G. E. Busch, K. S. Greve, G. L. Olson, R. P. Jones, and P. M. Rentzepis, *Appl. Phys. Lett.*, **27**, 450 (1975).
75. D. J. Bradley, in *Ultrashort Light Pulses*, S. L. Shapiro, Ed., Springer-Verlag, New York, 1977, p. 41.
76. M. E. Mack, *Appl. Phys. Lett.*, **15**, 166 (1969); C. Lin, T. K. Gustafson, and A. Dienes, *Opt. Commun.*, **8**, 210 (1973).

77. A. N. Rubinov, M. C. Richardson, K. Sala, and A. J. Alcock, *Appl. Phys. Lett.*, **27**, 358 (1975); B. Fan and T. K. Gustafson, *Appl. Phys. Lett.*, **28**, 202 (1976).

78. W. H. Glenn, M. J. Brienza, and A. J. DeMaria, *Appl. Phys. Lett.*, **12**, 54 (1968); B. H. Soffer and J. W. Linn, *J. Appl. Phys.*, **39**, 5859 (1968); T. R. Royt, W. L. Faust, L. S. Goldberg, and C. H. Lee, *Appl. Phys. Lett.*, **25**, 514 (1974); L. S. Goldberg and C. A. Moore, *Appl. Phys. Lett.*, **27**, 217 (1975); T. R. Royt and C. H. Lee, *Appl. Phys. Lett.*, **30**, 332 (1977).

79. (a) C. V. Shank and E. P. Ippen, in *Dye Lasers*, F. P. Schäfer, Ed., Springer-Verlag, New York, 1973, p. 137; (b) C. K. Chan and S. O. Sari, *Appl. Phys. Lett.*, **25**, 403 (1974); N. J. Frigo, T. Daly, and H. Mahr, *IEEE J. Quantum Electron.*, **QE-13**, 101 (1977); H. Mahr, *IEEE J. Quantum Electron.*, **QE-12**, 554 (1976); H. Mahr and M. D. Hirsch, *Opt. Commun.*, **13**, 96 (1975); R. K. Jain and J. P. Heritage, private communication.

80. A. Dienes, E. P. Ippen, and C. V. Shank, *Appl. Phys. Lett.*, **19**, 258 (1971).

81. D. J. Kuizenga, *Appl. Phys. Lett.*, **19**, 260 (1971).

82. (a) C. V. Shank, E. P. Ippen, and A. Dienes, *Digest of Tech. Papers, VII International Quantum Electronics Conf. Montreal, 1972*, IEEE, New York, 1972, p. 7; (b) E. P. Ippen, C. V. Shank, and A. Dienes, *Appl. Phys. Lett.*, **21**, 348 (1972).

83. (a) F. O'Neill, *Opt. Commun.*, **6**, 360 (1972); (b) J. P. Letouzey and S. O. Sari, *Appl. Phys. Lett.*, **23**, 311 (1973).

84. C. V. Shank and E. P. Ippen, *Appl. Phys. Lett.*, **24**, 373 (1974).

85. (a) E. B. Treacy, *Phys. Lett.*, **28A**, 34 (1968); *Appl. Phys. Lett.*, **14**, 112 (1969); (b) E. P. Ippen and C. V. Shank, *Appl. Phys. Lett.*, **27**, 488 (1975).

86. I. S. Ruddock and D. J. Bradley, *Appl. Phys. Lett.*, **29**, 296 (1976).

87. E. P. Ippen and C. V. Shank, *Opt. Commun.*, **18**, 27 (1976).

88. J. P. Heritage and R. K. Jain, private communication.

89. Z. A. Yasa, *J. Appl. Phys.*, **46**, 4895 (1975); Z. A. Yasa and O. Teschke, *Appl. Phys. Lett.*, **27**, 446 (1975); Z. A. Yasa, A. Dienes, and J. R. Whinnery, *Appl. Phys. Lett.*, **30**, 24 (1977).

90. E. P. Ippen and C. V. Shank, private communication.

91. S. Kielich, in *Dielectric and Related Molecular Processes*, Vol. 1, Specialist Periodical Reports, The Chemical Society, London, 1972, p. 192.

92. For an outline of the theory regarding orientational relaxation, see K. B. Eisenthal, *Acc. Chem. Res.*, **8**, 118 (1975), and references therein; the theory behind the concept of photoselection was well described by Albrecht and may be useful to those who require additional background: A. C. Albrecht, *J. Mol. Spectrosc.*, **6**, 84 (1961).

93. T. J. Chuang and K. B. Eisenthal, *Chem. Phys. Lett.*, **11**, 368 (1971).

94. A. Einstein, *Ann. Phys.*, (Leipzig), **19**, 371 (1906); translated into English in *Investigations on the Theory of Brownian Movement*, Dover, New York, 1956, and P. Debye, *Polar Molecules*, Dover, London, 1929, p. 83.

95. G. R. Fleming, J. M. Morris, and G. W. Robinson, *Chem. Phys.*, **17**, 91 (1976).

96. H. E. Lessing and A. von Jena, *Chem. Phys. Lett.*, **42**, 213 (1976), and references therein.

97. D. Magde and M. W. Windsor, *Chem. Phys. Lett.*, **24**, 144 (1974).

98. T. Förster and G. Hoffman, *Z. Phys. Chem.*, *Neue Folge* (Frankfurt), **75**, 63 (1971).

99. L. J. E. Hofer, R. J. Grabenstetter, and E. O. Wiig, *J. Am. Chem. Soc.*, **72**, 203 (1950).

100. G. Oster and Y. Nishijima, *J. Am. Chem. Soc.*, **78**, 1581 (1956).

101. B. Bushaw, Masters Thesis, Chemistry Department, Washington State University, 1976.

102. D. Cremers and M. W. Windsor, manuscript in preparation.

103. R. Olinger, U. Schindewolf, A. Gaathon, and J. Jortner, *Ber. Bunsenges. Phys. Chem.*, **75**, 690 (1971).

104. J. H. Baxendale and P. Wardman, *Nature*, **230**, 449 (1971); *J. Chem. Soc., Faraday Trans. 1*, **69**, 584 (1973).

105. M. J. Bronskill, R. K. Wolff, and J. W. Hunt, *J. Chem. Phys.*, **53**, 4201 (1970).

106. P. M. Rentzepis, R. P. Jones, and J. Jortner, *Chem. Phys. Lett.*, **15**, 480 (1972).

107. P. M. Rentzepis, R. P. Jones, and J. Jortner, *J. Chem. Phys.*, **59**, 766 (1973).

108. C. D. Jonah, *Rev. Sci. Instrum.*, **46**, 62 (1975).

109. G. A. Kenney-Wallace and C. D. Jonah, *Chem. Phys. Lett.*, **39**, 596 (1976).

110. W. J. Chase and J. W. Hunt, *J. Phys. Chem.*, **79**, 2835 (1975).

111. G. A. Kenney-Wallace and C. D. Jonah, *Chem. Phys. Lett.*, **47**, 362 (1977).

112. J. B. Birks, *Photophysics of Aromatic Molecules*, Wiley, New York, 1970, pp. 166, 187.

113. R. M. Hochstrasser and T. Y. Li, *J. Mol. Spectrosc.*, **41**, 297 (1972).

114. P. M. Rentzepis, *Photochem. Photobiol.*, **8**, 579 (1968).

115. P. M. Rentzepis, *Science*, **169**, 239 (1970).

116. P. M. Rentzepis, J. Jortner, and R. P. Jones, *Chem. Phys. Lett.*, **4**, 599 (1970).

117. D. Huppert, J. Jortner, and P. M. Rentzepis, *J. Chem. Phys.*, **56**, 4826 (1972).

118. E. Drent, G. Makkes van der Deijl, and P. J. Zandstra, *Chem. Phys. Lett.*, **2**, 526 (1968).

119. P. Wirth, S. Schneider, and F. Dorr, *Chem. Phys. Lett.*, **42**, 482 (1976).

120. J. P. Heritage and A. Penzkofer, *Chem. Phys. Lett.*, **44**, 76 (1976).

121. E. P. Ippen, C. V. Shank, and R. L. Woerner, *Chem. Phys. Lett.*, **46**, 20 (1977).

122. G. Eber, S. Schneider, and F. Doerr, *J. Photochem.*, **7**, 91 (1977).

123. E. C. Lim, in *Excited States*, Vol. 3, E. C. Lim, Ed., Academic Press, New York, 1976, and references therein.

124. R. S. Becker, *Theory and Interpretation of Fluorescence and Phosphorescence*, Wiley, New York, 1969, and references therein.

125. M. A. El Sayed and M. Kasha, *Spectrochim. Acta*, **15**, 758 (1959).

126. V. L. Ermolaev and I. P. Kotlyar, *Opt. Spectrosc. USSR*, **9**, 183 (1960).
127. Y. Hirata and I. Tanaka, *Chem. Phys. Lett.*, **41**, 336 (1976).
128. V. Sundstrom, P. M. Rentzepis, and E. C. Lim, *J. Chem. Phys.*, **66**, 4287 (1977).
129. Y. Hirata and I. Tanaka, *Chem. Phys. Lett.*, **43**, 568 (1976).
130. P. M. Rentzepis, *Science*, **169**, 239 (1970).
131. A. Nitzan, J. Jortner, and P. M. Rentzepis, *Chem. Phys. Lett.*, **8**, 445 (1971).
132. R. M. Hochstrasser, H. Lutz, and G. W. Scott, *Chem. Phys. Lett.*, **24**, 162 (1974).
133. R. W. Anderson, Jr., R. M. Hochstrasser, H. Lutz, and G. W. Scott, *Chem. Phys. Lett.*, **28**, 153 (1974).
134. D. Magde, B. A. Bushaw, and M. W. Windsor, *IEEE J. Quantum Electron.*, **QE-10**, 394 (1974).
135. M. Gouterman, in *The Porphyrins*, D. Dolphin, Ed., Academic Press, New York, in press.
136. M. Gouterman, F. P. Schwarz, P. D. Smith, and D. Dolphin, *J. Chem. Phys.*, **59**, 676 (1973).
137. G. P. Gurinovich and B. M. Jagarov, in *Luminescence of Crystals, Molecules and Solutions*, F. Williams, Ed., Proc. Intern. Conf. Luminescence, Leningrad, Aug. 1972, Plenum, New York, 1973.
138. J. B. Callis, M. Gouterman, and J. D. S. Danielson, *Rev. Sci. Instrum.*, **40**, 1599 (1969).
139. A. D. Kirk, P. E. Hoggard, G. B. Porter, M. G. Rockley, and M. W. Windsor, *Chem. Phys. Lett.*, **37**, 199 (1976).
140. G. B. Porter, in *Concepts of Inorganic Photochemistry*, A. W. Adamson and P. D. Fleischauer, Eds., Wiley, New York, 1975, p. 37, and references therein.
141. E. V. Zinato, in *Concepts of Inorganic Photochemistry*, A. W. Adamson and P. D. Fleischauer, Eds., Wiley, New York, 1975, p. 143, and references therein.
142. S. C. Pyke, M. W. Windsor, G. B. Porter, and A. D. Kirk, 32nd Annual Northwest Regional Meeting of the American Chemical Society, Portland, Oregon, June 15-17, 1977, Abstract #156.
143. A. Laubereau, D. von der Linde, and W. Kaiser, *Phys. Rev. Lett.*, **28**, 1162 (1972).
144. A. Laubereau, *Chem. Phys. Lett.*, **27**, 600 (1974).
145. S. F. Fischer and A. Laubereau, *Chem. Phys. Lett.*, **35**, 6 (1975).
146. A. Laubereau, L. Kirschner, and W. Kaiser, *Opt. Commun.*, **9**, 182 (1973).
147. K. Spanner, A. Laubereau, and W. Kaiser, *Chem. Phys. Lett.*, **44**, 88 (1976).
148. P. R. Monson, S. Patumtevapibal, K. J. Kaufmann, and G. W. Robinson, *Chem. Phys. Lett.*, **28**, 312 (1974).
149. D. Ricard, W. H. Lowdermilk, and J. Ducuing, *Chem. Phys. Lett.*, **16**, 617 (1972).
150. G. Mourou and M. M. Malley, *Chem. Phys. Lett.*, **32**, 476 (1975).
151. E. P. Ippen and C. V. Shank, *Opt. Commun.*, **18**, 27 (1976).

152. A. Penzkofer, W. Falkenstein, and W. Kaiser, *Chem. Phys. Lett.*, **44**, 82 (1976).

153. W. Kaiser and A. Laubereau, in *Proceedings of the 2nd International Conference on Laser Spectroscopy, Meqine, France*, S. Haroche, J. C. Pebay-Peyroula, T. W. Hänsch, and S. E. Harris, Eds., Springer-Verlag, Berlin, 1975, p. 380.

154. T. J. Chuang, G. W. Hoffman, and K. B. Eisenthal, *Chem. Phys. Lett.*, **25**, 201 (1974).

155. R. M. Noyes, *J. Chem. Phys.*, **22**, 1349 (1954); *J. Am. Chem. Soc.*, **78**, 5486 (1956).

156. R. M. Noyes, *Z. Elektrochem.*, **64**, 153 (1960).

157. R. W. Anderson, Jr., and R. M. Hochstrasser, *J. Phys. Chem.*, **80**, 2155 (1976).

158. D. A. Weirsma and J. Kommandeur, *Mol. Phys.*, **13**, 241 (1967).

159. T. Shida and A. Kira, *J. Phys. Chem.*, **73**, 4315 (1969).

160. D. A. Weirsma, J. H. Lichtenbelt, and J. Kommandeur, *J. Chem. Phys.*, **50**, 2794 (1969).

161. K. K. Smith, J. Y. Koo, G. B. Schuster, and K. J. Kaufmann, *Chem. Phys. Lett.*, **48**, 267 (1977).

162. N. J. Turro, P. Lechtken, A. Lyons, R. R. Hautala, E. Carnahan, and T. J. Katz, *J. Am. Chem. Soc.*, **95**, 2035 (1973).

163. P. Lechtken and N. J. Turro, *Angew. Chem. Int. Ed.*, **12**, 314 (1973).

164. N. J. Turro and P. Lechtken, *J. Am. Chem. Soc.*, **94**, 2886 (1972).

165. A. J. Campillo and S. L. Shapiro, in *Ultrashort Light Pulses*, S. L. Shapiro, Ed., Springer-Verlag, New York, 1977, p. 317.

166. D. Holten and M. W. Windsor, *Ann. Rev. Biophys. Bioeng.*, **7**, 189 (1978).

167. A. Yu. Borisov and M. D. Il'ina, *Biochim. Biophys. Acta*, **305**, 364 (1973).

168. H. Merkelo, J. H. Hammond, S. R. Hartman, and Z. I. Derzko, *J. Luminescence*, **1/2**, 502 (1970).

169. M. Seibert and R. R. Alfano, *Biophys. J.*, **14**, 269 (1974).

170. A. J. Campillo, R. C. Hyer, V. H. Kollman, S. L. Shapiro, and H. D. Sutphin, *Proc. Natl Acad. Sci. USA*, **74**, 1997 (1977).

171. A. J. Campillo, R. C. Hyer, T. Monger, W. W. Parson, and S. L. Shapiro, *Proc. Natl. Acad. Sci. USA*, **74**, 1997 (1977).

172. G. S. Beddard and G. Porter, *Biochim. Biophys. Acta.*, **462**, 63 (1977).

173. A. J. Campillo, V. H. Kollman, and S. L. Shapiro, *Science*, **193**, 227 (1976).

174. D. Mauzerall, *J. Phys. Chem.*, **80**, 2306 (1976).

175. J. C. Hindman, R. Kugel, A. Svirmickas, and J. J. Katz, *Proc. Natl. Acad. Sci. USA*, **74**, 5 (1977).

176. M. D. Hirsch, M. A. Marcus, A. Lewis, H. Mahr, and N. Frigo, *Biophys. J.*, **16**, 1399 (1976).

177. W. Yu, P. P. Ho, R. R. Alfano, and M. Seibert, *Biochim. Biophys. Acta*, **387**, 159 (1975).

178. R. S. Knox, in *Bioenergetics of Photosynthesis*, R. Govindjee, Ed., Academic Press, New York, 1974.

179. G. F. W. Searle, J. Barber, L. Harris, G. Porter, and C. J. Tredwell, *Biochim. Biophys. Acta.*, **459**, 390 (1977).
180. R. K. Clayton, *Ann. Rev. Biophys. Bioeng.*, **2**, 131 (1973).
181. S. C. Straley, W. W. Parson, D. Mauzerall, and R. K. Clayton, *Biochim. Biophys. Acta*, **305**, 597 (1973).
182. K. J. Kaufmann, P. L. Dutton, T. L. Netzel, J. S. Leigh, and P. M. Rentzepis, *Science*, **188**, 1301 (1975).
183. W. W. Parson, R. K. Clayton, and R. J. Cogdell, *Biochim. Biophys. Acta*, **387**, 265 (1975).
184. M. G. Rockley, M. W. Windsor, R. J. Cogdell, and W. W. Parson, *Proc. Natl. Acad. Sci. USA*, **72**, 2251 (1975).
185. J. Fajer, D. C. Brune, M. S. Davis, A. Forman, and L. D. Spaulding, *Proc. Natl. Acad. Sci. USA*, **72**, 4956 (1975).
186. J. Fajer, M. S. Davis, D. C. Brune, L. D. Spaulding, D. C. Borg, and A. Forman, *Brookhaven Symp. Biol.*, **28**, 74 (1977).
187. D. Holten, M. W. Windsor, W. W. Parson, and M. Gouterman, *Photochem. Photobiol.*, in press.
188. G. E. Busch, M. L. Applebury, A. A. Lamola, and P. M. Rentzepis, *Proc. Natl. Acad. Sci. USA*, **69**, 2802 (1972).
189. T. Kobayashi, H. Ohtani, T. Yoshizawa, and S. Nagakura, paper presented at 26th IUPAC Conference, September 1977, Tokyo, Japan.
190. R. Lozier, R. Bogomolni, and W. Stoeckenius, *Biophys. J.*, **15**, 955 (1975).
191. T. Yoshizawa, in *Handbook of Sensory Physiology*, Vol. VIII/1, H. J. A. Dartnall, Ed., Springer-Verlag, New York, 1972, p. 146.
192. R. R. Alfano, W. Yu, R. Govindjee, B. Cecher, and T. G. Ebrey, *Biophys. J.*, **16**, 541 (1976).
193. K. J. Kaufmann, P. M. Rentzepis, W. Stoeckenius, and A. Lewis, *Biochem. Biophys. Res. Commun.*, **68**, 1109 (1976).
194. C. V. Shank, E. P. Ippen, and R. Bersohn, *Science*, **193**, 50 (1976).
195. S. L. Shapiro, A. J. Campillo, V. H. Kollman, and W. B. Goad, *Opt. Commun.*, **15**, 308 (1975).
196. D. Holten, M. W. Windsor, W. W. Parson and J. P. Thornber, *Biochim. Biophys. Acta*, **501**, 112 (1978).

Chapter **VI**

HIGH-TEMPERATURE TECHNIQUES

John L. Margrave
Robert H. Hauge

1   INTRODUCTION

High-temperature science is an ancient field, dating back to the discovery and utilization of fire. Metallurgical processing of ores to obtain mercury, lead, and iron; the fusion of sand with other oxides to form glass, tile, and bricks; the utilization of solar heating for dehydrating biomass or for extracting salt from sea water; fire assaying of minerals; and the distillation of alcoholic beverages may be cited to demonstrate that this is not a new field. On the other hand, it was not until a coherent periodic table of the elements was constructed, an appreciation of the chemical significance of oxygen in combustion was developed, and efficient means for generating electrical current were perfected that the groundwork was available for the establishment of the high-temperature field.

Methods for generating and utilizing high temperatures have evolved over the past 75 years from primarily "brute force" techniques represented by the huge coke-fired furnaces of the steel industry, the electrically heated tanks of molten glass, or the gas-fired reactors and kilns of the refractory and cement industries to a high technology utilizing the most sophisticated of modern electronic devices, new materials, and scientific expertise. Not only can one attain high temperatures over large volumes for long times but one can also utilize controlled pulses of energy to heat individual atoms or molecules selectively to desired temperatures for very short periods of time. It is the goal of this review to present a variety of methods that will allow one to generate and measure high temperatures in systems of interest to scientists and engineers.

In 1979 it is routinely possible to generate temperatures of tens of thousands of degrees in moles of material over periods of hours at low or high pressures. At the forefront of technology it is possible to generate temperatures of hundreds of millions of degrees in micromoles for picosecond periods of time in high-vacuum systems. Obviously the techniques for generating and measuring high temperatures will span a wide range of disciplines and devices. The modern high-temperature scientist who makes use of these methods has available to him a greater range of temperature than has any scientist in history. One can document the synthetic achievements and detailed understanding of molecular and atomic phenomena that have been possible through the application of high-temperature techniques.

Several reviews of this field have been published and are cited as additional resources [1–7]. In addition to reviews of techniques, there have been many reviews of the various high temperature properties and processes (see Refs. 8–11). There are many other publications in specialized areas as well. An international assortment of journals is devoted especially to the high-temperature field [12–18]. Several books devoted to high-temperature science have also been published [19–25], and extensive tabulations of scientific data and engineering parameters for use by high-temperature scientists are available [26, 27, 489, 491].

2   THE GENERATION OF HIGH TEMPERATURES

A system is at "high temperatures" when it contains more energy of some type (electronic, transitional, vibrational, or rotational) than it did at some arbitrary reference, for example, a "lower temperature." The usual reference temperature is in the 15° to 25°C (59° to 77°F) range based on the classical centigrade or Fahrenheit temperature scales. Modern scientists prefer to use the absolute zero as reference temperature (0°K = -273.16°C = -459.7°F). Thus the arbitrary choice of a reference temperature determines what will be a high temperature. It is necessary to utilize a combination of thermodynamics, statistical mechanics, and quantum mechanics to appreciate fully the subtle ways in which one may generate and utilize high temperatures in modern technology.

The attainment of uniform temperatures above 1000°C over an extended period of time is primarily limited by the availability of construction materials. Such temperatures are usually generated in the laboratory by electrical means, although where weather is cooperative the sun as a source of radiant energy may be utilized directly. Electrically powered heating devices for these temperature ranges have been favored over those using chemical methods for simplicity of design, ease of operation, controlled atmosphere, and the close control of temperature that is possible. The gas-fired furnace, which is typical of the chemical method, has found wide application in industry because of economics, but current energy problems are changing this picture. There is a new surge of interest in solar augmentation of gas-fired furnaces, in the use of coal or lignite, and in other novel approaches to high temperatures.

Resistance Furnaces

Resistance-wound furnaces are popular in most laboratories because of the simplicity of design, low cost, and the efficient temperature control.

Melting points, thermal conductivities, thermal diffusivities, and the rates of evaporation, oxidation, and grain growth are some of the limiting properties that have to be considered in the choice of resistor materials from currently available elements, alloys, and compounds, including certain oxides, borides, carbides, nitrides, silicides, etc. Of course, the atmosphere desired may also limit the choice of resistor materials.

At temperatures above 1000°C one can use Kanthal, platinum, iridium, molybdenum, tantalum, or tungsten [28–30]. Kanthal and platinum resistors are useful and satisfactory up to 1400°C, while molybdenum, because of its ductility, is preferred over the other refractory metals for temperatures between 1500° and 2000°C. The advantages, limitations, and practical aspects of molybdenum-wound tube and tungsten mesh-type furnaces have been discussed [31]. Small tubes of refractory metals are easily interchanged as sources for metal vapors in sputtering, chemical vapor decomposition, etc. [32–35]. Iridium can be used up to ~2500°K [36, 37]. For high-vacuum use, refractory materials to support the resistor may be unsatisfactory due to the formation of gaseous products by reaction between the metal and the refractory support, and one usually tries to design self-supported heater elements. For typical designs of resistance furnaces, see Fig. 6.1.

Furnaces that can be used up to 3000°C with vacua of the order of 10^{-6} mm Hg or pressures up to 100 atm and an inert or reducing gas atmosphere have been built with heater tubes consisting of either a seamless tantalum tubing or a split tungsten tube with metal foil radiation shields that, in turn, are surrounded by a water-cooled mild steel pressure shell. Tungsten heater tubes and radiation shields may be used when hydrogen or ammonia atmospheres are used in the chamber.

Granular carbon or mixed silicon carbide and carbon resistors can be used up to 1800°C. The fact that the contact resistance of solid particles is higher than the bulk resistance of the material itself makes the granular resistor furnace work. Hollow silicon carbide tube heaters are widely used in air up to 1700°C [38].

At 1700°C and above, all metal carbides oxidize appreciably, and the more volatile elements (Si, Ti, B, etc.) begin to sublime out. Above 2000°C, one is getting into the range of melting points; and above 2500°C, all metals and even carbon sublime at appreciable rates. TaC, ZrC, and HfC are among the most stable and highest-melting carbides and thus offer the ultimate in heater properties since they are essentially metallic conductors. Commercial TaC shapes have been prepared, or one may convert a metal shape into a carbide by packing it in graphite powder and heating.

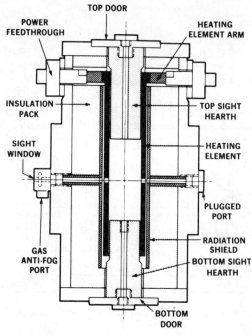

POWER FEEDTHROUGH

TOP DOOR

HEATING ELEMENT ARM

INSULATION PACK

TOP SIGHT HEARTH

SIGHT WINDOW

HEATING ELEMENT

PLUGGED PORT

GAS ANTI-FOG PORT

RADIATION SHIELD

BOTTOM SIGHT HEARTH

BOTTOM DOOR

Fig. 6.1. Composite resistance furnace capable of reaching 2700°C.

Transition element sulfides, nitrides, silicides, and borides are also metallic conductors and potential heater materials. Most interest has been shown in borides and silicides, since the nitrides tend to decompose at fairly low temperatures and the sulfides oxidize to produce SO_2 or SO_3. The borides and silicides are protected from oxidation by adherent films of B_2O_3 or SiO_2, and heater units of siliconized Mo or W wire ($MoSi_2$ or WSi_2 layers) and of sintered $MoSi_2$ (Kanthal) are available commercially [39, 40].

Graphite tube furnaces cut to form a spiral for increased electrical resistance are widely used. A horizontal graphite tube furnace thermally insulated with carbon black can be operated routinely up to 3000°C. The graphite heating element can be mounted vertically for more efficient vacuum use, and various workers have used carbon tube furnaces for a variety of high-temperature research [41–44]. One of the major limitations of resistance-heated furnaces arises from the necessity of a physical contact to deliver the electric power to the heaters. Water-cooled hollow copper rings are often used to connect the power supply to a carbon, tantalum, or tungsten tube furnace which operates above 2000°C.

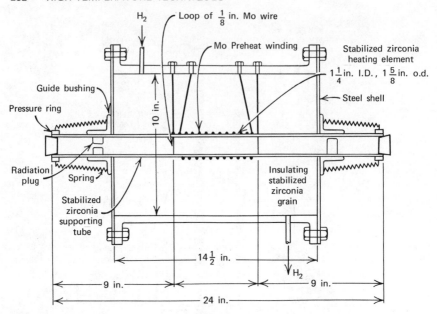

Fig. 6.2.   Diagram of a stabilized zirconia resistor tube furnace, hydrogen atmosphere.

Refractory oxide resistor-type furnaces for temperatures up to 2000°C have been described in the literature [45]. One can use, for example, an oxide resistor made of 85% zirconia and 15% of either yttria or ceria, a mixture which shows increased electrical conductivity with increasing temperatures (see Fig. 6.2).

The inherent problem in operating oxide resistor-type furnaces is the difficulty of controlling the current through the various resistors which have negative temperature coefficients. In addition, preheating of these oxides is necessary since the oxides are poor electrical conductors at low temperatures. Thus in the ZrO_2 furnace described by Davenport et al. [46], a molybdenum winding was used as the preheating element at low temperatures. At higher temperatures the resistance of the zirconia tube was lower than that of molybdenum, thus effectively short circuiting the molybdenum coil. A carbon rod may be used as a starter and then withdrawn as the oxide begins to conduct.

Induction Heating Apparatus

Induction-heated furnaces can generate high temperatures in various atmospheres, above the upper limiting temperatures for simple resistance furnaces, and they are widely used for high-temperature work from 1000°

to 3000° [47]. The upper limits of temperatures are determined by the stability of susceptor materials, that is, their rates of sublimation and/or melting points and by the efficiency of electrical coupling. An induction heating system consists of a water-cooled copper work coil which carries the alternating current acting as the primary of a transformer while a susceptor placed in the center of the concentric work coil serves as the secondary. One can heat inductively with 60 Hz or other low frequencies, but systems that operate at 10,000 Hz, 450,000 Hz, or even several MHz are used in various special applications.

In an all-glass system under vacuum, the work coil is ordinarily placed outside a quartz or a vycor tube which forms part of the vacuum system. An alternate method is to put the work coil outside the cooling jacket, when one is used. For an all-metal system under vacuum, the work coil is placed inside the system.

The main requirement for a susceptor is that it be a conductor. It may also serve as a crucible, and the heating of a nonconducting sample will take place by conduction and radiation. The most common susceptor material presently used is graphite. Where a carbon atmosphere cannot be tolerated, other susceptor materials may be used, such as the refractory metals with a ceramic liner (MgO, Al_2O_3, ZrO_2). Direct coupling with the sample is best whenever it is experimentally feasible. The design, construction, and applications of susceptors in the heating of metallic and nonmetallic parts are discussed in the engineering and metallurgical literature.

High-frequency generating devices include (1) the motor generator set [see Ref. 48, pp. 65–96], (2) the spark-gap converter (see Ref. 48, pp. 52–65), and (3) the vacuum tube oscillator (see Ref. 48, p. 96; 116 and 49). The frequencies range from 0.5 to 10 kHz for motor generator sets, from 10 to 100 kHz for spark-gap converters, and from 100 kHz to 100 MHz for vacuum tube oscillators. It is possible to attain 90% power transfer efficiency with large motor-generator sets compared to less than 50% for vacuum tube oscillators. Spark-gap generators are mainly of historical interest. For a detailed discussion on high-frequency generators see Curtiss or Jordan [50, 51].

For heating metals and good conductors, the lower-frequency devices (< 500 kHz are best; but for poorer conductors, one finds oscillators that produce power at 5 to 50 MHz more desirable. A dual-frequency unit is a particularly versatile type of laboratory instrument. One of the major differences between low-frequency and high-frequency heaters is described by the "skin effect." The eddy currents induced in the susceptors tend to concentrate nearer the surface for the higher-frequency sources.

High-frequency oxide induction furnaces for use in air at temperatures

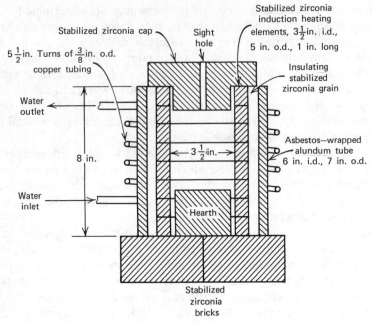

Fig. 6.3. Diagram of a stabilized zirconia induction furnace, air atmosphere.

in excess of 2200°C have been constructed with poorly conducting ceramic rings made of yttrium oxide-stabilized ZrO_2 as the susceptor elements. Unstabilized ZrO_2 grains were used for the thermal insulators. Excellent control of temperature in the furnace chamber up to a maximum of 2350°C was reported. Davenport et al. [46] and Duwez et al. [52] have also reported work with such furnaces. A typical furnace is shown in Fig. 6.3.

One of the problems in using induction heaters is to get the heat into the desired region. The "skin effect" accomplishes this in certain metallurgical processes where a thin layer needs to be heated to provide surface hardening, a fused oxide layer, etc. In some early work, Babat and Losinsky [53] developed the eddy current concentrator that made it possible for the magnetic field of a large coil to be coupled efficiently with a small piece of material serving as the secondary. At Los Alamos, the U.S. Bureau of Standards, and other laboratories, induction heaters/current concentrators have been used, for example, in a study of the vaporization behavior of niobium carbide. An overall view of a typical apparatus is shown in Fig. 6.4 [54, 55].

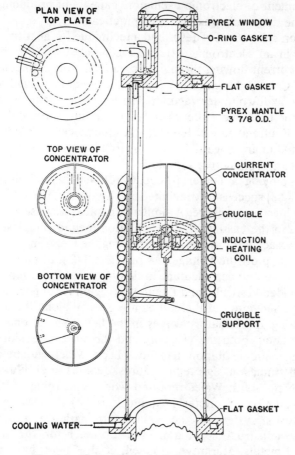

PLAN VIEW OF
TOP PLATE

PYREX WINDOW
O-RING GASKET

FLAT GASKET

PYREX MANTLE
3 7/8 O.D.

TOP VIEW OF
CONCENTRATOR

CURRENT
CONCENTRATOR

CRUCIBLE

INDUCTION
HEATING
COIL

BOTTOM VIEW OF
CONCENTRATOR

CRUCIBLE
SUPPORT

FLAT GASKET

COOLING WATER

Fig. 6.4. Concentrator for induction heating.

Electronic oscillators that operate in the microwave frequency region are now widely used for diathermy and home cooking devices since the permanent dipole movement of water interacts strongly with these electromagnetic fields. With polar gases one can create microwave discharges with ions, atoms, molecules, and electrons at temperatures in the 5000°K range [56–60].

Levitation Heating

Electromagnetic levitation for heating and melting metals without the use of crucibles was claimed as early as 1923 in various patents, but until the development of electronic oscillators capable of producing kilowatts of power the approach was not very practical.

Wroughton et al. [63, 64] described a technique for heating and melting substances in an electromagnetic field generated by a high-frequency alternating current flowing through two coaxial coils connected in series opposition. Careful design of the coil arrangement is needed to minimize electrical breakdown under vacuum. The levitation meltings of 10-g samples of Sn and brass were successfully carried out in air, and of Al under vacuum using a frequency of 10 kHz, and solid pieces weighing up to 550 g were levitated in air. Sagardia and Ross [65] have reported the levitation melting of large conductive samples.

While employing a heater frequency of 400 kHz, Polonis and co-workers [66–68] successfully levitated 5 to 8 g of Al, Ti, Fe, Ni, and Ti–Fe alloys in an inert atmosphere. A similar apparatus was used by Comenetz and Salatka in the casting of 10-g levitation-melted samples. Casting of molten Ti was accomplished by superheating the melts and by reducing or suddenly cutting off the voltage. As much as 140 g of molten aluminum has been levitated under vacuum. In floating-zone levitation, the sample bar is suspended vertically and the molten region is supported and stirred by the levitating action of the coil as it is passed along the length of the bar. By setting up oscillatory waves in the liquid metal, one can deduce information about both the viscosity and the surface tension [69].

Electromagnetic levitation has now been widely adapted to physicochemical studies at high temperatures. Jenkins et al. [70] at the University of New South Wales reported work concerning (a) the kinetics of direct oxidation of solutes in liquid iron, (b) the development of a levitation–fusion system for the rapid determination of gases in metals, and (c) the construction of a microcalorimeter for the study of reactive liquid metals. Margrave, Bonnell, et al. [71–78] have reported an extensive series of calorimetric measurements on Pt, Ti, V, Fe, Co, Ni, Cu, Mo, Zr, and Pd; the schematic apparatus is shown in Fig. 6.5. An adaptation of the technique in which hot, levitated samples are dropped into liquid argon, reminiscent of the Bunsen ice calorimeter, has been reported [79]. Scientists in the USSR have also applied the levitation method to studies of refractory metals [80–84]. Frost et al. [85–87] at the General Electric Space Services Laboratory have described several special types of coil designs (see Fig. 6.6), in some of which they can accomplish levitation and melting of elemental tungsten.

Recently there has been an interest in other means of levitation (acoustic, Bernoulli, or in space, where the microgravity environment prevails) and supplementing these nonelectromagnetic approaches with electron beam heating or laser heating [85–91].

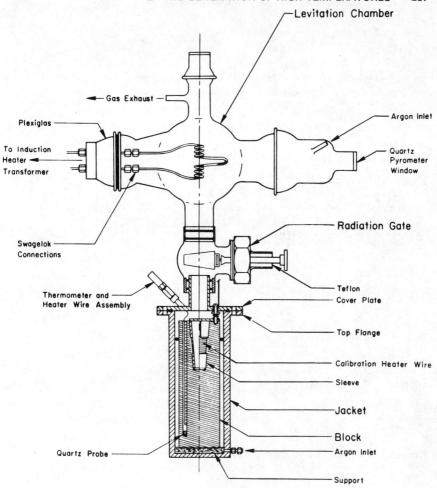

Fig. 6.5. Schematic of levitation calorimeter.

Chemical Methods

Exothermic chemical reactions such as the oxidation (with O_2, O_3, Cl_2, F_2, OF_2, H_2O_2, $LiClO_4$, NO_2, etc.) of various fuels (hydrocarbons, H_2 metals, boron hydrides, etc.) or the use of energetic species such as NH_4NO_3, $CH_3C_6H_2(NO_2)_3$ (TNT), etc., which contain both fuel and oxidizer in the same molecule, can produce temperatures in the 2000° to 6000°K range. The important requirements for obtaining high flame temperatures are that the reactants be relatively unstable or weakly bonded and that the reaction products be very stable. Various authors have discussed

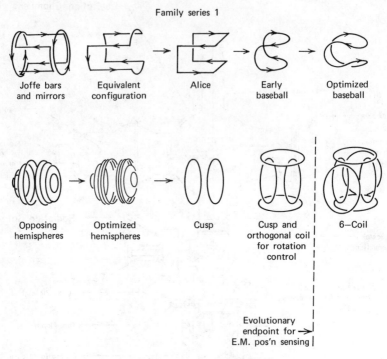

Family series 1

Joffe bars and mirrors — Equivalent configuration — Alice — Early baseball — Optimized baseball

Opposing hemispheres — Optimized hemispheres — Cusp — Cusp and orthogonal coil for rotation control — 6–Coil

Evolutionary endpoint for → E.M. pos'n sensing

Family relationships amongst various coil types

Fig. 6.6. Special levitation coil designs.

Table 6.1 Flame Temperature Predictions

Reactants	Major Products[a]	Flame Temperature[a] (°K)
$C_2H_2 + \frac{5}{2}O_2$	CO_2, H_2O	3300
$2H_2 + O_2$	H_2O	2800
$2Al + \frac{3}{2}O_2$	Al_2O_3	3900
$H_2 + F_2$	HF	4300
$2HCN + O_2 + F_2$	HF, CO, N_2	3950
$C_4N_2 + 2O_2$	CO, N_2	5261[b]
$C_4N_2 + \frac{4}{3}O_3$	CO, N_2	5516[b]
$C_2N_2 + O_2$	CO, N_2	4640

[a]At 1 atm total pressure.
[b]Calculated.

the thermodynamic and kinetic considerations in the production of high temperatures by chemical processes [92, 93]. In Table 6.1 are shown some thermodynamically predicted flame temperatures based on available bond energies at 1 atm total pressure assuming equilibrium. Table 6.2 shows the compositions of the flame gas in some typical energetic reactions at various pressures.

Common choices for producing high temperatures are the hydrocarbon–air or hydrocarbon–oxygen flames, with the hottest being the oxyacetylene flame with a temperature of 3140°C. Grosse and his coworkers have studied several combinations by which high temperatures may be produced, including $H_2 + F_2$, $Al + O_2$, $C_2N_2 + O_2$, $C_4N_2 + O_2$, and $C_4N_2 + O_3$ [94]. In the consideration of metal–gas reactions, one should note whether ignition occurs below or above the melting point. Aluminum, for example, ignites after it melts and yields a burning pool of molten aluminum that floats on top of the aluminum oxide formed during the combustion. Magnesium vapors can be fed to a torch and burned in the same way as in an oxyacetylene torch.

Flames and electrical discharges can produce sustained ionization in a gas, with fairly large concentrations of ions and electrons often above the equilibrium expectations, as shown by mass-spectrometric and microwave attenuation studies. Neutral species such as C_2 and OH can also be produced in excited states with nonequilibrium distributions over available energy levels. The successul production of electrical energy by direct extraction of electrons from cesium- or potassium-seeded flames while the combustion products are extremely hot (2000°K and above), followed by further energy production via more conventional heat exchange to a working fluid such as steam or a liquid metal, is of course the goal of magnetohydrodynamics researchers [95–98]. This technique appears especially promising for the energy-technology oriented to the efficient use of our coal resources, as shown in Fig. 6.7.

Electric Arcs and Discharges

In order to heat a gas to temperatures of 5000° to 8000°K one usually uses the low-intensity, air-cooled electric arc at 1 atm, the theory and operation of which have been discussed by Finkelnburg and others [99–101]. Above 4000°C all solids and liquids have appreciable vapor pressures and molecular species are extensively dissociated. The atoms have started to lose their outer electrons and become singly ionized. At still higher temperatures the ions lose more electrons, and this resulting high-temperature mixture of ions, electrons, and some neutral particles is referred to as a plasma. One can also build high-intensity arcs [102–106] in which extra energy is dissipated by gross electrode sublimation, but the

Table 6.2 Composition of Flame Gases

For the Reaction $C_4N_2(g) + 2O_2(g) \rightarrow N_2(g) + 4CO(g) + 254.6$ kcal

P (atm)	1.0	10.0	40.8 (600 psi)
T (°K)	5261	5573	5748
Mole Fraction			
N	0.02979	0.01988	0.01387
O	0.01675	0.01079	0.00757
O_2	0.000002	0.000006	0.00001
NO	0.00003	0.00008	0.00011
CO_2	0.00004	0.00013	0.00027
C	0.01683	0.01101	0.00796
CO	0.75780	0.77224	0.78011
N_2	0.17875	0.18587	0.19010
Total	0.99999	1.00000	1.00000

For the Reaction $C_4N_2(g) + \frac{4}{3}O_3 \rightarrow N_2(g) + 4CO(g) + 299.9$ kcal

P (atm)	1.0	10.0	40.8 (600 psi)
T (°K)	5516	5936	6100
Mole Fraction			
N	0.05311	0.03915	0.02711
O	0.02947	0.02215	0.01476
O_2	0.00001	0.00001	0.00002
NO	0.00006	0.00015	0.00020
CO_2	0.00004	0.00013	0.00027
C	0.02958	0.02245	0.01527
CO	0.72553	0.74396	0.76171
N_2	0.16220	0.17199	0.18066
Total	1.00000	0.99999	1.00000

For the Reaction of $2HCN + O_2 + F_2$

		T (°K)	3950	4400	4890
		P (atm)	1	10	100
Mole Fraction					
HCN	⎫		0.500	0.500	0.500
O_2	⎬ Reactants		0.250	0.250	0.250
F_2	⎭		0.250	0.250	0.250
HF	⎫		0.234	0.265	0.302
F_2			0.000	0.000	0.000
F			0.120	0.098	0.073
H_2	⎬ Products		0.005	0.008	0.009
H			0.109	0.083	0.055
CO			0.354	0.364	0.375
N_2	⎭		0.177	0.182	1.187

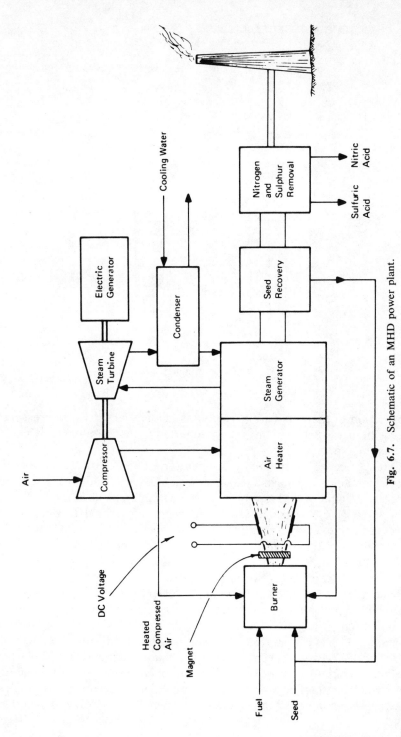

Fig. 6.7. Schematic of an MHD power plant.

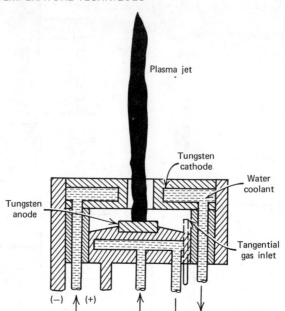

Fig. 6.8. Schematic of a dc plasma arc.

constriction of the conduction volume is ambient air at 1 atm pressure. Applications of such arcs are widely reported in the literature, especially for analytic spectroscopy.

For electrical generation of gas temperatures up to 20,000°K in a variety of gases one uses the plasma jet arc in which the arc region is constricted either by a liquid or a gas to improve the energy transfer efficiency [106, 107] (see Fig. 6.8). Plasma arcs have been constructed to operate at powers of five to several hundred kilowatts and have been used for welding, cutting, crystal growing, steel refining, and chemical syntheses. The limiting factor in arc design and operation is that of nozzle corrosion and erosion by the high-velocity exit gases.

Reed [108] has pioneered a technique for getting a stable plasma in a practically static gas at atmospheric pressure using an inductively coupled plasma torch. The power source was a 10-kW rf heating unit with a frequency of 4 MHz, and the temperatures obtained ranged as high as 14,000° to 19,000°K. A schematic diagram of an induction-coupled plasma torch is shown in Fig. 6.9 [108]. Gas is supplied at one end of the quartz tube, which has an rf coil of a few turns around it; the rf generator couples and transfers energy first into a graphite or metal starter rod and then into

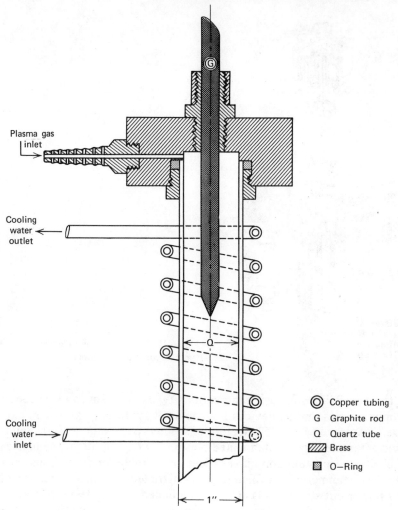

Fig. 6.9.   Schematic of an induction-coupled rf plasma arc.

the plasma. Once the breakdown potential of the gas is lowered sufficiently by the conduction of the gases in the area of the hot rod, the starter rod can then be slowly removed. Stabilization of the plasma is enhanced by vortex stabilization as the gas is fed into the tube tangentially causing it to flow spirally down the walls and form a low-pressure region in the middle of the tube. Argon is usually used to start rf plasmas, but argon mixtures with air, helium, hydrogen, nitrogen, oxygen, etc., and various other gases may be substituted once the plasma is operating.

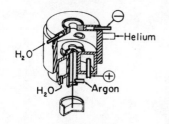

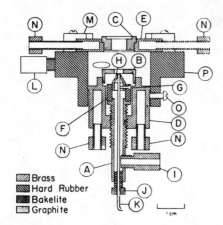

Brass
Hard Rubber
Bakelite
Graphite

Fig. 6.10. Schematic of a plasma source for spectroscopic applications.

The possibilities for using the various types of plasma arcs as spectroscopic sources and as devices for chemical syntheses have been extensively explored [109–115]. Water-constricted plasma generators tend to yield a complex and contaminated plasma; and, in addition, the high rate of electrode consumption makes them undesirable for continuous operation. The commercially available gas-constricted plasma arc units require less power but still give plasmas contaminated by the electrode material. Especially useful spectroscopic sources are the inductively heated plasma, which may be started with Ta, W, or Mo to avoid carbon contamination and the plasma jet. The lower energy and wide range of operating gases have led to many applications as a source of molecular, atomic, and/or ionic spectra as illustrated in Fig. 6.10 [116–118].

Among the important practical applications of plasma techniques are (1) the "plasma etching" methods of the electronics industry which allow selective removal of Si or SiO_2 or Si_3N_4 from silicon wafers which serve as the base of an integrated circuit module [119, 120] (see Fig. 6.11) (typical etchant gases are CF_4 or CF_4/O_2 mixtures which in the rf discharge yields CF_3^+, CF_2^+, COF_2^+, F^+, and F^- atoms as the active species); and (2) the

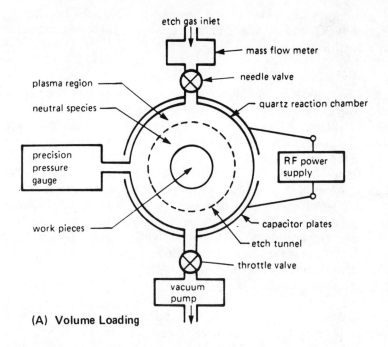

(A) Volume Loading

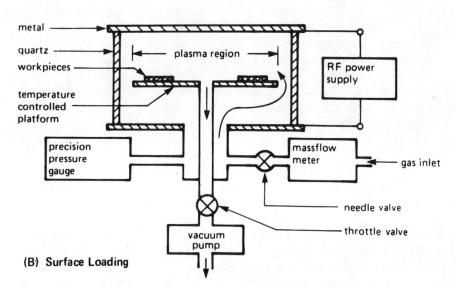

(B) Surface Loading

Fig. 6.11. Schematic of a plasma etching apparatus.

plasma ashing furnaces which utilize an oxygen plasma for burning organic matter to CO_2 and H_2O on a quantitative basis, leaving an inorganic residual ash [121].

Imaging Furnaces and Solar Heating

The development of imaging furnaces as high-temperature devices involves the use of optical systems to collect, focus, and concentrate an available radiant energy flux onto a small area with simultaneous enhancement of flux density. Typical sources of radiant energy presently available are the sun, the crater of a carbon arc, the filament of a tungsten lamp, plasma torches, etc. [122–124]. Maximum temperatures attainable with these systems are of the order of 10^3 to $10^{5°}$K. Some typical optical systems for imaging furnaces are shown in Fig. 6.12. Solar furnaces normally use large parabolic reflectors to concentrate the sun's radiation which is very diffuse into a solar image, while smaller parabolic or spherical mirrors and lenses may be used for the formation of an image of the carbon arc crater or tungsten filament. Other commonly used laboratory imaging sources are the high-current carbon arc or high-pressure Hg or Xe lamps [125–128].

The forerunner of the high-current carbon arc was developed by Beck [129] using a positive carbon electrode, cored with cerium fluoride and oxide, and operating at 100 to 400 A/cm^2 current density. Finkelnburg [130], in later investigations, found that the cerium compound core was not necessary and that other kinds of positive carbons could be employed provided that the anodic current density was high. Optical systems that can be used in a carbon arc image furnace are discussed by Null and Lozier [131]; temperatures of the order of 4000°K are obtained.

The sun is a convenient yet unreliable source for high temperatures. It is obvious from the radiation laws that no temperature higher than the surface temperature of the sun (5800°K) can be achieved with a solar imaging device. The highest temperature actually reached has been 4100°K on a 4-cm^2 area using the CRNS megawatt furnace in France. To achieve the 1-MW capacity of this furnace, Trombe and associates [132–134] designed an installation consisting of a parabolic reflector which focuses the energy from 63 steerable heliostats with a total area of 2835 m^2 (Fig. 6.13). At the solar insolation value of 950 W/m^2 they were able to achieve 1 MW total power with a maximum heat flux of 1600 W/cm^2 at the center. This was accomplished at a 1970 price of $2 per watt, a relatively expensive furnace.

To achieve really high temperatures it is necessary to have a radiation intensity several thousand times that of the sun and the best way this can be achieved is to use a parabolic reflector. This reflector can

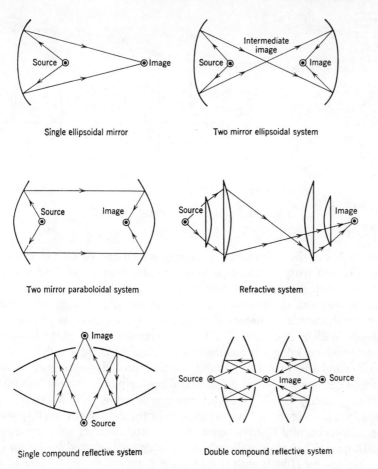

Fig. 6.12.   Optical systems for imaging furnaces.

be a single surface, a surface of multiple mirrors reflecting parallel light onto a large fixed parabola as with Trombe, or an effective parabola of many separate flat mirrors aimed at the same point as with Hildebrandt and others [135]. One is faced with the problem of getting a very small focal point with a very large mirror. With a large mirror like Trombe's there is a vast difference in heat flux at different points; and to achieve a target diameter of 40 cm, the average temperature drops to less than 3000°C using only 75% of the reflected light.

At present the only significant high-temperature uses made of solar devices have been for research and for the production of high-purity fused ceramics. The primary advantage of solar energy for high-temperature

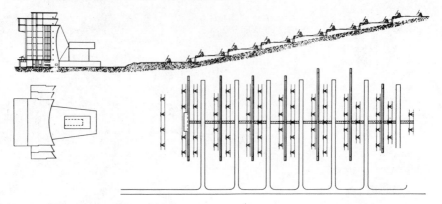

Fig. 6.13.   Solar furnace, Odeillo, France.

research lies in the abundance of energy which can be obtained without contamination from containers, fields, or the heat source. In some experiments, for example, nuclear weapons effects tests and thermodynamic properties, the required absence of electric charges or currents, flame, or shock waves means that only solar sources or large lasers can be used. With a capacity of 150 lb/hr Trombe et al. [136] have been able to produce kilogram castings of purified Al_2O_3 in 1 min.

To expand the research capabilities and provide testing of receiver-absorber designs for future commercial power plants, a new Solar Thermal Test Facility (STTF) was put into operation in New Mexico in 1978 [137, 138] (see Fig. 6.14). An outgrowth of the 5-MW test facility will be the construction of a 10-MW commercial central receiver solar power plant in California. This plant will test the concepts developed by Baum [139] and modified by Hildebrandt [140] for the competitive production of electricity in commercial quantities. Current planning is for the test facility to be operational in 1980, with a demonstration plant planned around 1985 to operate at a capital cost of $1700 per kW. Should the operating characteristics match the expectations, it should be possible to build large power plants powered by the sun at a price competitive with coal and nuclear plants.

Other large test facilities for solar energy research include the "street light" design now in operation at the Georgia Institute of Technology [141] and a large fixed parabola with movable heliostats which bring parallel light rays to the parabola along with a "Venetian blind" adjustable slit system for controlling the amount of sunlight incident on the parabola at White Sands, New Mexico [142]. Of course, there are more than a hundred searchlight mirrors which have been adapted for high-temperature solar research [143].

Fig. 6.14.   Solar thermal test facility, Albuquerque, New Mexico.

The current applications of solar energy to the high temperature field are limited (heating water/dissociating water, etc.); but with recent energy shortages and restrictions in the use of natural gas, it appears that solar energy might be commercially attractive in the production of lime, cement, CaC_2, CO, and H_2 through various high-temperature cycles [144] (Table 6.3). Solar augmentation of gas-fired furnaces appears to be practical for cement kilns, glass tanks, ore-roasting facilities, etc.

Table 6.3 Some High-Temperature Applications of Solar Energy to Chemical Processes

---

Desalination of water
Reduction of iron ore
Fractionation of rare earths
Crystallization of refractory oxides
Pyrolysis of limestone
Water–gas reaction
Preparation of calcium carbide
Manufacture of cement

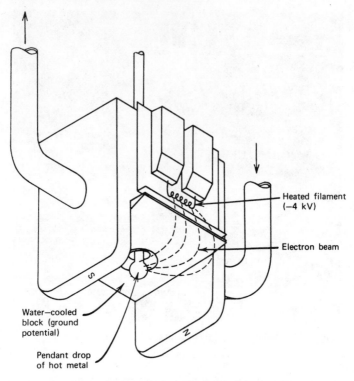

Heated filament
(−4 kV)

Electron beam

Water—cooled
block (ground
potential)

Pendant drop
of hot metal

Fig. 6.15.   Electron bombardment vaporization source.

Electron Bombardment

A conducting sample may be heated by directing a beam of electrons moving at high velocities in high vacuum under the influence of high voltage into the sample [145]. This is essentially resistance heating, and a simple and inexpensive power supply plus voltage control will allow one to regulate temperatures over a wide range. Although usually used for the evaporation of refractory substances, electron bombardment heating has found application in the preparation and refining of metals and refractory materials as a means of augmenting induction heating and in situations where very local heating is desired. Because the electrons may be magnetically focused to concentrate the heat on small areas, the temperatures attainable are limited mainly by the stability of the sample and its container (see Fig. 6.15) [146]. Since the source of the electrons is not in contact with the sample, the introduction of impurities into the sample is minimized, so that this method is especially useful in semiconductor work where extreme purity is required. Several commercial electron beam heating units are now available [147–166].

Shock Waves

Shock waves offer unique advantages for chemical studies since they allow the investigation of reactions at very high temperature in an environment that is kept at a much lower temperature [167–169]. Kinetic studies of complicated reactions at high temperatures are feasible since the initial heating process often does not necessarily make use of a chemical reaction as a heat source (for example, an inert driver gas) and the heating may take place in a very short time compared to the characteristic relaxation times of the system under consideration.

A shock tube for gas reactions basically consists of a tube divided in two sections separated by a diaphragm which can be made of metal, cellophane, mylar, or some other strong thin material. The high-pressure section usually contains a "driver" gas at a pressure of 1 to 10 atm, while the low-pressure section, containing the sample, is operated at a pressure of 0.1 mm to 1.0 cm Hg. When the diaphragm is broken, the resulting shock wave through the gas produces an almost instantaneous temperature jump due to the accompanying compression and heating that follows. Since the compression takes place in fewer than 20 collisions, one has an effective system for studying high-temperature equilibria and/or kinetics as, for example, ionization and dissociation phenomena [170, 171].

The propagation of shock waves through solids also produces heating, and commercially feasible methods for production of diamonds [172–175] and borazon [176] in explosive shocks have been developed as shown in Fig. 6.16.

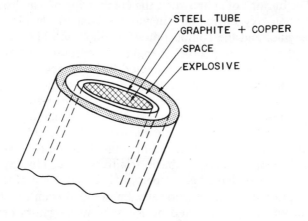

STEEL TUBE
GRAPHITE + COPPER
SPACE
EXPLOSIVE

Fig. 6.16.   Arrangement for shock synthesis of diamonds.

The shock wave through the air following a nuclear explosion causes both mechanical shock and thermal shock while gas temperatures in the range of $10^{3°}$ to $10^{6°}K$ are generated. Underground confinement of nuclear explosives leads to shock waves through the earth and pressures of at least 1,000,000 atm (1 megabar) at temperatures of 1000° to 3000°K [177–179].

Chemical Explosions and Exploding Wires

Some of the important methods for generating high temperatures (and simultaneously high pressures), although for very short periods of time, utilize chemical explosives or exploding wires [180, 181]. Although these methods may seem at first to be ill controlled and likely to yield ambiguous physical conditions, the sophistication of modern physics, chemistry, and engineering is such that one can select explosive materials and appropriately shape and formulate composite charges in order to define rather precisely the regions that will be exposed to high temperatures and/or high pressures. In Table 6.4 are listed some of the currently used high-explosive chemical formulations along with appropriate descriptive information. The fact that most explosive systems involve auto-oxidation–reduction processes within a given molecule means that the times for propagation of explosive waves will be of the order of the times for atomic motions in molecules. Therefore chemical explosive phenomena tend to fall in the microsecond and longer periods of time rather than times shorter than microseconds.

The line is thin between an unconfined explosion and a confined explosion where the confining barrier serves as an orientation system for directing the mechanical and thermal forces generated by the explosion process. The burning of gunpowder, the combustion of a hydrocarbon–air or hydrocarbon–oxygen mixture in the combustion chamber of an automobile engine, and the combustion of alcohol with N_2O_4 or hydrogen peroxide in the chamber of a rocket engine all represent partially confined explosions with the forces being directed through mechanical and other means into channels whereby useful work can be accomplished. The chemical side of high-temperature explosives is concerned with questions of explosive design, that is, the preparation and isolation of metastable molecules which will be ultimate explosive materials. Much effort has gone into this area of science and technology over the past 50 years, yet few breakthroughs beyond the basic trinitrotoluene, nitroglycerin, and the related molecules have been reported [182–184]. One of the ultimate chemical explosive materials currently known is hexanitrobenzene, the synthesis of which was reported by Russian investigators before 1970. Although details of this synthesis and information about the properties

Table 6.4 Modern High Explosives

Notation	Chemical Name	Remarks[a]
RDX	Cyclotrimethylene-trinitramine	S-1; very powerful
HMX	Cyclotetramethylene-tetranitramine	S-1; very powerful
PETN	Pentaerythritol tetranitrate	S-1; very powerful
HNB	Hexanitrobenzene	S-1; new material
NG	Nitroglycerin	S-1; sensitive liquid
TNT	Trinitrotoluene	S-1; common sensitizer
AN	Ammonium nitrate	S-2; low cost
ANFO	Ammonium nitrate plus fuel oil	S-2; low cost
SBA(AN + TNT)	Slurry blasting agents	S-2; low cost
NG + { sodium or ammonium nitrates + combustibles }	Dynamite	S-2; low cost
———	Picric acid and derivatives	S-2
———	Lead azide	P; blasting caps and fuzes
———	Mercury fulminate	P; blasting caps and fuzes
———	Lead styphnate	P; blasting caps and fuzes
———	Diazodinitrophenol	P; blasting caps and fuzes
———	Nitromannite	P; blasting caps and fuzes

[a]P = Primary explosive, extremely sensitive; S-1 = secondary explosive, fairly sensitive; S-2 = secondary explosive, less sensitive than S-1 materials.

and uses for hexanitrobenzene are unavailable, it seems clear that this material is one of the most energetic chemical explosives that can be prepared and isolated in significant quantities. More unstable compounds are known—the fulminates [see Ref. 185, p. 449], the sensitive peroxides (see Ref. 185, p. 457), nitrogen triiodide [see Ref. 185, p. 456], and other nitro compounds—but they are so unstable that handling is a problem and practical usage is not to be expected.

The chemical environment created by an explosive compound is not usually simple nor is it a synthetically attractive environment for preparation of pure materials. $CO_2/H_2O/N_2$/etc., however, form a sufficiently inert environment to allow one the option of diamond synthesis or cubic BN synthesis without problems of excessive contamination in the product (see Refs. 172–176). Explosive cladding, a process in which one metal is essentially welded onto another as a thin sheet, represents a metallurgical application of this high-temperature/high-pressure alternative approach [186]. This is a unique solution to one of the most difficult kinds of forming processes, to attach a thin layer of one metal as a protective film on a substrate metal—platinum on iron, nickel on iron, etc.

In the quest for high-energy content available in a chemically defined system and in short periods of time, one discovers that exploding wires offer many advantages [187]. By discharging a bank of capacitors in a very short time through a small metallic conductor, one can raise the metal to temperatures far beyond the normal sublimation and boiling points in periods of time so short that the atoms still maintain a general geometry reminiscent of the original conducting wire. In other words, one can create a conducting rod of gaseous plasma by melting and vaporizing the metal without the metal atoms deflecting appreciably under the influence of gravity in periods of the order of microseconds or less. Such techniques have been known for a long time, and exploding wires are commonly used as high-temperature initiators for chemical processes and/or for generating gaseous atoms to undergo synthetic interactions with a surrounding atmosphere [188]. Temperatures of the order of 20,000° to 50,000°K have been achieved by exploding wires, and high-precision studies of the energy balance in exploding wire systems have allowed investigators to make estimates of fundamental properties of not only the original conducting solid over a range of temperature but also of the heat of fusion, heat capacity, and other properties of the liquid metal, as well as properties of the gaseous plasma which is ultimately formed [189–193]. Among the interesting investigations of this sort are those of Levedev et al. [194, 195] whose studies of exploding tungsten wires have indicated the heat of fusion of tungsten to be 61.5 kilojoules/mole and allow one to estimate the heat capacity of liquid tungsten as 73 joules/(deg)(mole), in considerable disagreement with levitation studies [490].

Lasers

Probably no device has opened up more avenues of unique high-temperature (excited state) chemistry than the development of a great variety of lasers over the past two decades [196–201]. The ruby laser, providing coherent red light in the form of very-high-energy pulses of

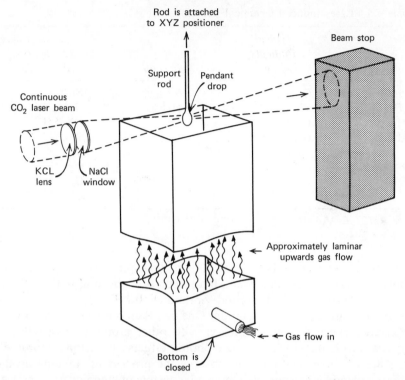

Fig. 6.17. Schematic of apparatus for laser heating of samples in a defined gaseous atmosphere.

short duration [202, 203], was followed by the YAG (yttrium–aluminum garnet) laser [204], which provided high-energy pulses at 1.06 microns and then the carbon dioxide gas laser [205–208], providing high energy at the 10.6-micron wavelength, and these have been followed by several generations of new lasing systems and/or new concepts for varying the optical frequencies, including the almost continuously variable wavelength output from modern dye lasers [209–211]. One can now find a laser system that operates in almost any desired region of the optical spectrum from the UV up to the far infrared and, of course, even in the microwave region where one has the maser devices [212] which were the forerunners of lasers. Surface temperatures of 3000° to 4000°K are easily obtained with lasers; and in terms of Planck's radiation law and energy fluxes, one has radiant energy/(cm^2)(sec) at levels corresponding to $10^{9°}$ to $10^{12°}$K, for very short periods of time (see Fig. 6.17).

With this option of optically introducing energy into atomic or molecular systems by taking advantage of absorptions in very specific

Table 6.5   Laser-Induced Chemical Reactions[a]

Molecule	Typical Radicals Produced	Molecule	Typical Radicals Produced
C_2F_3Cl	C_2*	SiF_4	$SiF*$
C_2H_4	$C_2,* C_2$	HDCO	HD, CO
CH_3CN	$CN,* C_2*$	NH_3	NH_2, H
CH_3NO_2	$CN,* CH,* C_2*$	SF_6	$SF_5,* SF_4,$
	NH, O_2*		SF_3
CF_2Cl_2	$CCl*$		
BCl_3	$BCl*$	$C_2H_2Cl_2$	$C_2,* CH,* C_2,$
			CH

[a]See Ref. 196, Vol. III.

wavelength regions, it is possible to create essentially unique sets of "hot" atoms or molecules which have been brought to a first electronic excited state or to a first excited vibrational state or to a first excited rotational state by laser irradiation. One can then observe the relaxation of excited species as they interact with other atoms or molecules in physical or chemical ways. Single-photon effects or multiple-photon effects are possible, and one can observe the process of stepping up the vibrational ladder in a complex molecular system or stepping up the series of electronic energy levels toward the ionization potential in an atomic species. The various approaches to isotopic separation that have been recently announced in the flood of publications on laser-induced chemistry [213–225] are representative of the unique opportunities that have become available in this field, as summarized in Table 6.5.

Besides using laser irradiation for the selective excitation of atoms and molecules, methods have also been developed for using laser radiation to heat one gas that can then serve as a heat transfer medium for heating a second material that does not have absorptions in the particular range of the available lasers. In this way one can construct "laser furnaces" or "laser reactors" in which traditional chemistry can be performed based on ordinary thermodynamic equilibrium types of arguments and classical thermodynamic concepts of temperature. For instance, one can use a CO_2 laser to excite a gas-filled reactor that contains SF_6, SiF_4, CF_4, or some other stable compound that has strong absorptions which fall in the infrared region close enough to the CO_2 laser radiation output so that effective heating of these gases is possible. One can then use the relaxation of excited SiF_4, CF_4, or SF_6 to heat argon or some other gas and then

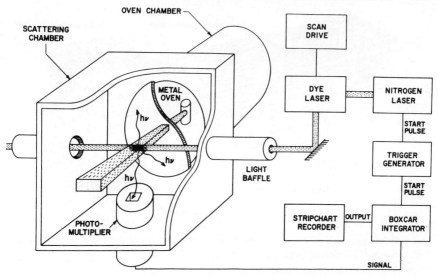

Fig. 6.18. Schematic of apparatus used to study metal–gas reaction products via laser-induced fluorescence.

ultimately transfer this energy to a chemically reactive system in the range of 1000° to 2000°K range [226–229].

A combination of molecular beam techniques, using atoms or molecules from high-temperature sources, followed by laser excitation and fluorescence of product species has allowed considerable expansion of our spectroscopic knowledge of high-temperature molecules [230], as illustrated in Fig. 6.18.

The ultimate applications of lasers to high-temperature generation is the initiation of thermonuclear fusion, and the SHIVA–NOVA experiments currently in progress [231] at the Lawrence Livermore Laboratory are illustrated in Fig. 6.19. It is believed that temperatures in excess of $10^{8\circ}$K can be attained for at least 100 microseconds by this approach.

Nuclear Fission and Fusion

The ultimate processes for generating high temperatures and unique interactions of particles are, of course, nuclear fission and fusion. In these processes one not only rearranges electrons but also converts nuclei either by splitting (fission) or joining (fusion). In these interchanges of nucleons one has evolution of energy on a scale at least 1000 times greater than the tradition exothermic chemical processes. Thus high temperatures in nuclear systems mean $10^{6\circ}$ to $10^{10\circ}$K, corresponding to particle energies of 10^3 to 10^7 eV. It is certainly the case that "chemistry" in any

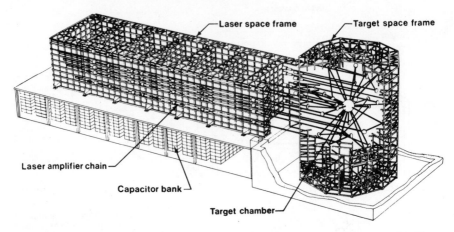

Fig. 6.19.   Schematic of Shiva-Nova high-energy laser facility at the University of California, Lawrence Livermore Laboratory, Livermore, California.

traditional sense is not important at these extreme temperatures. They are truly astronomical since it is through fission/fusion processes that the energies of stars and the sun are produced.

It is a challenge to scientists and engineers when one asks for controlled nuclear fission or controlled nuclear fusion processes. An explosion can be accomplished with the proper reactants when the reactions are sufficiently exothermic and spontaneous by standards of chemical or nuclear thermodynamics. The problem of containing a reaction which can produce particle temperatures of $10^{8\circ}$K has been solved in a variety of ways, ranging from the nuclear reactors developed by Fermi and co-workers in the early days of nuclear fission [467–470] to the modern techniques of proposed fusion reactions [471, 472, 473]. One approach makes use of scattering or moderating media such as graphite, paraffin, or light atoms to interact with the high-energy products of the nuclear reaction. Also there are particular isotopes of cadmium or boron that can absorb the neutrons formed in the reaction to form stable heavy nuclei. Through proper design, it is possible to carry out a controlled fission process in a modern nuclear reactor as shown in Fig. 6.20 [474]. The nuclear reaction temperature may range as high as $10^{8\circ}$K, but with suitable confinement, neutron moderating, and/or absorption one can reduce the temperature to only a few hundred degrees K at the extreme edge of the nuclear pile. Thus normal materials of construction such as graphite and various refractory metals serve satisfactorily for nuclear reactors in spite of the very high temperatures in the core of the system [475]. At present the major problem in the production of power by fission concerns the handling of nuclear wastes, that is, the spent fuel elements and the fission products. Storage in metal tanks, solidification in blocks of concrete, dumping in deep caves or wells, or dispatching waste containers into space for eventual collision with the sun are among the suggestions for disposal of nuclear waste [476–478].

The fusion reaction, which when uncontrolled yields the hydrogen bomb, is attractive when controlled as an energy production system not only because of the high-energy yield per pound of fuel of material but also because the starting materials are isotopes of hydrogen or perhaps lithium while the product materials are stable isotopes of helium [479]. Thus very available raw materials—deuterium from the ocean or lithium from naturally occurring ores—can be reacted in one of several fusion processes (as shown in Table 6.6) to produce large amounts of energy and safe stable products.

Unfortunately the ignition temperatures for getting fusion reactions to occur are even higher than the typical fission reactions, and one has to be concerned not only about generating these very high temperatures but also about confining the fusion materials long enough for more energy to be produced than was required to get the system into the proper configuration for reaction. This problem is most usefully thought of in terms of a parameter that involves both the time at a high temperature (τ) and the number density (n) reacting at high temperature so that the product $n\tau$ is a maximum (see Fig. 6.21). It is apparent that to get particle temperatures in the range of 10^{8}°K one has only a few alternatives. For example, one may accelerate charged particles such as D^+ across potential differences of 10^6 volts and thus create particle temperatures in the range of 10^{10}°K [480–482]. In order to confine the reacting plasma—a mixture of ions and electrons—for long enough to have an appreciable opportunity of producing useful energy, one needs a magnetic confinement system, often called the magnetic mirror, which uses superconducting magnets. In this system one can fairly effectively deflect particles as they try to escape from the magnetic mirror except in the direction along the mirror axis. Actually leakage along the mirror axis can provide the method by which particles, that is, electric current, is extracted and energy produced [483, 484].

Another technique for producing particles at high enough temperatures to induce fusion utilizes laser irradiation methods [485, 486]. Since laser pulses have equivalent temperatures of 10^{10}° to 10^{12}°K, the major problem is simply to find a laser that operates at an appropriate wavelength so that the energy can be transferred to the deuterium or to the tritium systems and then to provide sufficient magnetic containment so that fusion can

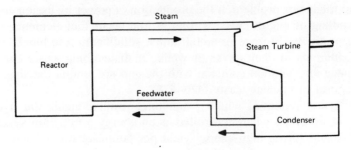

(a) Boiling Water Reactor (BWR)

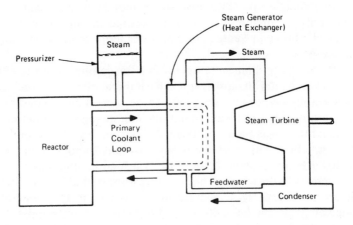

(b) Pressurized Water Reactor (PWR)

Fig. 6.20. (a) Boiling water reactor; (b) pressurized water reactor; (c) core assembly for a boiling water reactor.

occur. Some of the current ideas about construction of a fusion reactor are shown in Fig. 6.22. No successful energy-producing fusion reactor has yet been operated, but the principles appear to be clearly developed and through the combination of high-energy ion bombardment or laser excitation with magnetic confinement and a system of heat-absorbing and heat-transferring gases (the lithium blanket and metal heat exchangers shown in the schematic) [479, 487, 488] scientists and engineers hope to be able to operate successful fusion reactors within the next 10 to 20 years. This will be a major achievement, and yet the almost infinite source of energy which fusion can provide makes the dedicated efforts and cost worthwhile.

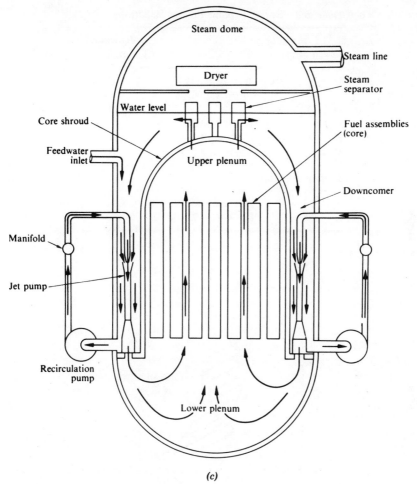

(c)

Fig. 6.20. *continued*

Attaining High Temperatures in High-Pressure Systems

The problem of simultaneous achievement of high temperatures and high pressures creates a difficult experimental situation since most materials become softer and therefore less satisfactory as anvils or retainers in high-pressure devices. The long-time interest in synthesis of diamonds called the attention of many experimentalists to the problem of attaining high temperatures—1000° to 2000°K—while at the same time confining a system at pressures of 10,000 to 100,000 atm. Professor P. W. Bridgman succeeded in obtaining very high pressures at or near room temperature

Table 6.6 Some Typical Fusion Reactions

Fuel Cycles	Ignition Temperatures
D + T → ⁴He + n + 17.6 MEV	5×10^7 K
D + D ⟨ ³He + n + 3.2 MEV / T + H + 4.0 MEV	3×10^8 K
D + ³He → ⁴He + H + 18.3 MEV	5×10^8 K
H + ⁶Li → ³He + ⁴He + 4.0 MEV	9×10^8 K

Tritium Breeding

$n + {}^6Li \rightarrow T + {}^4He + 4.8 \text{ MEV}$

$n + {}^7Li \rightarrow T + {}^4He + n - 2.5 \text{ MEV}$

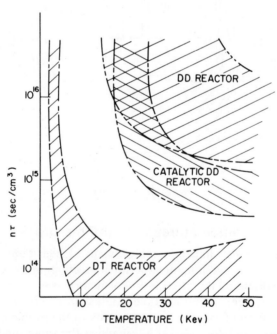

Fig. 6.21. Lawson criterion relating plasma density (h) and burn time (τ) to the operating temperature (T) of a thermonuclear reactor.

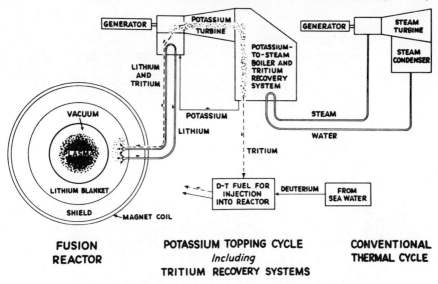

Fig. 6.22. Suggested design for a nuclear fusion power plant.

[232], and various experimenters were able to extend his methods to a few hundred degrees centigrade. However the major breakthrough in the high-pressure/high-temperature regime was made by scientists at the General Electric laboratory, especially H. T. Hall [233–236], who personally designed and first operated the high-pressure/high-temperature device in which synthetic diamonds were prepared. The Hall device, called the "belt," made use of simple resistive heating as current was passed through a thin-walled metal tube in the center of which the sample was heated and compressed (see Fig. 6.23). Various other heated sample techniques for the range 1000° to 2000°K and 50,000 to 10,000 atm are now in use [237–240].

If one combines an explosive with some degree of confinement, then it is also possible to generate high temperatures at high pressures. Apparatus for the shock synthesis of diamonds [172–175] and of borazon is illustrated in Fig. 6.16.

3 MEASUREMENT OF HIGH TEMPERATURES

Temperature-measuring techniques have necessarily become more sophisticated as new temperature-generating techniques have been developed. Ideally one seeks to define an absolute temperature scale based on thermodynamic and statistical mechanics and then to utilize this absolute scale to establish a convenient laboratory reference scale with which

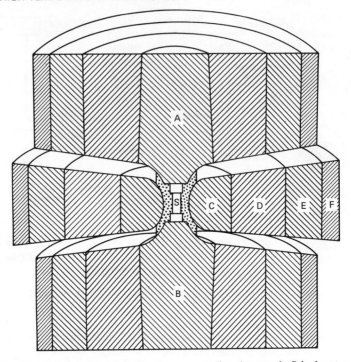

Fig. 6.23.   The Hall belt. The dotted region surrounding the sample S is the pyrophyllite gasket; A and B are tapered anvils of cobalt-bonded tungsten carbide; C, D, E, and F are shrunk on steel rings to provide support.

practicing scientists can compare and calibrate their laboratory or plant thermometers. Since the range of temperatures to be measured can extend from the temperatures of boiling He, O_2, H_2O, or S to the temperatures of the gaseous ions generated in a nuclear fireball, it is not surprising that no single measuring device or method is universally applicable. Rather, one can develop thermodynamic reference scales based on ideal gas properties and on the Planck radiation law. From these through calibration and interpolation one can develop "practical" temperature scales and from these, a "laboratory" temperature scale for day-to-day use in performing experiments [241]. Laboratories such as the United States Bureau of Standards, the British Physical Laboratory, the French Bureau of Weights and Measurements, and their German, Japanese, and Russian analogs have provided a continuing international interest and concern with consistent, precise, and practical temperature scales for use in industry and in research. Through cooperative efforts expert international committees have devised temperature scales with specified limits on precision and accuracy for use by practicing scientists and engineers.

Table 6.7 International Temperature Scales[a]

Defining Fixed Points	IPTS-68[b] °C	IPTS-68[b] °K	EPT-76[c] °K
sp Cadmium			0.519
sp Zinc			0.851
sp Aluminum			1.179_6
sp Indium			3.414_5
bp Helium (1std atm)			4.222_1
sp Lead			7.199_9
tp Equilibrium Hydrogen	−259.34	13.81	13.804_4
bp Equilibrium Hydrogen (25/76 std atm)	−256.108	17.042	17.037_3
bp Equilibrium Hydrogen	−252.87	20.28	20.273_5
tp Neon (2.7 nmole of ^{21}Ne and 92 nmole of ^{22}Ne per 0.905 mole of ^{20}Ne)			24.559_1
bp Neon	−246.048	27.102	27.102
tp Oxygen	−218.789	54.361	
bp Oxygen	−182.962	90.188	
tp Water	+0.01	273.16	
bp Water	100	373.15	
fp Zinc	419.58	692.73	
fp Silver	961.93	1235.08	
fp Gold	1064.43	1337.58	
Some Secondary Fixed Points			
fp Tin	231.9681	505.1181	
fp Lead	372.502	600.652	
bp Sulfur	444.674	717.824	
fp Antimony	630.76	903.89	
fp Aluminum	660.37	933.52	

[a]tp = triple point; bp = boiling point; fp = freezing point; sp = superconducting transition point.
[b]International Practical Temperature Scale.
[c]Proposed changes to IPTS-68[b], ref. 242.

One of the convenient international temperature scales is presented in Table 6.7.

The concept of temperature that is used to characterize systems in thermal equilibrium and its representation in terms of the temperature scales have been the object of study in many laboratories, the subject for symposia and reviews, and the special concern of several international committees of physicists and chemists [243–247]. Two temperature scales are presently in use—the Absolute Thermodynamic Temperature scale and the International Practical Temperature scale.

The Absolute Thermodynamic Temperature scale is based on the triple

Table 6.8   Comparison of Conventional Units With SI Units

Quantity	Typical Value	
	Conventional	*SI*
Atomic size	1 Å	10^{-10} m; 0.1 nm; 100 pm
Liquid density	1 g/cm^3	10^3 kg/m^3; 1 kg/dm^3
Pressure	1 atm	1.01325×10^5 Pa
	1 mm Hg (torr)	132.7 Pa
Concentration	1 mole/liter	10^3 moles/m^3; 1 mole/dm^3
Surface tension	20 erg/cm^2	0.02 J/m^2
Dipole moment	1 debye	3×10^{-30} Cm
Polarizability	1 A^3	9×10^{-42} A^{2}s^{4}kg^{-1}
Energy	1 kcal/mole	4.184 kJ/mole

point of water, which was assigned the temperature 0.01°C or 273.16°K (Kelvin) by international agreement in 1954. The original basis was the melting point of ice in equilibrium with water saturated with air (0°C). Thermodynamic temperatures are usually determined by gas thermometry below the gold point (1063°C) and have actually been measured up to ~2000°C. Normally, at T > 1063°C, the optical pyrometer reading from a blackbody is used with Planck's radiation equation to define temperatures.

An International Practical Temperature Scale (IPTS) was first adopted in 1927, and it is now revised. It was devised to define conveniently and precisely any attainable temperature, and it is close to the thermodynamic temperature scale. The IPTS-1968 is based on six fixed points: (a) oxygen point—liquid oxygen in equilibrium with its vapor at 1 atm = −182.98°C; (b) triple point of water—equilibrium among ice, liquid water, and water vapor = +0.01°C; (c) steam point—equilibrium between liquid water and its vapor at 1 atm = 100°C; (d) sulfur point—equilibrium between liquid sulfur and its vapor at 1 atm = 444.66°C; (e) silver point—equilibrium between solid silver and liquid silver at 1 atm = 961.55°C; and (f) gold point—equilibrium between solid gold and liquid gold at 1 atm = 1064.24°C. All of these defined reference points on the various temperature scales are subject to revision as one converts to the new SI unit system. For example, at 1 kilopascal, the boiling point of water is approximately 97.5°C.

It will apparently require the time and efforts of many workers in the high-temperature/high-pressure fields to develop consistent and convenient, practical reference points in the framework of the new unit system [248–252] (see Table 6.8). The gas thermometer has provided an important bridge between thermodynamic and practical temperature scales and is still used to transfer the low-temperature calibration points to the radiation-based temperature calibrations. Various laboratory radiation sources are available for calibration of optical pyrometers; and at temperatures above 10,000°K, plasma arc devices will be useful in providing laboratory temperature standards [253–254].

Thermocouples

In most laboratories the thermocouple, properly used, is the most versatile and rugged temperature-sensing element. Its unique value lies in the fact that it indicates directly the true temperature of its junction. Materials have been available for some time that are stable at temperatures up to 3000°C in neutral atmospheres or under vacuum and to 2200°C in oxidizing atmospheres [255–258]. Wires are available in various diameters to satisfy either the most delicate or the most rugged temperature-measuring requirements, either bare or in metal sheaths with swaged ceramic insulation. Of particular interest to the high-temperature technologist is the commercial availability [259] of refractory thermocouple wires: tungsten/rhenium, tungsten/tungsten–rhenium 26%, and iridium/iridium–rhodium 40%. The platinum–rhodium 10% thermocouple is the most reliable thermocouple presently in use, with a reproducibility of ±0.1°C up to about 1100°C for short periods of time. By having both legs of the thermocouple made of platinum–rhodium alloys, routine measurements are possible up to 1800°C. For example, a platinum–rhodium 20%/platinum–rhodium 40% thermocouple in an alumina protection sheath has been used at 1800°C with a maximum change of 10°C from the calibration. Sometimes a protective metal cladding such as Ta is used.

Among the thermocouples made up of pairs of refractory metals, the tungsten/iridium combination is reported to have an accuracy better than ±6°C at 2100°C, and iridium/rhodium–iridium 40% thermocouples have been used up to 2100°C. In both cases an inert atmosphere is the most satisfactory for long life; the use of a protective tube also helps to prolong their usefulness in air. Molybdenum/rhenium and tungsten/rhenium thermocouples that are useful up to about 1900° and 2700°C, respectively, have been described. Some typical emf data for thermocouple pairs are shown in Fig. 6.24.

Nonmetallic thermocouples such as silicon carbide/graphite, graphite/doped graphite, silicon carbide/doped silicon carbide, and graphite/tungsten have found some use in the steel industry. Calibrations are necessary for each one, but changes have been found to be small up to 100 hr. The tungsten/graphite thermocouple is good up to 2400°C in a reducing atmosphere [260]. Shepard and co-workers have described calibration of a graphite/boronated graphite thermocouple up to 3100°C [261]. Other authors have discussed thermoelements of metal carbides, metal borides,

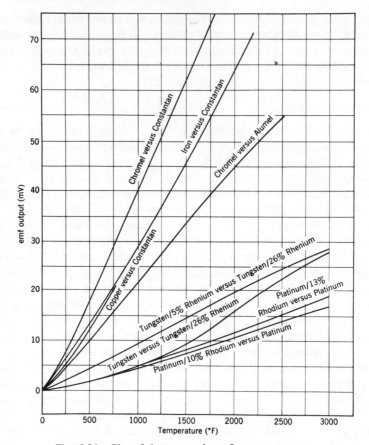

Fig. 6.24. Plot of thermocouple emf vs. temperature.

and various oxides [262]. These latter materials should be of special interest in oxidizing atmospheres.

There are several comprehensive reviews of thermocouple materials which present emf–temperature tables for the refractory metal couples, extensive tables of physical properties, use and performance data, a discussion of sheathed thermocouples, and extensive bibliographies [263–265].

Resistance Thermometry

One can determine an unknown temperature by comparing the resistance of a calibrated wire at the unknown temperature to its resistance at the ice point (usually done with a platinum resistance thermometer)

[266–271]. This is the most accurate of all currently available thermometric techniques. One can specify absolute temperatures to at least $\pm0.001°K$ and changes of temperature to $10^{-5}°K$, or better, if suitable calibrations are available. Although most platinum resistance thermometry is restricted to the temperature range below 900°K, there have been attempts to extend the useful range of this method. For example, the use of a gas-tight platinum protection sheath helps to maintain the stability of the normal-type platinum resistance thermometer. Other metals may be used in resistance thermometers (Cu, Ni, W, etc.), but Pt is especially appropriate because of its chemical intertness and the availability of wires, sheets, etc., of high purity.

In addition to the metallic conductors, which show an approximately linear resistance increase with temperature, one may use semiconductors with their negative temperature coefficients of resistance as resistance thermometers. Traditional thermistors find their main applications at temperatures up to 300°C, but there is no basic reason why higher-temperature applications are not possible [272]. For example, UO_2 and ZrO_2 become good conductors at high temperatures although they are essentially insulators at room temperature and there have been efforts to use them as the working elements of a resistance thermometer [273].

Various constructions using protective tubes of platinum or dense alumina and platinum wires supported on sapphire discs or recrystallized alumina tubes to support four coils of platinum wire inside an alumina tube have been used for temperature measurements up to 1063°C. A resistance thermometer with the platinum wire fused into a ceramic matrix has been used up to 1600°C [274], and a tungsten resistance thermometer enclosed in a quartz capillary has been recommended for temperatures to 1000°C [275]. Several companies have available a variety of nonstandard and custom-engineered resistance probes for aerodynamic, surface, high-pressure, etc., temperature measurement [276].

Optical Pyrometry

It is convenient to monitor the temperature of a hot object by optical radiation methods because of the continuous spectral radiation emitted as explained by the Planck radiation theory for blackbodies [277–279] (see Fig 6.25). Brightness pyrometers yield the apparent temperature of the surface of a hot body by comparison with a calibrated pyrometer lamp. The conventional, monochromatic optical pyrometer utilizes the spectral radiation around 6500 Å as defined by a red filter and the observing eye or photo cell [280–283]. For temperatures above the gold point (1063°C), the optical pyrometer is used to indicate temperatures on the International Practical Temperature Scale. The calibration of optical pyrometers is car-

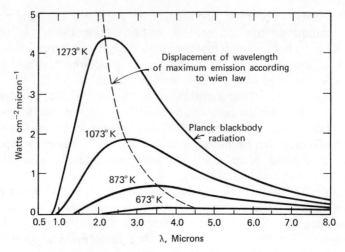

Fig. 6.25. Plot of radiation emitted by a blackbody as a function of wavelength and temperature.

ried out by the U.S. National Bureau of Standards and its counterpart in other countries, usually by intercomparison with tungsten strip lamps, which are convenient, reproducible sources of radiant energy. Accurate calibrations in the range from 800° to 1500°C ($\pm 1°$), from 1500° to 2000°C, ($\pm 2°$), and at 2800°C ($\pm 6°$) are routine. In all surface temperature measurements a correction for the nonideal surfaces is required, that is, an emissivity correction is needed.

In a typical optical pyrometer, one either varies the temperature of the reference tungsten filament until it matches the brightness of the experimental hot body or one attenuates the radiation from the hot body to get a match, or both. In all cases, the optical pyrometer is a blackbody steradiance meter with the scale calibrated relative to the steradiance of the gold point blackbody. The success of the optical pyrometer is due to the fact that steradiance ratios are more easily and more accurately determined than are absolute intensities. Moreover brightness can be matched with good precision by eye or with photosensitive devices.

The brightness temperature of a tungsten reference filament is usually limited to a maximum of 1350°C in order to preserve its calibration. To measure higher brightness temperatures, the light from a high-temperature source is attenuated to bring it to within the brightness range of a tungsten filament operated below 1350°C by inserting neutral absorbing screens or rotating sectors into the path of the target image as needed [284, 285].

When an optical pyrometer is sighted upon a nonblackbody with a spectral emissivity less than unity, the net effect is a less intense object brightness image. Molten steel, for example, generally has an emissivity of about 0.4 at 0.65 micron, and many oxides have emissivities around 0.2 to 0.3. The true temperatures for materials with known emissivities at 0.65 micron are conveniently determined from the extensive tabulation of corrections by Poland, Green, and Margrave [286].

One can usually approximate a blackbody in an object whose true temperature is desired. A 10° V-groove in a solid, a V-bend in foil, a small hole in a spherical cavity, a long cylindrical tube with a small hole at the center, or a hole with a length-to-diameter ratio greater than about 5 drilled into a solid are all satisfactory blackbodies. Thus if any of these small probes can be put into the high-temperature nonblackbody, emissivity can be determined as a function of temperature, or the true temperature may be read directly by sighting into the blackbody cavity [287]. If blackbody conditions are achieved experimentally, no details of the interior are observable since each element of the surface being sighted upon is radiating with equal intensity. Little is currently known about the emissivities of liquid metals except at the melting points [288, 289].

High-precision photoelectric pyrometers have been designed at the National Bureau of Standards [280–283, 290]. Basically, the experimental arrangement is similar to that used in other optical pyrometers. In one design, three interchangeable pyrometer lamps are included—one working standard and two reference standards. A combination of interference filters and the usual pyrometer red glass are used to isolate a 0.6530-micron band with a half-width of 0.0110 micron. The phototube alternately sees the source and the tungsten filament at 30-sec intervals and the output is read out on a recorder. Equal output of the source and the pyrometer lamp is taken as a radiance match. Precision at 1063°C is said to be better than 0.02°C. A schematic is presented in Fig. 6.26.

Other photoelectric pyrometers have been designed for operation at 0.4500 micron [291–294], and another recording pyrometer uses a photosensitive transistor with a response time of the order of 0.02 sec [295].

Since Planck's law requires the intensity ratio of blackbody radiation at two wavelengths to be a unique function of temperature, it can accordingly be used to measure blackbody temperature [277–279]. Among the advantages claimed for two-color pyrometry are (1) no thermal contact; (2) reasonable variations in intensity due to smoke or absorptance in the source optical path do not change the indicated temperatures; (3) limited variations in target area do not change the indicated temperature; and (4) short-term radiance changes of the source tend to cancel and make the temperature indication relatively independent of surface effects or surface

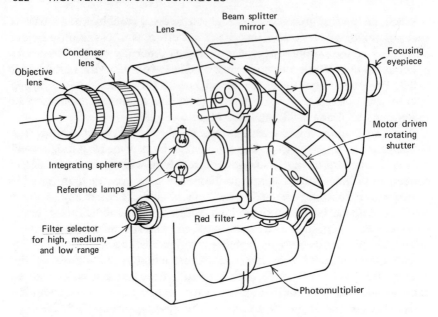

Fig. 6.26. Automatic brightness pyrometer for recording and controlling temperatures in the 700° to 3500°C range.

condition. Several commercially available two-color pyrometers embody the principles outlined above [296–305].

Hornbeck has discussed the problems associated with the design of a three-color pyrometer with the assumption that emissivity varies linearly with wavelength, but the work was carried only sufficiently far to prove the feasibility of the idea [306].

Currently available infrared detectors (PbS, PbSe, PbTe, InSb, GaAs, etc.) make it possible, in principle, to use Planck's law to determine the gold point on the thermodynamic scale by using an infrared mono-chromatic pyrometer calibrated at the sulfur or zinc points with a gas thermometer and thus to interrelate the temperature scales. Several commercial units are available [307–310].

An infrared radiometer suitable for temperature measurements from ambient to 3600°C has been described by Astheimer and Wormser [311]. This instrument used a thermistor detector, sensitive to about 50 microns, in conjunction with a glass filter which shut out light below 1.8 micron. Response time is reported to be 10 msec, sensitivity 0.1°C, and the absolute accuracy of the instrument calibration ±2.0°C.

Several portable instruments are available commercially for the temperature range of 0 to 850°C. Another commercial infrared pyrometer [312] uses a thermistor bolometer and has a spectral bandpass of 1 to 12 microns with optics for targets as small as 1 mm^2. An operating range from 15 to 5500°C above ambient, 0.25 sec response, and 1% accuracy are claimed.

Infrared devices which use InSb or GaAs detectors are also commercially available [313]. These instruments have the fastest responses of any known detectors, being less than 1 μsec, and can be used for recording, indicating, and controlling fast temperature changes near 25°C; their range can also be extended to 3500°C.

Commercially available optical pyrometers [314] are well suited to electronic instrumentation and are being utilized to solve temperature control problems to an increasing extent. Units have been designed for sighting between the turns of an induction heating coil, for use with an $\frac{1}{8}$-in.-diameter sapphire rod lightpipe for small openings, and other specialized applications.

Tingwaldt [315] has described a novel method for determining the thermodynamic temperature of a heated metal under nonblackbody conditions. When the metal surface is illuminated with radiation from a blackbody source and viewed at an oblique angle with a standard disappearing filament optical pyrometer, the reflection of the blackbody image from the surface of the metal is unpolarized only when the blackbody and the metal have equal temperatures. Thus if a polarizer is inserted in the beam entering the pyrometer, the metal and the blackbody are at the same temperature when the rotation of the polarizer produces no change in the brightness match.

Spectroscopic Methods

The temperatures of gases at ultrahigh temperatures are normally measured by spectroscopic methods. The underlying principles and techniques are given in detail by Broida [316], by Gaydon and Wolfhard [317, 318], and others [319] for temperatures below 4000°C. Experimental techniques for temperatures above 4000°C are discussed in the works of Lochte–Holtgreven, Finkelnburg and Maecker, and others [320–322].

Although it is much easier to measure the thermodynamic temperature of solids and liquids at their surfaces than it is to measure the thermodynamic temperature of gases, it is paradoxical that the theoretical understanding of the properties of gases as a function of temperature is more complete than of solids and liquids. There are accordingly many more methods to determine gas temperatures. It is customary to refer to several kinds of temperature for gases, for example, total temperature, static temperature, vibrational temperature, translational temperature, rotational temperature, electron temperature, stagnation temperature, etc. This is merely a manifestation of the dependence of gas temperature on environmental factors and the various degrees of freedom which one uses to describe the properties of gases. Either one must indicate all of the various gas temperatures or else bring the gas to thermodynamic equilibrium and measure temperature in the conventional sense.

In a flame front, the temperature gradient in a gas may be of the order of $10^{5\circ}$K/mm, and the interesting phenomena may occur in the order of 10^4 collisions; or the cooling or heating rate of the gas may be as high as $10^{6\circ}$K/sec. Under such conditions, the thermodynamic properties of the gas may lag the heat transfer process; hence there is a need to know the temperatures characteristic of particular energy modes in order to describe correctly the changes occurring in the gaseous system.

Gas temperature measurements are either accomplished by radiation methods [8, 323] or by the insertion of a probe of some sort to determine a local property of the gas which is temperature dependent and from which the temperature may be derived either by calibration or by theory [324]. Radiation methods require a large target area which is viewed through a layer of semitransparent gas whose properties must also be considered when determining temperature. Probes may be used to determine the gas steam temperature point by point; but heat exchange occurs between the probe and the gas and between the probe and the gas duct and must either be eliminated or corrected for in order to obtain accurate temperatures.

Gas temperatures may be measured with a thermocouple up to the temperature limit of the thermocouple by designing an environmental shield which permits the junction to indicate the true temperature of the gas. The design of the environmental shield is determined principally by the size of the probe required for a particular measurement as well as the velocity, heat capacity, molecular weight, pressure, and temperature of the gas and the wall temperature of the confining duct [325–327].

The rate of flow of a gas through a restriction is a function of the molecular weight, orifice area, total temperature, total pressure, and the static pressure. The maximum discharge rate of a gas through an orifice, venturi, or nozzle will increase as the pressure drop across the orifice increases until the throat velocity equals the velocity of sound. A further increase of pressure drop or decrease in discharge pressure will not change the discharge rate. Thus temperature may be determined by measuring upstream pressure provided that an equation of state, molecular weight, and the ratio of C_p/C_v are known as a function of temperature. Such probes require a rather large amount of gas but may be operated at either sonic or subsonic velocity. A comprehensive account of the theory and historical development of this probe with an extensive bibliography is available [328, 329].

Emission and absorption spectra of gases permit one simultaneously to measure temperature and to identify the constituents which comprise the gas mixture [330, 331]. These techniques have provided almost all of our knowledge about electric arcs and flames. The special advantage of the spectrographic method is that no perturbation is introduced into the system being studied, that is, the system serves as its own thermometer.

Deriving accurate temperatures from spectroscopic observations requires considerable insight and experience because of the complications brought about by the presence of several species, namely, molecules, atoms, positive and negative ions, and electrons, which may be created by various physical processes to yield different distributions of energy over the various degrees of freedom (i.e., translation, vibration, rotation, and electronic) which may not be in equilibrium with each other and which may not be Boltzmann in character. The usual approach is to assume equilibrium as predicted by the Boltzmann equation and to calculate separate electronic, vibrational, or rotational temperatures in chemical flames [332] (see Fig. 6.27) or to assume that the Saha equation holds and utilizes the modified Holtsmark theory of Stark-effect broadening of spectral lines to derive temperatures in partially ionized gases (plasmas) [333–335]. Deviations from ideality and coulombic effects may also be considered quantitatively by a Debye–Hückel type of treatment [336].

The sodium line reversal method has been used more extensively than any other optical method of gas temperature measurement, and techniques to achieve time resolutions of the order of a few tenths of a microsecond have been described [337, 338]. In addition to measuring temperature, Bundy and Strong [339] also estimated pressure and gas velocity from the line broadening and Doppler shift, respectively, of the Na D-lines. The precision of the line reversal method is estimated to be about 5°K at 1500°K and approximately 10°K at 2300°K. Photoelectric self-balancing pyrometer line reversal temperatures are conveniently obtained with a recording line reversal pyrometer (see Fig. 6.28). The characteristic lines of Li, K, Rb, and Tl [339] as well as the 2.7-μ band of H_2O and the 4.4-μ band of CO_2 [340, 341] have also been used for line reversal studies. The CH and the C_2 bands present in the reaction zone of flames possess abnormal excitation and cannot be reversed and are therefore not usable for temperature measurements by this method.

There are observable changes in the emission spectrum of a gas which may be used to determine temperatures to $10^{6°}$K. Line broadening, line intensity, and continuum intensity measurements have been most used for

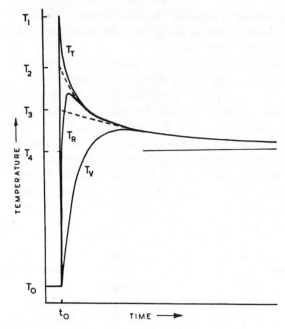

Fig. 6.27. Effective translational, rorational, and vibrational temperatures behind a shock front, showing successive relaxation processes.

the measurement of these high temperatures. The method to be used can only be determined after a comprehensive analysis of the nature of the spectral emission of the particular system whose temperature is sought. It is usually possible to find a suitable optical probe either by adding a small amount of H_2, He, or other indicator gas, or by selecting an atomic emission line of one of the components or a natural impurity that is in an optically clear region of the spectrum and is not affected by self-absorption.

As the temperature of a gas is increased above 4000°K, the spectroscopic changes which occur may be described as follows: (1) rotation–vibration transitions disappear below $10^{4\circ}$K; (2) with increasing temperature, atomic lines broaden, some to a far greater extent than others; (3) all lines first increase in intensity and then decrease in intensity with increasing temperature; (4) the intensity of background continuum radiation increases in proportion to line intensities with increasing temperature; (5) atomic lines are slightly shifted toward the red at higher temperatures; (6) the continuum series limit becomes diffuse and shows an apparent shift toward the red. All of these changes are related to temperature, and equilibrium thermodynamics may be used along with ion and electron

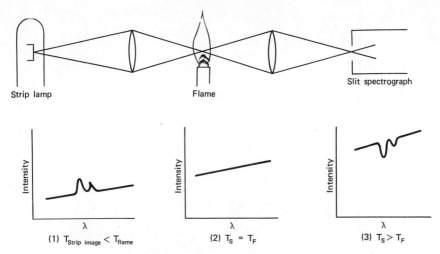

Fig. 6.28. Principle of the line reversal method for gas temperature measurement.

concentrations to compute the gas temperatures by using the Saha equation. It is usually assumed that the particles comprising the plasma possess Maxwell velocity distributions and Boltzmann population distributions among the excited states.

Line broadening of the H_β line at 4861 Å has been widely used for temperature measurements in astrophysics, shock tubes, and arc plasmas [342–344]. The other lines of the Balmer series, H_α, H_γ, and H_δ, have also been studied, but the H_β line is more convenient experimentally and is most often used to measure temperatures in the range of 7000° to 30,000°K. Higher temperatures may be measured by choosing a line corresponding to a higher energy level, that is, a lower principal quantum number. Typical contours of Holtsmark-broadened lines and the derived temperatures are given in Fig. 6.29.

The theoretical and historical development of line broadening has been covered in comprehensive fashion by Breene [345]. In the original Holtsmark theory of Stark line broadening, the radiating atom is supposed to be always surrounded by a fluctuating microfield due to the presence of plasma ions and dipole and quadrupole interactions. With this assumption, Holtsmark calculated as a function of time the probability of the existence of a field strength F at the radiating atom. The probability is normalized in terms of a normal field strength F_0 which depends only upon the number of ions per cc and is given by the equation

$$F_0 = 2.61eN^{2/3}$$

where e is the charge of the electron and N is the number of ions per cc.

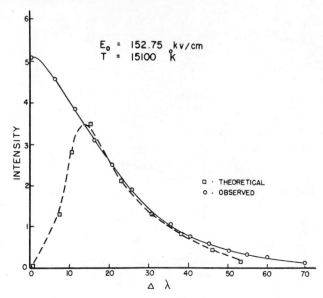

Fig. 6.29. Profile fittings of H_α line broadened in a high-temperature plasma.

In the Holtsmark theory the field strength probability distribution is independent of the ion density; and, as the line broadens, its intensity increases and the line shape does not change. Schmaljohan [346] using the Holtsmark theory has calculated the theoretical intensity profiles of the H_α, H_β, H_γ, and H_δ Balmer lines in terms of the distance from the center of the emission line, $\Delta\lambda$, and the normal field strength F_0. Thus in order to measure temperature, it is only necessary to assume values of F_0 and calculate the intensity of the H_β line that has the same half-width as the experimentally observed line. From this value of F_0, N may be calculated and, in turn, the temperature may be determined by application of the Saha equation [347].

$$\frac{N_+ N_c}{N_0} = f(T)$$

where N_+, N_c, and N_0 are the numbers of positive ions, electrons, and neutral atoms. If there are several atomic species contributing to the ion and electron concentrations, then one can write a Saha equation for each component and solve the set of equations simultaneously with the help of Dalton's law and equations for charge and mass balance.

Various authors [348–350] have calculated corrections to be applied to the Holtsmark theory in order to include the effect of electrons on line broadening and other parameters for various plasma devices and various gases. The agreement between theory and observed line profiles is good. The accuracy of this method of determining electron concentrations and temperatures is $\pm 50°K$ at $15,000°K$, and a given temperature is reproducible in a wall-stabilized arc device to $\pm 10°K$.

4 THE UTILIZATION OF HIGH TEMPERATURES

In a review of high-temperature techniques it is clearly a requirement that one not only be concerned with generation but also with applications of high temperatures. Some of these are obvious and many of them are quite historical, predating our modern concepts and theories of physical and chemical phenomena. Keeping warm, cooking food, lighting the way at night, manufacturing and using fireworks, and purifying metals such as mercury, gold, silver, and copper were among the ways in which high temperatures have been utilized over the past 2000 years. Molding bricks and blowing glass, using gun powder, preparing iron alloys, and fire assays were among the applications of high temperatures that were common in the Middle Ages. Nevertheless, the growth of applications really began with the isolation of oxygen and the basic understanding that combustion required a fuel and an oxidizer. After this, the development of batteries for producing electric current and finally the construction of high voltage/high current electrical machines made it possible for scientists and engineers in any laboratory to have temperatures from a few hundred to tens of thousands of degrees at their fingertips. With this unique opportunity, how have scientists utilized high temperatures in our modern world?

High-Temperature Reactions

Many high-temperature syntheses can be accomplished by simply mixing the reactants (or compounds that will generate the reactants) at room temperature and then heating the mixture. Two factors determine whether a particular synthetic approach is practical: (1) the reaction must be thermodynamically possible, and (2) the rate of formation of product must be appreciable and must exceed its rate of decomposition. At high temperatures, reaction rates are generally fast so that thermodynamic factors are somewhat more important than kinetic factors. In electric arcs, however, one has species and thermodynamic potential to form almost any molecule, and kinetics will determine which results.

The general situation has been summarized in the three ''laws'' of high-temperature chemistry: (1) at high temperatures everything reacts with everything; (2) reactions go faster the higher the temperature; and (3) the products might be any combination of the available atoms [351].

One may prepare many alloys and binary compounds such as carbides, borides, silicides, etc., by simply mixing the elements. Further purification to yield a particular stoichiometry is sometimes necessary. The preparation of metals and carbides such as Fe, Ni, Si, SiC, etc., by the high-temperature reduction of appropriate oxides with carbon represents slightly more complex reaction systems.

The production of CH(g) and CN(g) by heating H_2 or N_2 in a carbon tube forms a basis for high-temperature syntheses of C_2H_2 and C_2N_2 [352–354]. A mixture of CaF_2 + carbon yields $CF_2(g)$ at high temperatures, and these radicals combine to form C_2F_4 in a good yield [355]. In the case of $N_2 + O_2$, in contrast, it is the high-temperature species itself (NO) that is desired. Since NO is metastable at room temperature, a rapid quench is needed after the molecule has been synthesized to a temperature low enough so that the spontaneous decomposition rate is negligible [356].

In other cases, however, one reactant may be stable only at very high temperatures and even then at low pressures, while a desired reactant or product may be stable only at low temperatures. In such cases a tube furnace or Knudsen cell may serve as the source of the high-temperature species, which can effuse out of the hot zone and either contact another reactant in a cooler gas phase or else be condensed into a low-temperature matrix where controlled reactions can be carried out. Typical of such latter cases are the generation of H, O, and F atoms, etc., in an electric discharge or by photolysis and subsequent reaction of these species in a matrix to yield HO_2, OH, H_2O_2, XeF_4, etc. Skell and associates have utilized C_2, C_3, C_4, and C_5 molecules, etc., generated in a high-temperature furnace for the synthesis of new organic molecules (see Fig. 6.30). Also, they have studied the reaction of metal atoms (Fe, Pt, Mo, etc.), with various organic species [357, 358]. (There is also a long series of subsequent publications by Skell and associates.) Margrave, Timms, and associates [359] studied reactions of SiF_2, $SiCl_2$, SiO, BF, and other high-temperature molecules with C_2H_4, C_2H_2, benzene, BF_3, and a variety of other species. It is apparent that the low-temperature reactions of high-temperature molecules offer almost an infinity of opportunities for new syntheses, as shown in Table 6.8.

Another active area of high-temperature synthesis covers the chemistry of coal when subjected to pyrolysis, solvent extraction, or hydrogenation to produce the gaseous or liquid hydrocarbons now needed to supplement the dwindling supplies of crude oil and natural gas [360–364].

High-Temperature Electrochemical Syntheses

Many pure elements are prepared at high temperatures by fused salt electrolysis, including aluminum, magnesium, sodium, and chlorine on an

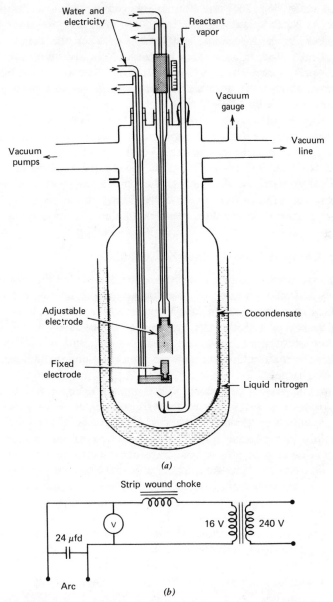

Fig. 6.30. Apparatus for cryochemistry on carbon vapor: (a) apparatus, (b) electrical circuit to control carbon arc.

industrial scale [365, 366] and lithium, beryllium, calcium, fluorine, and tantalum in batch quantities [367]. Alloys such as potassium–sodium, lithium–sodium, lead–sodium, and lead–potassium are easily prepared electrolytically [368]. Calcium (or barium) alloys and aluminum-calcium alloys as well as other alloy systems such as zirconium with beryllium, manganese, aluminum, uranium, or titanium have also been prepared from fused salt media [369–371].

Refractory borides were synthesized by fused salt electrolysis in 1927 by Andrieux from a mixture of boric oxide, metallic oxide, and fluoride, and he was later able to synthesize 17 other borides, ten of which were prepared for the first time [372–374]. Extensive studies of the electrochemical preparation of silicides, phosphides, carbides, arsenides, antimonides, and sulfides were made by Andrieux and associates [375–378] who also prepared single and complex oxides (ferrites, ferrates, etc.) [379].

Electric Arc and Plasma Syntheses

Plasma arc devices can provide both a heat source for the processing of chemicals and also a source of potentially reactive species for synthetic reactions. Typical applications of plasma arcs for syntheses have included the preparation of endothermic nitrogen compounds (NO, HCN, C_2N_2, etc.), other endothermic molecules such as C_2H_2, and various polymeric films whose formation from metastable monomers can be facilitated by an electric discharge.

A typical dc plasma generator [380] is shown in Fig. 6.8. The anode is made of thoriated tungsten and the tungsten cathode is water-cooled. Power is supplied by a 600-A dc welder. Gases or fluidized solids may be injected into the plasma from a ring attached to the bottom of the generator, and cold fingers or cold traps are used to collect samples. To prevent air from entering the system, a positive pressure is maintained; the power inputs vary from 10 to 50 kW. Various metal nitrides have been prepared in plasma arcs of this type. Cyanogen and nitric oxide have been prepared, but in low yield [381, 382].

Acetylene has been synthesized from methane in an argon plasma at 80% yield [383]. Three methods have been tried: (1) feeding methane into the "flame" of the argon plasma jet, which was the most successful; (2) feeding powdered carbon into a hydrogen plasma jet; and (3) using a methane plasma jet. There have also been attempts to utilize plasma arc devices for synthesis of N_2H_4 [384], fluorocarbons [385, 386], and HCN [387, 382]. The induction-heated plasma arc has been used for synthetic chemistry, especially for processes that require oxidizing atmospheres, and for production of the fluorocarbon CF_x [389, 390].

High-intensity electric arcs have been utilized for a variety of metallurgical refining processes [391, 392], including the growth of single crystals, and for preparation of C_2F_4 [393]. Unfortunately the power efficiency in terms of moles per kilowatt does not make plasma syntheses competitive with our present high cost of energy.

Other applications of low-intensity electric arcs, microwave discharges, glow discharges, etc., have been discussed by Jolly [394, 395], Kanaan and Margrave [396], and others [397, 398].

Biochemists have an interest in plasma syntheses since one can prepare simple amino acids in electric discharges through NH_3, H_2O, and CO_2 [399–401]. Recently a plasma process which produced HCN has been suggested as a source of primordial amino acids, proteins, and other basic molecules of living organisms. The major difficulties lie in the techniques for quenching the potentially reactive ions, atoms, and molecules in such a way that the most desirable products are formed. It seems likely that isolation of these unstable dissociation products on suitably reactive substrates could be one solution to this problem, and this idea has recently been invoked as part of a cycle leading to the production of biologically active systems [402].

Syntheses in Flames

The synthesis of organic compounds by use of the free radicals present in flames has been reported [403–405]. A stream of aqueous organic reactants is rapidly passed through the inner cone of a flame and presumably presents a cold wall of solvent that continuously extracts the free radicals from the flames. These free radicals may then react with the organic solutes. Various flames (methane–oxygen, hydrogen–oxygen, etc.) have been utilized to convert aqueous acetic acid to glycolic acid, succinic, hydroperoxyacetic, and chloracetic acids in the presence of suitable solutes.

When one introduces metal halides into flames, there is both mass-spectrometric and optical evidence for a variety of gaseous hydroxides—LiOH, NaOH, KOH, MgOH, CaOH, SrOH, BaOH, CuOH, InOH, GaOH, etc. [406, 407]. Gaseous H_3BO_3, HBO_2, $(HBO_2)_3$, (BOX), $(BOX)_3$, etc., are produced in B-O-X-H flames [408]. The condensation and fractionation of products from such inorganic flame systems would seem to be an important type of process to try in developing this synthetic tool for the future.

Crystal Synthesis and Growth

The growth of large crystals of a variety of high-temperature materials has become a major part of modern industrial use of high-temperature techniques. Thus single crystals of silicon, SiO_2, Al_2O_3, $LiNbO_2$, diamond, cubic BN, ruby, doped garnets, ferrites, rutile, CdS, ZnS, ZnSe, and many other pure metal compounds and even minerals from hydrothermal processes are currently being prepared for industrial applications. This area is truly interdisciplinary, involving the methods of physics, chemistry, and metallurgy along with sophisticated aspects of engineering. For useful reviews of the field see Laudise [409–411] and others [412, 413].

Solid-State Crystal Growth

In the solid state the growth of one grain at the expense of another characterizes the preparation of crystals. If the process is allowed to proceed until all other grains are consumed, a single crystal is formed. This method is especially advantageous where segregation of solute atoms is to be avoided, such as in alloy systems. In addition, contamination and defect structures in the crystal are minimized. Chalmers [414] has discussed three methods by which single crystals can be formed without fusion: (a) critical strain followed by annealing, (b) growth after straining in a very sharp temperature gradient, and (c) growth during phase transformation. The spontaneous growth of grains of an unworked or slightly cold-worked polycrystalline metal during annealing is referred to as grain growth and has been used to grow tungsten and molybdenum single crystals. A phase transformation process is characterized by nucleation, growth, and composition and structural changes. It is possible to produce single crystals in some instances by controlled cooling or by thermal cycling through the transition point.

Recrystallization is a special type of phase transformation that also involves nucleation and growth. Depending on the metal and on the degree of cold working, annealing above a certain temperature causes the old grains to be replaced by new ones. The original composition and structure are retained, but the orientation is changed. An incubation period is followed by growth of new grains at an approximately constant rate until the stable state characteristic of the processes is formed. Another type of crystal growth from the solid state combines the use of high pressures along with high temperatures as, for example, in the recrystallization of graphite into diamond or hexagonal BN into cubic BN. One can get direct conversion to sintered polycrystals or single crystals without the use of a catalyst by brute force techniques [415–419], but the industrial conversions of graphite to single crystal diamond and hexagonal BN to cubic BN proceed through an intermediate solution phase using a transition metal (Ni, Ta, etc.) in the case of diamond [420, 421] or a nitride-forming metal (Li, Mg, etc.) in the case of BN [422]. Interesting practical uses for synthetic carbonado, the sintered polycrystalline form of diamond, are being developed, and large pieces ($\frac{1}{4}$ in. thick × $\frac{1}{2}$ in. in diameter) have been prepared [415–419].

Liquid-State Crystal Growth

Solidification from the molten state is a widely used technique for the preparation of metal crystals. There are three desirable conditions for bringing a single crystal nucleus into existence: (1) seeding of the melt with an available single crystal, (2) keeping to a minimum the volume of the metal in the temperature range in which nucleation is favorable, and (3) restricting the number of initial nuclei formed and allowing only one of these to grow.

Among the approaches to crystal growth from the liquid phase are (1) the moving temperature differential methods of Czochralski and Bridgman [423–425], (2) the zone-refining and crystal growth methods of Pfann [426], (3) the Verneuil technique [427–430], and the growth of single-crystal thin films [431, 432].

In the Czochralski method a seed crystal on the end of a rod is slowly withdrawn from the melt. In the Bridgman method one uses a mold with a pointed bottom to induce nucleation. The melt in the mold is gradually passed through a temperature differential to induce unidirectional growth. Various adaptations of the above techniques are given in the literature [431, 432]. Purification of a metal by the repeated passage of a narrow molten zone through the sample was first reported by Pfann and is now widely used in industry [433–435] (see Fig. 6.31). The success of zone melting depends on the fact that the impurities in the sample either increase or decrease the freezing point of the solvent. The impurities that lower the freezing point of the solvent concentrate in the liquid phase and are swept to one end with the repeated passage of the molten zone. For impurities that raise the freezing point, stepwise purification during the passage of the molten zone does not occur, and such impurities tend to segregate in the solid phase. One can use induction heating of the sample by a horizontal graphite susceptor to produce a narrow molten zone. If containers are used, one can still achieve further purification of the initial sample by several approaches. One method consists of heating the sample to its melting point while contained in a water-cooled crucible, which is coupled to the induction coil of a high-frequency generator [436]. The resistivity of the material being melted increases rapidly with increasing temperature, while the resistivity of the water-cooled copper-tube crucible remains low. In this way it has been possible to melt a sample whose melting point is more than 2000°C above that of the crucible, without contamination or reaction by the crucible. This procedure minimizes diffusion from the crucible into the melt since the molten metal does not wet the crucible. Samples of any form and shape can be used in this method, and alloys of refractory metals can easily be prepared.

Another method for avoiding container contamination uses induction heating of a vertical rod to create a narrow floating liquid zone which is held intact by the high surface tension of the liquid [437, 438]. Thus with a silicon rod the molten phase can be passed up and down a vertically held rod. Single crystals can be grown by this method if the process is carefully controlled. One can also levitate a sample in vacuum and boil out the impurities [439, 440].

Electron bombardment can be used to produce a floating zone [441]. The purification takes place in three ways: (a) by outgassing on vacuum fusion, (b) by vaporization of impurities, and (c) by zone refining. The anode is the sample and the cathode is a circular tungsten filament from which electrons are generated. Pressures below 5×10^{-4} torr are required. The maximum diameter of the specimen depends on the surface tension and density of the sample in the liquid state. Single crystals of tungsten, molybdenum, and niobium have been grown and purified to a considerable degree when zone melted under vacuum.

The Verneuil apparatus for growing crystals consists essentially of a power feed, a heat source, and a crystal moving device (see Fig. 6.32). Verneuil, in 1904, used this method for the preparation of rubies by flame fusion. Suitable feed powders must be used, for example, for oxides, fluffy powders characterized by a loose metastable crystal structure [442, 443]. This technique has been widely applied, and the literature reports single-crystal boules of various transition element oxides by fusing the powders in an oxyhydrogen flame [444].

Hydrothermal Synthesis of Crystals

A completely different approach to crystal synthesis has been developed by scientists at the Bell Telephone Laboratories [445, 446], Roy et al [447], and others [448] for preparation of synthetic crystals of large size and high quality. To get synthetic quartz, a $Na_2SiO_3–H_2O$ flux mixture is heated above the critical temperature of water and then allowed to cool according to a carefully controlled schedule. A small seed of SiO_2 can be used to produce crystals as large as a football. A variety of anhydrides and hydrated minerals can be obtained as single crystals by this high-temperature/high-pressure technique [449, 450].

Growth from the Vapor Phase

The growth of crystalline materials at high temperatures from the vapor phase has only recently become significant, although organic chemists

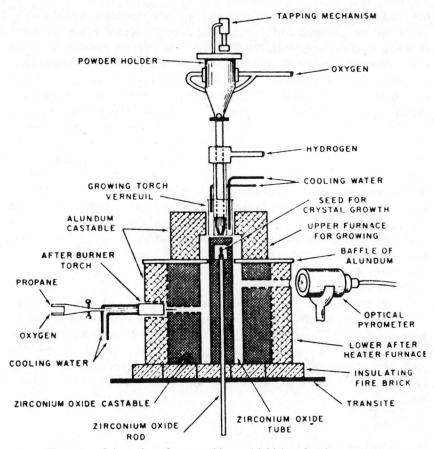

Fig. 6.31. Schemation of zone melting and initial perforation apparatus.

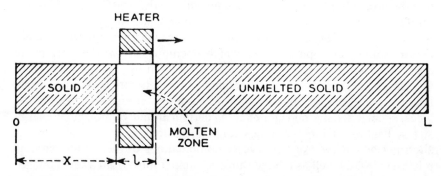

Fig. 6.32. Apparatus for growing single crystals via the Verneuil technique.

have long used the technique. In a few cases, the starting material may be vaporized or sublimed and a crystalline deposit formed when the vapor strikes a cold surface [451], but more often a reactive mixture is passed over a hot wire to initiate reaction, and the material to be deposited is nucleated from the vapor on a suitable seed crystal. Such techniques (chemical vapor deposition, CVD) are widely used for the preparation of ceramics, alloys, and oriented crystals [452–454] and for the deposition of thin films [455, 456]. There is also a considerable interest in the growth of whiskers from the vapor [457–458] which can be formed into threads and woven into refractory or conducting "fabrics." High strength, ductility, and anisotropy have been observed for whiskers. Single crystals of CdS [459] have been grown from crude CdS under an atmosphere of H_2S or argon at a pressure of 1 atm or less, for periods ranging from four to seven days at temperatures up to 1250°C. Crystals of GaAs, InSb, etc., have been grown by gradually lowering the temperature in an evacuated reaction tube until nucleation occurs [460]. A cold finger may be employed to control the nucleation, structure, and morphology of single crystals grown from the vapor by sublimation.

Ductile titanium can be deposited on an electrically heated filament by the thermal decomposition of TiI_4 vapor [461], silicon crystals can be prepared from $SiCl_4 + H_2$ [462], and there are several reports of the growth of diamonds on seed crystals by thermal pyrolysis of methane [463, 464]. Other metals may be deposited by evaporation, sputtering, exploding wires, or by thermal decomposition of various carbonyls [465, 466].

5 CONCLUSIONS

In summary, it is apparent that much work and development have gone into the area of high-temperature science during the past 25 years and that this field has grown from several specialized and narrow types of activities with no real coherence as a discipline to a clearly defined interdisciplinary research and development field which encompasses workers from the areas of chemistry, physics, ceramics, metallurgy, and a variety of applied engineering fields. The techniques for generating and measuring high temperatures are diverse and offer options in terms of heating devices, specialized methods for exciting certain atomic or molecular energy states, heating of large volumes for long periods or small volumes for short periods of time or the reverse combinations of these time–volume relationships, as well as a huge range of temperature from a few hundred to over $10^8°K$. Measurements over such a wide range of temperatures obviously require sophisticated instrumentation, and techniques have been developed that make use of chemical and physical properties, including the radiation properties of matter as described by Planck's law. Although measuring techniques are excellent, reliable standards for reference purposes are scarce, and it is especially difficult to find temperature standards that apply to the range 2000°K and above.

New research efforts in this field are clearly justified. In a closely related area, further studies of the emissivities of solids and liquids as a function of temperature and wavelength would provide needed information for application of optical pyrometer methods to temperature measurements. There is a continuing need for more versatile thermocouple combinations which are usable at the higher ranges of temperature and especially for those combinations which would be stable in oxidizing environments.

From the viewpoint of new materials, one has to regard the high-temperature regime as unique in providing an infinity of new species and opportunities. In some cases one can achieve temperature/pressure combinations that are unknown in nature. With so many opportunities for preparing new materials, it is not surprising that the list of new solids and liquids as well as new gaseous molecules has grown exponentially over the past two decades. If one includes ions in addition to neutral species or chemical reactants, then it is safe to say that chemists have barely explored the fringes of the islands of stability that define the neutral or other possible combinations of atoms. The techniques of plasma chemistry combined with gas chromatography, mass spectrometry, optical spectroscopy, esr spectroscopy, and other specialized methods such as matrix isolation have already shown that there are hundreds of new and reactive molecules that can be prepared. Synthetic chemistry at high temperatures has barely been explored.

ACKNOWLEDGMENT

Research in high-temperature science is supported at Rice University by the National Science Foundation, the U.S. Army Research Office, the National Aeronautics and Space Administration, and the Robert A. Welch Foundation. We are especially appreciative of the excellent assistance of Mrs. Donna Montez who helped prepare the final manuscript. We also wish to acknowledge the patience and consideration of the editor, Dr. Bryant Rossiter, without which this chapter might never have been completed.

References

1. J. L. Gole, *Ann. Rev. Phys. Chem.*, **27**, 525 (1976).
2. R. P. Benedict, *Fundamental Temperature, Pressure and Flow Measurements*, 2nd ed., Wiley, New York, 1977.
3. J. W. Hastie, R. H. Hauge, J. L. Margrave, *Ann. Rev. Phys. Chem.*, **21**, 475 (1970).
4. J. L. Margrave, *High Temperature Chemistry*, American Chemical Society, Washington, D.C., 1967.
5. R. G. Bautista and J. L. Margrave, "High Temperature Technique," in *Technique of Inorganic Chemistry*, H. B. Jonassen and A. Weissberger, Eds., Wiley, New York, 1965, pp. 65–135.
6. R. J. Thorn, *Ann. Rev. Phys. Chem.*, **17**, pp. 83–118 (1966).
7. L. Brewer, "Principles of High Temperature Chemistry," in *Proc. R. A. Welch Foundation Conf. on Chemical Research*, VI. *Topics in Modern Inorganic Chemistry*, 1962, p. 47.
8. "Characterization of High Temperature Vapors and Gases," *Proc. of 10th Materials Research Symposium, U.S. Natl. Bureau of Standards*, J. W. Hastie, Ed., Sept. 18–22, 1978, Washington, D.C.
9. For example, the Materials Science and Technology Series, Academic Press, New York, includes volumes on "Diffusion in Solids," "Epitaxial Growth," "Advanced Materials in Catalysis," "Solid Electrolytes," "Martensitic Transformation," etc.
10. S. Takeuchi, Ed., *The Properties of Liquid Metals*, Halsted Press, Wiley, New York, 1973.
11. S. Z. Beer, *Liquid Metals*, Marcel Dekker, New York 1972.
12. J. L. Margrave, Ed., *High Temperature Science*, The Humana Press, Clifton, N.J., Vol. 1 (1969) et seq.
13. A. E. Sheindlin, Ed., *High Temperature (USSR)*, Plenum, New York, Vol. 1 (1963) et seq.
14. Internationale Revue Des Hautes Temperatures et Des Refractaires. Masson, S.A., 120, Boulevard, Saint-Germain-F 75280, Paris CEDEX06 France.
15. E. Fitzer, Ed., *High Temperatures–High Pressures*, Pion, London, Vol. 1, 1969, et seq.
16. The Journal of Solar Energy Sciences and Engineering, Published by the Association for Applied Solar Energy, Suite 202, Central Plaza Bldg., 3424 N. Central Ave., Phoenix, Arizona, Vol. 1 in 1957.
17. Journal of the High Temperature Society, Tokyo, Japan, Vol. 1, 1975, et seq.
18. In addition, there are journals devoted to the various applied technologies which utilize high temperatures like the iron and steel industry, the glass and ceramic industry, the electrochemical industry, the mining and metallurgy industries, the nuclear energy field, the various areas of solid-state applications to modern electronics, etc.
19. J. W. Hastie, *High Temperature Vapors*, Academic Press, New York, 1975.

20. "High Temperature Metal Halide Chemistry," *Proc. of Symposium at Electrochemical Soc. Meeting, Atlanta, Georgia*, Oct. 1977, The Electrochemical Society, P.O. Box 2071, Princeton, N.J. 08540, 1978. D. Hildenbrand.

21. L. Eyring, Ed., *Advances in High Temperature Chemistry*, Vols. 1–4, Academic Press, New York, 1967–1971.

22. J. L. Margrave, Ed., *The Characterization of High Temperature Vapors*, Wiley, New York, 1967.

23. "High Temperature Chemistry-Current and Future Problem," *Proc. of Conf. at Rice Univ., Houston, Texas*, Jan. 26–27, 1966, J. L. Margrave, Ed., N.A.S., N.R.C., N 1470, 1967.

24. "Metallurgy at High Pressures and High Temperatures," *Proc. of Symposium of Metallurgical Soc. Conf.*, Feb. 25–26, 1963, Gordon and Breach, New York, 1964.

25. "Condensation and Evaporation of Solids," *Proc. of Int. Symp. on Condensation and Evaporation of Solids*, Dayton, Ohio, Sept. 12–14, 1962, E. Rutner, P. Goldfinger, and J. P. Hirth, Eds., Gordon and Breach, New York, 1964.

26. JANAF Tables, 2nd ed., D. R. Stull, Prophet H. and M. W. Chase, NSRDS-NBS, 37, 1971, and subsequent revisions and additions.

27. L. V. Gurvich, et al., *Thermodynamic Properties of Individual Substances*, USSR Acad. of Sciences, Inst. of Combustible Materials, Vol. 1 and 11, 1962, and subsequent revisions and additions.

28. J. G. Beck, *Solid State Technol.*, **20**, 59 (1977).

29. M. C. Davidson and L. R. Holland, Space Sciences Laboratory Preprint No. 78-106, Marshall Space Flight Center, Alabama, 35812; *Rev. Sci. Instrum.*, 49, 1156 (1978).

30. Among the industrial suppliers of resistance furnaces are ABAR, 905 Pennsylvania Blvd., Feasterville, PA 19047; Astro Industries, 606 Olive St., Santa Barbara, CA 93101; BTU Engineering, Esquire Rd., North Billerica, MA 01862; Centorr Associates, Route 28, Suncook, NH 03275; GCA/Vacuum Industries, 34 Lindon St., Somerville, MA 02143; C.I. Hayes, 800 Wellington Ave., Cranston, RI 02910; Labtronics, 26 Valley Rd., Port Washington, NY 11050; Columbia Scientific Industries, Box 9908, Austin, TX 78766; Harrop Laboratories, 3470 E Fifth Ave., Columbus, OH 43219; L&L Special Furnace, Box 938, Chester, PA 19016; Tem-Pres, 1526 William St., State College, PA 16801; Varian Associates, 611 Hansen Way, Palo Alto, CA 94303; Aremco Products, Box 429, Ossining, NY 10562.

31. J. A. Kitchener and J. O'M. Bockris, *Discuss. Faraday Soc.*, **4**, 91 (1948).

32. J. A. Stockdale, L. Schumann, H. H. Brown, Jr., and B. Bederson, *Rev. Sci. Instrum.*, **48**, 938 (1977).

33. G. A. Ozin and M. Moskovits, "Techniques of Matrix Cryochemistry," in *Cryochemistry*, pp. 9–60, G. A. Ozin and M. Moskovits, Eds., Wiley, New York, 1976.

34. L. I. Maissel and G. Reinhard, Eds., *Handbook of Thin Film Technology*, McGraw-Hill, New York, 1970, pp. 36–49.

35. J. W. Hastie, R. H. Hauge, and J. L. Margrave, *J. Chem. Phys.*, **51**, 2648 (1969).
36. M. B. Panish, *J. Chem. Phys.*, **34**, 1079 (1961).
37. H. Jehn, R. Volker, and M. I. Ismail, *Plat. Metals Rev.*, **22**(3), 92 (1978).
38. J. R. Soulen and J. L. Margrave, *Rev. Sci. Instrum.*, **31**, 68 (1960).
39. Fansteel Met. Corp., One Tontalum Place, North Chicago, Illinois 60064.
40. Kanthal Corp., Woster Street, Bethel Ctr, Stamford, Connecticut 06801.
41. G. D. Bell, M. H. Davis, R. B. King, and P. M. Routly, *Astrophys. J.*, **129**, 437 (1959).
42. K. H. Olsen, P. M. Routly, and R. B. King, *ibid.*, **130**, 688 (1959).
43. L. Brewer, P. Gilles, and F. Jenkins, *J. Chem. Phys.*, **16**, 797 (1948).
44. T. Y. Kometani, Ed., *Graphite Furnace Absorption Spectroscopy*, Franklin Inst. Press, Philadelphia, 1978.
45. S. M. Lang and R. F. Geller, *J. Am. Ceram. Soc.*, **34**, 193 (1951).
46. W. H. Davenport, S. S. Kistler, W. M. Wheildon, and O. J. Whittemore, Jr., *J. Am. Ceram. Soc.*, **33**, 333 (1950).
47. Commercial suppliers of industry furnaces include Ajax Magnethermic, 1745 Overland Ave., Warren, OH 44482; Angstrom, Box 248, Belleville, MI 48111; Atomergic Chemetals, 100 Fairchild Av., Plainview, NY 11803; GCA/Vacuum Industries, 34 Lindon St., Somerville, MA 02143; Ipsen Industries, Box 6266, Rockford, IL 61125; Leco, 3000 Lakeview Av., St. Joseph, MI 49085; Lepel, 59-21 Queens Midtown Expy, Maspeth, NY 11378; Lindberg, 2450 W Hubbard St., Chicago, IL 60612; Radio Frequency, 50 Park St., Medfield, MA 02052; Varian Associates, 611 Hansen Way, Palo Alto, CA 94303.
48. E. May, *Industrial High Frequency Electric Power*, Wiley, New York, 1950.
49. J. W. Cable, *Induction and Dielectric Heating*, Reinhold, New York, 1954.
50. F. W. Curtiss, *High Frequency Induction Heating*, 2nd ed., McGraw-Hill, New York, 1950.
51. E. C. Jordan, in *Electric Power, Electrical Engineers Handbook*, 4th ed., H. Pender and W. A. Del Mor, Eds., Wiley, New York, 1949 p. 18.
52. P. Duwez, F. H. Brown, Jr., and F. Odell, *J. Electrochem. Soc.*, **98**, 356 (1951).
53. G. Babat and M. Losinsky, *J. Appl. Phys.*, **11**, 816 (1940).
54. R. J. Fries, *J. Chem. Phys.*, **37**, 320 (1962).
55. R. F. Walker, J. Efimenko, and N. L. Lofgren, *Planetary Space Sci.*, **3**, 24 (1961).
56. F. K. McTaggart, *Plasma Chemistry in Electrical Discharges*, Elsevier, New York, 1967.
57. Cober Electronics, 7 Gleason Av., Stamford, CT 06902.
58. Despatch Industries, Box 1320, Minneapolis, MN 55440.
59. Sharp Electronics, Box 389, Paramus, NJ 07652.
60. Varian Associates, 611 Hansen Way, Palo Alto, CA 94303.
61. O. Muck, Ger. Pat. 422,004 (Oct. 30, 1923).
62. W. V. Lovell, U.S. Pat. 2,400,869 (May 28, 1946); U.S. Pat. 2,566,221 (Aug. 28, 1951).

63. D. M. Wroughton, E. C. Okress, P. H. Brace, G. Comenetz, and J. C. R. Kelly, *J. Electrochem. Soc.*, **99**, 205 (1952).

64. R. T. Begley, G. Comenetz, P. A. Flinn, and J. W. Salatka, *Rev. Sci. Instrum.*, **30**, 38 (1959).

65. S. G. Sagardia and H. U. Ross, *Diss. Abst. Intl.*, **38B**(10), 4941 (1978).

66. D. M. Polonis, R. G. Butters, and J. G. Parr, *Research* (London), **7**, 272 (1954).

67. D. H. Polonis and J. G. Parr, *Trans. AIME*, **200**, 1148 (1954).

68. G. Comenetz and J. W. Salatka, *J. Electrochem. Soc.*, **105**, 673 (1958).

69. H. Soda, A. McLean, and W. A. Miller, *Met. Trans.*, **9B**, 145 (1978).

70. A. E. Jenkins, B. Harris, and L. Baker, Paper presented at the Symposium on Metallurgy at High Pressures and High Temperatures, Metallurgic Society, AIME Annual Meeting, Dallas, Texas, 1963.

71. D. W. Bonnell, Ph.D. Thesis, Rice University, Houston, Texas, 1972.

72. J. L. Margrave, "Liquid Metals at High Temperatures," in *Proceedings of the International Conference on Phase Transitions at High Temperatures*, Odeillo, France, Sept. 27–30, 1971, pp. 71–77.

73. P. W. Wilson, J. A. Treverton, and J. L. Margrave, *High Temp. Sci.*, **3**, 163 (1971).

74. J. A. Treverton and J. L. Margrave, *J. Phys. Chem.*, **75**, 3737 (1971).

75. J. A. Treverton and J. L. Margrave, *J. Chem. Thermodyn.*, **3**, 3737 (1971).

76. J. L. Margrave, *High Temp.–High Press.*, **2**, 283 (1970).

77. J. A. Treverton and J. L. Margrave, ASTM Conf. on Thermophysical Properties, Boston, Mass., Oct. 1970.

78. A. K. Chaudhuri, D. W. Bonnell, L. A. Ford, and J. L. Margrave, *High Temp. Sci.*, **2**, 203 (1970).

79. H. P. Stephens, *High Temp. Sci.*, **6**, 157 (1974).

80. E. E. Shpilrain, V. A. Fomin, O. N. Kagan, G. F. Sokol, V. V. Kachalov, S. N. Ulyanov, *High Temp.–High Press.*, **9**, 49 (1977).

81. V. Ya. Chekhovskoi, A. E. Sheindlin, and B. Ya. Berezin, *High Temp.–High Press.*, **2**, 301 (1970); *ibid.*, **3**, 287 (1971); *ibid.*, **4**, 478 (1972).

82. B. Ya. Berezin and V. Ya. Chekhovskoi, *High Temp.* (USSR) (Engl. Trans.), **15**, 616 (1970).

83. A. A. Fogel, T. A. Sidorova, V. V. Smirnov, Z. A. Guts, and I. V. Korkin, *Izv. Akad. Nauk USSR, Metally*, **2**, 138 (1968).

84. V. Ya. Chekhovskoi and B. Ya. Berezin, *Bull. Izobr.*, **No. 17**, USSR Inventor's Certificate 218,492 (1968).

85. R. T. Frost, *EM Containerless Processing Requirements and Recommended Facility Concept*, General Electric Company, Final Report to NASA, NAS-8-29680, May 13, 1974.

86. R. T. Frost, *Free Suspension Processing Systems for Space Manufacturing*, General Electric Company, Final Report to NASA, NAS-8-26157, June 15, 1971.

87. R. T. Frost, *Unique Manufacturing Processes in Space Environment*, NASA Pub. ME-70-1, April 1970.

88. Anon., Aluminum Casting with Levitation Coil to Avoid Surface Imperfections, *Bus. Week*, Aug. 7, 1978, p. 60D.

89. *Ind. Res./Devel.*, Plane Tran—An Electromagnetically Levitated Train—Thru Evacuated Tunnels at 14,000 MPH, July 1978, p. 58.

90. S. Silverman and E. Passaglia, Eds., *Applications of Space Flight in Materials Science and Technology*, NBS Spec. Pub. No. 520, September 1978.

91. W. Oran, Ed., *Proceedings of Workshop on Microgravity Electromagnetic Levitation Facility*, NASA, Washington, D.C., May 1, 1978.

92. S. Gordon and B. H. McBride, *Computer Programs for Calculation of Complex Chemical Equilibrium Concentrations, Rocket Performance, Impulse, and Chapman-Jouget detonations*, NASA Sp-273, 1971.

93. M. Blander and C. C. Hsu, "Calculation of Equilibria between Gases and Condensed Phases" in Ref 95, pp. 60–66.

94. A. V. Grosse and A. D. Kirchenbaum, *J. Am. Chem. Soc.*, **77**, 5012 (1955).

95. Various papers in the *Proceedings of the Conference on High Temperature Science Related to Open-Cycle, Coal-Fixed MHD System, Argonne National Laboratory*, Argonne, Illinois, April 4–6, 1977.

96. R. Bunde, in *MHD Power Generation: Selected Problems of Combustion MHD Generators*, J. Raeder, Ed., Springer-Verlag, New York, 1975.

97. T. Sata and M. Yoshimura, *J. Mat. Sci. Soc.* (Japan), **15**(2), 87 (1978).

98. M. Hirabayashi and A. Ito, *Bull. Jpn. Inst. Metals*, **17**(4), 294 (1978).

99. W. Finkelnburg, *Z. Phys.*, **112**, 305 (1939).

100. W. Finkelnburg, *The High Current Carbon Arc, FIAT Final Report 1052-PB-81644*, Dept. of Commerce, Washington, D.C., 1947.

101. W. Finkelnburg, *J. Appl. Phys.*, **20**, 468 (1949).

102. C. Sheer, "Investigation of High-Intensity Arcs," in *Technique for Material Testing*, WADC TR 58-142, ASTIA 205,364, PB 161,262, November 1958.

103. V. Harris, J. Holmgren, S. Korman, and C. Sheer, *J. Electrochem. Soc.*, **106**, 874 (1959).

104. S. D. Korman and C. Sheer, Vitro Laboratory Technical Note No. 6, Aug. 22, 1958.

105. C. Sheer, The High Intensity Arc and its Applications to Process Chemistry, Vitro Laboratory Report, 1956.

106. H. Gerdien and A. Lotz, *Z. Techn. Phys.*, **4**, 157 (1923).

107. A. S. Kana'an and J. L. Margrave, "Chemical Reactions in Electrical Discharges" in *Advances in Inorganic and Radiochemistry*, H. J. Emeléus and A. G. Sharpe, Eds., Vol. 6, pp. 143–206 (1964).

108. T. B. Reed, *J. Appl. Phys.*, **32**, 821 (1961).

109. A. S. Kana'an, Ph.D. Thesis, Department of Chemistry, University of Wisconsin, 1963.

110. A. S. Kana'an, C. P. Beguin, and J. L. Margrave, "Spectroscopic Investigations of Some Halides in Plasma Torches," in *Symposium on Spectroscopy of Gases at High Temperatures*, Society of Applied Spectroscopy, Cleveland, Ohio, Sept. 28 & 29, 1964.

111. Advanced Kinetics, 1231 Victoria St., Costa Mesa, CA 92627.

112. International Plasma, Box 4136, Hayward, CA 94540.

113. LFE, 1601 Trapelo Rd., Waltham, MA 02154.

114. Technics, 5510 Vine St., Alexandria, VA 22310.

115. Tomorrow Enterprises, Box 85, Willow Creek, MT 59760.

116. R. M. Barnes and R. J. Winslow, *J. Phys. Chem.*, **82**, 1869 (1978).

117. R. M. Barnes, Ed., *Applications of Inductively Coupled Plasmas to Emission Spectroscopy*, Franklin Institute Press, Philadelphia, 1978.

118. H. U. Eckert, *High Temp. Sci.*, **6**, 99 (1974).

119. D. Kuppers, J. Koenigs, and H. Wilson, *J. Electrochem. Soc.*, **125**, 1298 (1978).

120. C. M. Melliar-Smith and C. J. Mogab, "Plasma Assisted Etching Techniques for Pattern Delineation," in *Thin Film Processes*, J. L. Vossen and W. Kern, Eds., Academic Press, New York, 1978.

121. R. S. Thomas, in *Techniques and Applications of Plasma Chemistry*, J. R. Hollahan and A. T. Bell, Eds., Wiley-Interscience, New York, 1974, p. 255.

122. T. S. Laszlo, "Image Furnace Techniques," in *Techniques of Inorganic Chemistry*, H. B. Jonassen, and A. Weissberger, Eds., Vol. V., Wiley-Interscience, New York, 1965, p. 194.

123. F. Kreith and J. F. Kreider, *Principles of Solar Engineering*, McGraw-Hill, New York, 1978.

124. T. Sakurai, "Solar Furnaces," in *Solar Energy Engineering*, A. A. M. Sayigh, Ed., Academic Press, New York, 1978, p. 233.

125. For example, movie projector lamps as sold by various commerical manufacturers.

126. B. J. Costello, *Western Electric Eng.*, **7**, 40 (1963).

127. C. Kooy and H. J. M. Louwenberg, *Phillips Tech. Rev.*, **23**, 161 (1962).

128. K. Kitazawa, K. Nagashima, T. Mizutani, K. Fueki, and T. Mukaibo, *J. Cryst. Growth*, **39**, 211 (1977).

129. F. Beck, Ger. Pat. 262,913 (1910).

130. W. Finkelnburg, *J. Appl. Phys.*, **20**, 468 (1949).

131. M. R. Null and W. W. Lozier, *Rev. Sci. Instrum.*, **29**, 163 (1958).

132. F. Trombe and A. Le Phat Vinh, *Rev. Int. Hautes Temp. et Refract.*, **10**, 199 (1973).

133. F. Trombe, A. Le Phat Vinh, and C. Royere, in *Proc. of NSF Int. Seminar on Large Scale Solar Energy Test Facilities*, Nov. 18 & 19, 1974, New Mexico State University.

134. C. Royere, in *Proc. Ann. Meeting, Solar Thermal Test Facilities Users Assoc.*, April 11 & 12, 1978, Golden, Colorado.

135. A. F. Hildebrandt and L. L. Vant-Hull, *Mech. Eng.*, **96**, Sept., 23 (1974).

136. F. Trombe, L. Grow, C. Royere, and J. F. Robert, *Rev. Int. Hautes Temp. et Refract.*, **10**, 205 (1973).

137. Organized in April 1977 to coordinate experiments of the U.S. Dept. of Energy high-temperature solar facilities in Albuquerque, New Mexico, and Atlanta, Georgia, and to conduct cooperative programs with the U.S. Army White Sands Solar Furnace and the CNRS Solar Furnace at Odeillo, France.

138. F. Smith, Solar Thermal Test Facilities, in *Proc. of Annual Meeting, Users Assoc.*, April 11 & 12, 1978.

139. V. A. Baum, R. R. Apparissi, and B. A. Garf, *Solar Energy*, **1**, 6 (1957).
140. L. L. Vant-Hull and A. F. Hildebrandt, *Solar Energy*, **18**, 31 (1976).
141. T. C. Brown, in *STTF Proceedings of Annual Meeting: Technical Sessions*, April 1978, p. 117.
142. R. Hays, in *STTF Proceedings of Annual Meeting: Technical Sessions*, April 1978, p. 111.
143. R. A. Wohlrede, *Exploring Solar Energy*, World Publishing, Cleveland, 1966, p. 76.
144. J. L. Margrave, in *STTF Proceedings of Annual Meeting: Technical Session*, April 1978, p. 284.
145. L. I. Maissel and G. Reinhard, Eds., *Handbook of Thin Film Technology*, McGraw-Hill, New York, 1970, pp. 50–55.
146. R. M. Silva, Ed., *2nd Electron Beam Processing Seminar*, June, Frankfurt am Main, Germany, Universal Tech. Corp., 1388 Research Park Dr., Dayton, Ohio 45432, 1972.
147. Airco Temescal, 2850 7th St., Berkeley, CA 94710.
148. Aydin Energy, 3180 Hanover St., Palo Alto, CA 94303.
149. Bertan Associates, 3 Aerial Way, Syosset, NY 11791.
150. Cober Electronics, 7 Gleason Av, Stamford, CT 06902.
151. D&W Industries, 1115 E Arques Av, Sunnyvale, CA 94086.
152. Del Electronics, 250 E Sandford Blvd, Mount Vernon, NY 10550.
153. Denton Vacuum, Cherry Hill Industrial Cntr, Cherry Hill, NJ 08003.
154. High Voltage Engineering, S. Bedford St., Burlington, MA 01803.
155. Nuclide AGV, Box 315, Acton, MA 01720.
156. Phys Electronics Indus, 6509 Flying Cloud Dr., Eden Prairie, MN 55344.
157. Radiation Dynamics, 316 S. Service Rd, Melville, NY 11746.
158. Sloan Tech, 535 E Montecito St, Santa Barbara, CA 93103.
159. Spellman High-Voltage, 7 Fairchild Av., Plainview, NY 11803.
160. ULVAC, Box 145, Colchester, VT 05446.
161. Varian Associates, 611 Hansen Way, Palo Alto, CA 94303.
162. Velonex, 560 Robert Ave., Santa Clara, CA 95050.
163. Zi-Tech, Box 26, Palo Alto, CA 94302.
164. Edwards High Vacuum Inc., 3276-TR Grand Island Blvd., Grand Island, NY 14072.
165. Perkin–Elmer Ultek, Inc., P.O. Box 10920, Palo Alto, CA 94303.
166. Thermionics Inc., 2285 Sutro St., P.O. Box 3711, Hayward, CA 94544.
167. R. A. Strehlow, in *Progress in High Temperature Physics and Chemistry*, Vol. 3, C. A. Rouse, Ed., Pergamon Press, New York, 1969, p. 1.
168. H. Oertel, *Stossroke*, Springer-Verlag, New York, 1966.
169. R. A. Strehlow, in *Progress in High Temperature Physics and Chemistry*, Vol. 2, C. A. Rouse, Ed., Pergamon Press, New York, 1968, p. 127.
170. C. G. Newman, M. A. Ring, and H. E. O'Neal, *J. Am. Chem. Soc.*, **100**, 5945 (1978).
171. R. H. Santoro and G. T. Diebold, *J. Chem. Phys.*, **69**, 1787 (1978).
172. D. S. De Carli and J. C. Jamieson, *Science*, **133**, 182 (1961).
173. D. S. De Carli and D. J. Milton, *Science*, **147**, 144 (1965).

174. G. R. Cowan, B. W. Dunnington, and A. H. Holtzman, U.S. Pat. 3,401,019 (Sept. 9, 1968).

175. L. C. Trueb, *J. Appl. Phys.*, **42**, 503 (1971).

176. A. N. Dremin, O. N. Breusov, V. T. Bavina, and S. V. Sergei, U.S. Pat. 4,014,979 (March 1977).

177. D. E. Grady, in *High Pressure Research*, M. H. Manghnani and S. Ahimoto, Eds., Academic Press, New York, 1977, pp. 389–438.

178. H. C. Rodean, *Nuclear Explosion Seismology*, TID-25572, NTIS, U.S. Dept. of Commerce. Springfield, Virginia 22151, 1971.

179. O. Dohlman and H. Israelson, *Monitoring Underground Nuclear Explosions*, Elsevier, New York, 1977.

180. F. D. Bennett, in *Progress in High Temperature Physics and Chemistry*, Vol. 2, C. A. Rouse, Ed., Pergamon Press, New York, 1968, p. 1.

181. W. G. Chace and H. K. Moore, Eds., *Exploding Wires*, Vols. 1–3, Plenum Press, New York, 1959–1964.

182. M. A. Cook, *The Science of High Explosives*, Reinhold, New York, 1958.

183. H. D. Fair and R. F. Walker, Eds., *Energetic Materials*, Vol. 1, Plenum Press, New York, 1977, p. 11.

184. J. N. Bradley, A. K. Butler, W. D. Capey, and J. R. Gilbert, *J. Chem. Soc., Farad. Trans. 1*, **73**, 1789 (1977).

185. N. I. Sax, *Handbook of Dangerous Materials*, Reinhold, New York, 1951.

186. DuPont "Detaclad" Explosive Products Div., Metal Cladding Oper., Coatesville, PA 19320.

187. G. R. Gathers, J. W. Shanes, and R. L. Brier, *Rev. Sci. Instrum.*, **47**, 471 (1976).

188. M. J. Joncich, in Progress in High Temperature Physics and Chemistry, Vol. 2, C. A. Rouse, Ed., Pergamon Press, New York, 1969, p. 231.

189. L. Baker, Jr., and B. L. Marchal, *Exploding Wires*, Plenum Press, New York, 1962, p. 207.

190. E. C. Cassidy, S. Abramowitz, and C. W. Beckett, NBS Report #109, Nov. 1968.

191. H. Fischer and L. C. Mansur, Eds., *Conference on Extremely High Temperatures*, Wiley, New York, 1958.

192. G. G. Ilin, E. M. Nurmatov, and J. S. Fishman, *High Temp.* (USSR), **13**, 266 (1975) (Engl. trans.).

193. I. S. Fishman and E. M. Nurmatov, *Teplotiz. Vys. Temp.*, **11**, 946 (1973).

194. S. V. Levedev, A. I. Sauvatimskii, and Yu. B. Smirnov, *Sov. Phys., High Temp.*, **9**, 578 (1971).

195. I. Ya. Dikhter and S. V. Lebedev, *High Temp.–High Press.*, **2**, 55 (1970).

196. C. B. Moore, *Chemical and Biochemical Applications of Lasers*, Vols. I–III, Academic Press, New York, 1974–1977.

197. J. D. Anderson, Jr., *Gasdynamic Lasers: An Introduction*, Academic Press, New York, 1976.

198. W. W. Duley, CO_2 *Lasers,* Academic Press, New York, 1976.

199. S. R. Skaggs, *High Temp. Sci.*, **9**, 197 (1977).

200. S. R. Leone, *J. Chem. Educ.*, **53**, 13 (1976).

201. L. S. Nelson, *Adv. High Temp. Chem.*, **4**, 171 (1971).
202. T. H. Maiman, *Nature*, **187**, 493 (1960).
203. J. Weber, Ed., *Lasers: A Collection of Reprints with Commentary*, Gordon and Breach, New York, 1968.
204. H. G. Danielmeyer, in *Lasers*, A. K. Levine and A. J. De Maria, Eds., Marcel Dekker, New York, 1976, pp. 1–67.
205. B. S. Patel, *Rev. Sci. Instrum.*, **49**, 1361 (1978).
206. L. S. Nelson, A. G. Whittaker, and B. Topper, *High Temp. Sci.*, **4**, 445 (1972).
207. A. G. Whittaker, P. L. Kinter, L. S. Nelson, and N. Richardson, *Rev. Sci. Instrum.*, **48**, 632 (1977).
208. L. S. Nelson, N. L. Richardson, K. Keil, and S. R. Skaggs, *High Temp. Sci.*, **5**, 138 (1973).
209. R. B. Green, *J. Chem. Educ.*, **54**, A365 (1977).
210. G. V. Shank, *Rev. Mod. Phys.*, **47**, 649 (1975).
211. F. P. Schafer, in *Topics in Applied Physics*, F. P. Schafer, Ed., Springer-Verlag, Berlin, 1973, pp. 1–85.
212. J. Weber, Ed., *Masers: Selected Reprints with Editorial Comment*, Gordon and Breach, New York, 1968.
213. M. C. Lin and M. E. Umstead, *J. Phys. Chem.*, **82**, 2047 (1978).
214. T. Cocca and S. Dakesian, *Solid State Tech.*, **21**, 63 (1978).
215. I. Glatt and A. Yogev, *J. Am. Chem. Soc.*, **98**, 7087 (1976).
216. A. Yogev and R. M. J. Benmair, *Chem. Phys. Lett.*, **46**, 296 (1977).
217. A. Yogev, R. M. J. Lovenstein, and D. Amar, *J. Am. Chem. Soc.*, **94**, 1091 (1972).
218. J. N. Butler, *J. Am. Chem. Soc.*, **84**, 1343 (1962).
219. D. Garcia and P. M. Keehn, *J. Am. Chem. Soc.*, **100**, 6111 (1978).
220. C. Cheng and P. M. Keehn, *J. Am. Chem. Soc.*, **99**, 5808 (1977).
221. *Laser Annealing of Semiconductors, Chem. Eng. Anon News*, 20 (Aug. 7, 1978).
222. G. Levin, D. Holloway, C. Mao, and M. Swarc, *J. Am. Chem. Soc.*, **100**, 5841 (1978).
223. R. G. Orth and R. C. Dunbar, *J. Am. Chem. Soc.*, **100**, 5949 (1978).
224. R. H. Ohse, J. F. Babelot, L. Cercignani, P. R. Kinsman, K. A. Long, J. Magill, and A. Scotti in "Application of Laser Pulsed Heating for the Study of High Temperature Vapors," in Ref. 8.
225. V. S. Letokhov, *Ann. Rev. Phys. Chem.*, **28**, 133 (1977).
226. E. Grunwald, D. F. Dever, and P. M. Keehn, *Megawatt Infrared Laser Chemistry*, Wiley, New York, 1978.
227. D. F. Dever and E. Grunwald, *J. Am. Chem. Soc.*, **98**, 5055 (1976).
228. M. P. Freeman, D. N. Travis, and M. F. Goodman, *J. Chem. Phys.*, **60**, 231 (1974).
229. J. T. Knudtson and E. M. Eyring, *Ann. Rev. Phys. Chem.*, **25**, 255 (1974).
230. R. N. Zare, in *Proc. of Symposium, at 173rd Meeting of the Amer. Chem. Soc. New Orleans, Mar. 21–23, 1977*, 1978, p. 50.

231. J. Trehnolme, in *Laser Interaction and Related Plasma Phenomena*, H. J. Schwarz and H. Horace, Eds., Plenum Press, New York, 1977, Part 4A, p. 1.

232. H. Brooks, F. Birch, G. Holton, and W. Paul, Eds., *Collected Experimental Papers of P. W. Bridgman*, Harvard University Press, Cambridge, Mass., Vols. 1–7, 1964.

233. H. T. Hall, U.S. Pat. 2,941,248 (June 21, 1960); U.S. Pat. 2,947,608 (Aug. 20, 1960).

234. F. P. Bundy, H. T. Hall, H. M. Strong, and R. H. Wentorf, *Nature*, **176**, 51 (1955).

235. H. T. Hall, *Rev. Sci. Instrum.*, **31**, 125 (1960).

236. H. T. Hall, *J. Chem. Educ.*, **38**, 383 (1961).

237. H. T. Hall, *Rev. Sci. Instrum.*, **29**, 267 (1964).

238. H. Liander and E. Lundblad, *Arkiv Kemi*, **16**, 139 (1969).

239. N. Kawai, M. Togoya, and A. Onodera, *Proc. Jpn. Acad.*, **49**, 623 (1973).

240. High-pressure anvil devices are available commercially from Cambridge Thermionic, 445 Concord Av, Cambridge, MA 02138; Harwood Engineering, 878 South St, Walpole, MA 02081; Leco, 3000 Lakeview Av, St. Joseph, MI 49085; Tem-Pres, 1526 William St, State College, PA 16801; and H. T. Hall, Inc. Provo, Utah.

241. *Temperature, Its Measurement and Control in Science and Industry*, American Institute of Physics, J. T. Tate, H. C. Wolfe, C. M. Herzfeld, and H. H. Plumb, Eds., Vols. I–IV, Reinhold, New York (1941–1972).

242. *New Temp. Scale (EPT-76) to Extend IPTS-68*, *Chem. Eng. News*, 21 (Aug. 7, 1978).

243. S. Hattori, *Rep. Natl. Res. Lab. Metrology* (Japan), **27, Suppl. No. 87**, 79 (1978).

244. G. Ruffino, *High Temp.–High Press.*, **9**, 253 (1977).

245. T. Ricolfi, and F. Lanza, *High Temp–High Press.*, **9**, 483 (1977).

246. S. R. Skaggs, *High Temp. Sci.*, **9**, 197 (1977).

247. M. Foex, *High Temp.–High Press.*, **9**, 269 (1977).

248. J. Terrien Dir., BIPM, *The International System of Units (SI)*, National Bureau of Standards Special Publication 330, U.S. Government Printing Office, Washington, D.C., 1977.

249. A. W. Adamson, *J. Chem. Educ.*, **55**, 634 (1978).

250. P. Kozma, *Meas. Control*, **12**(5), 112 (1978).

251. R. Freeman, *High Temp. Sci.*, in press (1979).

252. R. D. Freeman, *Bull. Chem. Thermodyn.*, **21**, 505 (1978).

253. Standard radiation sources are available from Land Instruments, Box 1623, Tullytown, PA 19007; Medtherm, Box 412, Huntsville, AL 35804.

254. K. C. Lapworth, R. C. Preston, D. Nettleton, and C. Brookes, "Standards and Measurements of High Temperatures at the National Physical Laboratory, United Kingdom," in *Proceeding of 3rd International Symposium on Plasma Chemistry, University of Limoges, France*, July 13–19, 1977.

255. F. R. Caldwell, *Thermocouple Materials*, U.S. National Bureau of Standards, Washington, D.C. Monograph, 40 1962, also in Ref. 241.

256. J. P. Evans and G. W. Burns, "Stability and Reproducibility of High Temperature Platinum Resistance Thermometers," in *Temperature, Its Measurement and Control in Science and Industry*, American Institute of Physics, C. M. Herzfeld, ed., Vol. III, Reinhold, New York, 1962.
257. G. L. Selman, *Plat. Met. Rev.*, **22**, (3) 90 (1978).
258. G. E. Glawe, *Rev. Sci. Instrum.*, **46**, 1107 (1975).
259. Commercial suppliers of thermocouples include ARI Industries, 9000 King St, Franklin Park, IL 60131; Acco, 929 Connecticut Av, Bridgeport, CT 06602; Ailtech, 534C Alla Rd, Los Angeles, CA 90066; Alnor Instrument, 7301 N Caldwell, Niles, IL 60648; Atkins Technical, 3314 SW 40th Blvd, Gainesville, FL 32608; BLH Electronics, 42 4th Av, Waltham, MA 02154; BTU Engineering, Esquire Rd, North Billerica, MA 01862; Bailey Instruments, 515 Victor St, Saddle Brook, NJ 07662; Beckman Instruments, 2500 Harbor Blvd, Fullerton, CA 92634; Burns Engineering, 10201 Bren Rd E, Minnetonka, MN 55343; Cambridge Instrument, 73 Spring St, Ossining, NY 10562; Conax, 2300 Walden Av, Buffalo, NY, 14224; Edmund Scientific, 7782 Edscorp Bldg, Barrington, NJ 08007; Edwards High Vacuum, 3279 Grand Island Blvd, Grand Island, NY 14072; En-cer, 16850 S State St, South Holland, IL 60437; Engelhard Industries, 70 Wood Av S, Iselin, NJ 08830; Fenwal, 400 Main St, Ashland, MA 01721; Foxboro, 38 Neponset Av, Foxboro, MA 02035; Glas-Col Apparatus, 711 Hulman St, Terre Haute, IN 47802; Harrop Laboratories, 3470 E Fifth Av, Columbus, OH 43219; Honeywell, 1100 Virginia Dr, Fort Washington, PA 19034; Hughes Aircraft, Centinela & Teale, Culver City, CA 90230; LFE, 1601 Trapelo Rd, Waltham, MA 02154; Leco, 3000 Lakeview Av, St. Joseph, MI 49085; Leeds & Northrup, Symneytown Pike, North Wales, PA 19454; Matthey Bishop, Malvern, PA 19355; Measurements & Control Sys, Gulton Indust Pk, East, Greenwich, RI 02818; Minco Products, 7300 Commerce Ln, Minneapolis, MN 55432; Omega Engineering, Box 4047, Stamford, CT 06907; Pyrometer Instrum, 234 Industrial Pkwy, Northvale, NJ 07647; RdF, 23 Elm Av, Hudson, NH, 03051; Scientific Manufacturing, 1399 64th St, Emerville, CA 94608; Simpson Electric, 853 Dundee Av, Elgin, IL 60120; Taylor Instrument, 95 Ames St, Rochester, NY 14601; Tem-Pres, 1526 William St, State College, PA 16801; Theall Engineering, Box 336, Oxford, PA 19363; Thermo Electric, 109 5th St, Saddle Brook, NJ 07662; Thermolyne, 2555 Kerper Blvd, Dubuque, IA 52001; C. W. Thornthwaite Assoc, Route 1, Elmer, NJ 08318; Thornton Associates, 87 Beaver St, Waltham, MA 02154; Wescor, 459 Main St, Logan, UT 84321; Weston Instruments, 614 Frelinghuysen Av, Newark, NJ 07114.
260. H. T. Clark, Iron and Steel Eng., **23**, 55 (1946).
261. R. D. Shepard, H. S. Patting, and R. D. Westbrook, *Bull. Am. Phys., Soc.*, **1**, 119 (1956).
262. P. A. Kinzie, "Thermocouple Temperature Measurement", Wiley-Interscience, New York (1973).
263. P. I. Roberts, *Plat. Met. Rev.*, **22**(3), 89 (1978).
264. W. F. Roeser, "Thermoelectric Thermometry," in *Temperature, Its Mea-*

surement and Control in Science and Industry, American Inst. of Physics, J. T. Tate, ed., Vol. I, Reinhold, New York, 1941.

265. W. F. Roeser and S. Lonberger, Natl. Bur. Std. Circ., 1958, p. 590.

266. H. F. Stimson, "Precision Resistance Thermometry and Fixed Points," in *Temperature, Its Measurement and Control in Science and Industry*, American Inst. of Physics, H. C. Wolfe, ed., Vol. II, Reinhold, New York, 1955.

267. E. H. McLaren, *Can. J. Phys.*, **35**, 78 (1957).

268. J. P. Evans and G. W. Burns, "Stability and Reproducibility of High Temperature Platinum Resistance Thermometers," in *Temperature, Its Measurement and Control in Science and Industry*, American Inst. of Physics, C. M. Herzfeld, ed., Vol. III, Reinhold, New York, 1962.

269. R. P. Benedict, *Fundamental Temperature, Pressure, and Flow Measurements*, 2nd ed., Wiley–Interscience, New York, 1977, p. 53.

270. D. J. Curtis and G. J. Thomas, *Metrologia*, **4**(4), 184 (1968).

271. J. P. Evans and S. D. Wood, "An Intercomparison of High Temperature Platinum Resistance Thermometers and Standard Thermocouples," *Metrologia*, **7**(3), 108 (1971).

272. There are many suppliers of thermistors: ARI Industries, 9000 King St, Franklin Park, IL 60131; Acco, 929 Connecticut Av, Bridgeport, CT 06602; Ambulatory Monitoring, 731 Saw Mill River Rd, Ardsley, NY 10502; American Optical, 14 Mechanic St., Southbridge, MA 01550; Analog Devices, Box 280, Norwood, MA 02062; Andonian Cryogenics, 26 Farwell St, Newtonville, MA 02160; Artronix, 1314 Hanley Industrial Ct., St. Louis, MO 63144; Athena Controls, 20 Clipper Rd, West Conshohocken, PA 19428; Atkins Technical, 3314 SW 40th Blvd., Gainesville, FL 32608; Bendix, 1400 Taylor Av., Montvale, NJ 07645; Calbio Research, 1515 Broadway, New York, NY 10036; Capintec, 136 Summit Av, Montvale, NJ 07645; Climatronics, 1324 Motor Pkwy, Hauppauge, NY 11787; Climet Instrum, 1320 W Conton, Redlands, CA 92373; Columbus Instruments, 950 N Hauge, Columbus, OH 43204; Conax, 2300 Walden Av, Buffalo, NY 14225; Consolidated Controls, 15 Durant Av., Bethel, CT 06801; Cox, 215 Park Av South, New York, NY 10003; Datametrics, 340 Fordham Rd, Wilmington, MA 01887; EG&G, 151 Bear Hill Rd, Waltham, MA 02154; En-cer, 16850 S State St, South Holland, Ill. 60473; Fischer & Porter, 295 Warminster Rd, Warminster, PA 18974; Foxboro, 38 Neponset Av, Foxboro, MA 02035; General Eastern, 36 Maple St, Watertown, MA 02172; General Electric, 40 Federal St, W Lynn, MA 01910; Gentran, 1290 Hammerwood Av, Sunnyland, CA 94086; Gilson Medical Electronics, Box 27, Middleton, WI 53562; Gould Measurement Systems, 2230 Statham Blvd, Oxnard, CA 93030; Gow-Mac Instrument, Box 32, Bound Brook, NJ 08805; Gulton Piezo Products, 212 Durham Av, Metuchen, NJ 08840; Kahl Scientific Instrument, Box 1166, El Cajon, CA 92022; LFE, 1601 Trapelo Rd, Waltham, MA 02154; Lafayette Instrument, Box 1279, Lafayette, IN 47902; Leeds & Northrup, Symneytown Pike, North Wales, PA 19454; Medtherm, Box 412, Huntsville, AL 35804; Minco Products, 7300 Commerce Ln, Minneapolis, MN 55432; Montedoro-Whitney, Box 1401, San Luis Obispo, CA 93401; Omega En-

gineering, Box 4047, Stamford, CT 06907; Quantem, Box 5141, Trenton, NJ 08638; RdF, 23 Elm Av, Hudson, NH 03051; Raytek, 325 E. Middlefield Rd, Mountain View, CA 94040; Relco Products, 5594 E Jefferson Av, Denver, CO 80237; Rosemount, Box 35129, Minneapolis, MN 55435; Scientific Instruments, 632 South F St, Lake Worth, FL 33460; Simpson Electric, 853 Dundee Av, Elgin, Ill. 80120; Taylor Instrument, 95 Ames St, Rochester, NY 14601; Thermo Electric, 109 5th St, Saddle Brook, NJ 07662; Transmed Sci, 860-C Capitolio Way, San Luis Obispo, CA 93401; Tri-R Instruments, 48 Merrick Rd, Rockville Centre, NY 11570; Victory Engineering, Victory Rd, Springfield, NJ 07081; Waters Instruments, 2411 7th St NW, Rochester, MN 55901; Wescan Instruments, 3018 Scott Blvd, Santa Clara, CA 95050; Weston Instruments, 614 Frelinghuysen Av, Newark, NJ 07114; Yellow Springs Instrument, Box 279, Yellow Springs, OH 45387.

273. T. C. Ehlert, M. S. Thesis, University of Wisconsin, 1958.

274. Weiller Inst. Corp., 2 Stone St., New York, NY 10004.

275. Macdonald Co. Inc., 33 University Road, Cambridge, MA 02138.

276. Commercial resistance thermometers are available from Acco, 929 Connecticut Av, Bridgeport, CT 06602; Andonian Cryogenics, 26 Farwell St, Newtonville, MA 02160; Burns Engineering, 10201 Bren Rd E., Minnetonka, MN 55343; Conax, 2300 Walden Av, Buffalo, NY 14225; Consolidated Controls, 15 Durant Av, Bethel, CT 06801; Cox, 215 Park Av, South, New York, NY 10003; Datametrics, 340 Fordham Rd, Wilmington, MA 01887; Edison Electronics, Grenier Field, Manchester, NH 03103; Engelhard Industries, 70 Wood Av S, Iselin, NJ 08830; John Fluke, Box 43210, Mountlake Terrace, WA 98043; Foxboro, 38 Neponset Av, Foxboro, MA 02035; Hewlett-Packard, 1501 Page Mill Rd, Palo Alto, CA 94304; Honeywell, 1100 Virginia Dr, Fort Washington, PA 19034; Kahl Scientific Instrument, Box 1166, El Cajon, CA 92022; Leeds & Northrup, Sumneytown Pike, North Wales, PA 19454; Mettler Instrument, Box 71, Highstown, NJ 08520; Minco Products, 7300 Commerce Ln, Minneapolis, MN 55432; Omega Engineering, Box 4047, Stamford, CT 06907; Precision Digital, 368 Hillside Av, Needham, MA 02194; Quantem, Box 5141, Trenton, NJ 08638; RdF, 23 Elm Av, Hudson, NH 03051; Relco Products, 5594 E Jefferson Av, Denver, CO 80237; Rosemont, Box 35129, Minneapolis, MN 55435; Science Associates, 340 Nassau St, Princeton, NJ 08540; Scientific Instruments, 632 South F. St, Lake Worth, FL 33460; Taylor Instrument, 95 Ames St, Rochester, NY 14601; Technical Hardware, Box 3609, Fullerton, CA 92634; Tekmar, Box 37202, Cincinnati, OH 45222; Thermo Electric, 109 5th St, Saddle Brook, NJ 07662; United Systems, Box 458, Dayton, OH 45401; Weston Instruments, 614 Frelinghuysen Av, Newark, NJ 07114; Yellow Springs Instrument, Box 279, Yellow Springs, OH 45387.

277. C. Kittel, *Thermal Physics*, Wiley, New York, 1969.

278. K. C. Lapworth, T. J. Quinn, and L. A. Allnutt, *J. Phys.*, **E3**, 116 (1970).

279. R. Von Seggern, *High Temp.–High Press.*, **9**, 283 (1977).

280. W. E. Forsythe, "Optical Pyrometry," in *Temperature, Its Measurement*

and Control in Science and Industry, American Inst. of Physics, J. T. Tate ed. Vol. I, Reinhold, New York, 1941.

281. D. R. Lovejoy, "Recent Advances in Optical Pyrometry," in Temperature, Its Measurement and Control in Science and Industry, Am. Inst. Phys., C. M. Herzfeld, ed. Vol. III, Reinhold, New York, (1963).

282. H. J. Kostkowski and R. D. Lee, "Theory and Methods of Optical Pyrometry," in *Temperature, Its Measurement and Control in Science and Industry*, American Inst. of Physics, C. M. Herzfeld, ed. Vol. III, Reinhold, New York, 1962; Natl. Bur. Std., U.S. Monograph, 1962, p. 41.

283. G. Ruffino, *Rev. Int. Hautes Temp. Refract.*, **12,** 172 (1975).

284. R. D. Lee, "The NBS Photoelectric Pyrometer of 1961," in *Temperature, Its Measurement and Control in Science and Industry*, American Inst. of Physics, C. M. Herzfeld, ed. Vol. III, Reinhold, New York, 1962.

285. E. K. Storms and B. A. Mueller, "A Very Accurate Pyrometer for General Laboratory Use," in Ref. 8, (1978).

286. D. E. Poland, J. W. Green, and J. L. Margrave, *Corrected Pyrometer Readings*, Nat. Bur. Std. U.S. Monograph No. 30, 1961.

287. R. P. Benedict, *Fundamental Temperature, Pressure and Flow Measurements*, 2nd ed., Wiley–Interscience, New York, 1977, p. 136.

288. D. W. Bonnell and J. L. Margrave, Rice University, unpublished work, 1972/1973.

289. J. L. Margrave, D. W. Bonnell, J. A. Treverton, and A. J. Valerga, "The Emissivities of Liquid Metals at their Fusion Temperatures," in *Proceedings of Fifth Symposium on Temperature Measurement, Washington, D.C.*, 1971.

290. L. H. Trieman, "A Precision Photon Counting Pyrometer," in *Temperature, Its Measurement and Control in Science and Industry*, American Inst. of Physics, C. M. Herzfeld, ed. Vol. III, Reinhold, New York, 1962.

291. Mikron Instrument Co., P. O. Box 211, Ridgewood, NJ 07451.

292. Barnes Engineering Co., 30 Commerce Road, Stamford, CT 06904.

293. Servo Corp. of America, 111 New South Road, Hicksville, NY 11802.

294. Radiation Electronics Corp., 8241 N. Kimball Ave., Skokie, IL 60076.

295. S. A. Elder, "A Completely Transistorized Recording Pyrometer," in *Temperature, Its Measurement and Control in Science and Industry*, American Inst. of Physics, C. M. Herzfeld, ed. Vol. III, Reinhold, New York, 1962.

296. B. B. Brenden, "An Infrared Radiation Ratio Pyrometer," in *Temperature, Its Measurement and Control in Science and Industry*, American Inst. of Physics, C. M. Herzfeld, ed. Vol. III, Reinhold, New York, 1962.

297. T. P. Murray and V. G. Shaw, *ISA J.*, **5,** 36 (1958).

298. D. L. Burk, *Instrum. Control Syst.*, **33,** 64 (1960).

299. E. C. Pyatt, *Brit. J. Appl. Phys.*, **5,** 264 (1954).

300. S. Ackerman, "Design of an Automatic Two Color Pyrometer," in *Temperature, Its Measurement and Control in Science and Industry*, American Inst. of Physics, C. M. Herzfeld, ed. Vol. III, Reinhold, New York, 1962.

301. V. G. Shaw, *Instrum. Control Syst.*, **33,** 58 (1960).

302. W. E. Hill, "Two-Color Pyrometry," in *Temperature, Its Measurement and Control in Science and Industry*, American Inst. of Physics, C. M. Herzfeld, ed. Vol. III, Reinhold, New York, 1962.

303. G. A. Hornbeck, "A High-Speed Ratio Pyrometer," in *Temperature, Its Measurement and Control in Science and Industry*, American Inst. of Physics, C. M. Herzfeld, ed. Vol. III, Reinhold, New York, 1962.

304. G. J. Hecht, "A Novel, Near Infrared, Two-Wavelength Pyrometer," in *Temperature, Its Measurement and Control in Science and Industry*, American Inst. of Physics, C. M. Herzfeld, ed. Vol. III, Reinhold, New York, 1962.

305. G. Ruffino, *Rev. Int. Hautes Temp. Refract.*, **12**, 187 (1975).

306. G. A. Hornbeck, "A High-Speed Ratio Pyrometer," in *Temperature, Its Measurement and Control in Science and Industry*, American Inst. of Physics, C. M. Herzfeld, ed. Vol. III, Reinhold, New York, 1962.

307. G. Ruffino and A. Rosso, *High Temp.–High Press.*, **2**, 227 (1970).

308. G. Ruffino, F. Reighini, and A. Rosso, *Temperature*, **4**, 531 (1972).

309. Ircon, Inc., 7555 N. Linden Av, Skokie, IL 60077.

310. Raytek, 325 E. Middlefield Road, Mountain View, CA 94040.

311. R. W. Astheimer and E. M. Wormser, *J. Opt. Soc. Am*, **49**, 179 (1959).

312. Servo Corp. of America, 111 New South Road, Hicksville, NY 11802.

313. Radiation Electronics Corp., 8241 N. Kimball Ave., Skokie, IL 60076.

314. Optical pyrometers are available from Barnes Engineering, 30 Commerce Rd, Stamford, CT 06904; Capintec, 136 Summit Av, Montvale, NJ 07645; Epic, 150 Nassau St, New York, NY 10038; Kollsman Instrum, Daniel Webster Hwy S., Merrimack, NH 03054; Land Instruments, Box 1623, Tullytown, PA 19007; Leeds & Northrup, Sunneytown Pike, North Wales, PA 19454; Medtherm, Box 412, Huntsville, AL 35804; Omega Engineering, Box 4047, Stamford, CT 06907; Photobell, 162 Fifth Av, New York, NY 10010; Pyrometer Instrum, 234 Industrial Pkwy, Northwale, NJ 07647; Willamson, 1152 Main St, Concord, MA 01742.

315. C. Tingwaldt, "New Optical Methods for the Determination of Thermodynamic Temperatures of Glowing Metals," in *Temperature, Its Measurement and Control in Science and Industry*, Vol. III, Reinhold, New York, 1962.

316. H. P. Broida, "Experimental Temperature Measurements in Flames and Hot Gases," in *Temperature, Its Measurement and Control in Science and Industry*, American Institute of Physics, H. C. Wolfe, ed., Vol. II, Reinhold, New York, 1955.

317. A. G. Gaydon, *The Spectroscopy of Flames*, Wiley, New York, 1957.

318. A. G. Gaydon and H. G. Wolfhard, "Flames, Their Structure, Radiation and Temperature," 3rd rev. ed., Chapman and Hall, London.

319. M. Lapp and C. M. Penney, Eds., *Laser Raman Gas Diagnostics*, Plenum Press, New York, 1974.

320. J. Richter, *Z. Astrophys.*, **61**, 57 (1965).

321. W. Lochte-Holtgreven, R. Schall, and F. Wecken, *Rep. Progr. Phys.*, **19**, 312 (1956).

322. W. Finkelnburg and H. Maecker, "Elektrische Bogen und Thermisches Plasma," in *Handbuch der Physik*, Vol. XXII, Springer-Verlag, Berlin, 1956.

323. H. R. Griem, *Plasma Spectroscopy*, McGraw-Hill, New York, 1964.

324. R. R. Dils and P. S. Follansbee, "Gas Temp. Measurements in Practical Combustors," in Ref. 8.

325. L. S. Hunter, G. Grunfelder, and L. H. Hoshall, "Moving Thermocouple Measurements of Heat Transfer in Hot Gases" in Ref. 8.

326. I. Warshawsky, "Pyrometry of High Velocity Gases," in *Proceedings of Sixth Symposium on Combustion*, Reinhold, New York, 1956.

327. L. N. Krause, G. E. Glawe, and R. C. Johnson, "Heat Transfer Devices for Determining the Temperature of Flowing Gases," in *Temperature, Its Measurement and Control in Science and Industry*, American Inst. of Physics, C. M. Herzfeld, ed., Vol. III, Reinhold, New York, 1962.

328. L. Fingerson and P. L. Blackshear, "Theory and Design of a Pneumatic Temperature Probe and Experimental Results Obtained in a High Temperature Gas Stream," in *Temperature, Its Measurement and Control in Science and Industry,* American Inst. of Physics, C. M. Herzfeld, ed., Vol. III, Reinhold, New York, 1962.

329. I. Warshawsky and P. W. Kuhns, "Review of the Pneumatic-Probe Thermometer," in *Temperature, Its Measurement and Control in Science and Industry*, Vol. III, Reinhold, New York, 1962.

330. A. G. Gaydon and W. B. Pearse, "The Identification of Molecular Spectra", 4th ed., Wiley-Interscience, New York (1976).

331. P. F. Hessin and A. G. Gaydon, "Estimation of carbon radical concentrations in fuel-rich acetylene-oxygen flames", p. 481–9, 12th International Combustion Symposium, Poitier, July, 1968. The Combustion Inst. Pitt. (1969).

332. K. C. Lapworth, L. A. Allnutt, and J. R. Pendlebury, *J. Phys.*, **D4**, 759 (1971).

333. J. Holtsmark, *Ann. Phys.*, **58**, 577 (1919).

334. J. Holtsmark, *Phys. Z.*, **25**, 73 (1924).

335. S. R. Seshodri, *Fundamentals of Plasma Physics*, Amer. Elsevier, New York, 1973.

336. E. M. Dewan, AFCRL-42, Air Force Cambridge Research Laboratory, Bedford, MA, May 1961.

337. L. H. Grabner, M. C. Drake, and J. W. Hastie, "A Comparison of Optical Flame Temperature Measurements; Na Line Reversal, Rotational and Vibrational Raman and OH Spectroscopy," in Ref. 8.

338. G. A. Hornbeck, "A High-Speed Ratio Pyrometer," in *Temperature, Its Measurement and Control in Science and Industry*, American Inst. of Physics, C. M. Herzfeld, ed., Vol. III, Reinhold, New York, 1962.

339. F. Bundy and H. M. Strong, in *Physical Measurements in Gas Dynamics and Combustion*, Vol. IX, Princeton University Press, 1954, pp. 343–86.

340. H. Kohn, *Ann. Phys.*, **44**, 749 (1914).

341. F. Henning and C. Tingwaldt, *Z. Phys.*, **48**, 805 (1928).

342. C. R. Vidal, J. Cooper, and E. W. Smith, *Astrophys. J., Suppl. 214,* **25,** 37 (1973).

343. P. Kepple and H. R. Griem, *Phys. Rev.,* **173,** 317 (1968).

344. H. Griem, "Spectral line Broodenig by Plasmas," Arcadia NY NY 1974.

345. R. G. Breene, *The Shift and Shape of Spectral Lines,* Pergamon Press, New York, 1961, p. 323.

346. P. Schmolhohann, State Examination work, 1936 (unpublished); see also A. Unsold, Physik der Sternatmospharen, 2nd. ed., chap. XI Springer-Verlag, Berlin (1955).

347. H. R. Griem, *Plasma Spectroscopy,* McGraw-Hill, New York, 1964.

348. R. C. Prestor, *J. Quant. Spectr. Radiant Transfer,* 18, 337 (1977).

349. R. C. Prestor, *J. Phys. B, Atom Mol. Phys.,* **10,** 523 (1977).

350. H. R. Griem, A. C. Kolb, and K. Y. Shen, *Phys. Rev.,* **116,** 6 (1959).

351. A. W. Searcy, in *Proceedings of an International Symposium on High Temperature Technology,* McGraw-Hill, New York, 1960.

352. H. W. Leutner and C. S. Stokes, *Ind. Eng. Chem.,* **53,** 341 (1961).

353. M. P. Freeman and J. F. Skrivan, *Hydrocarbon Process, Petrol Refiner,* **41**(8), 124 (1962).

354. R. F. Baddour and J. M. Iwasyk, *Ind. Eng. Chem., Process Design Develop.,* **1,** 169 (1962).

355. R. F. Baddour and B. R. Bronfin, *Chem. Eng. News,* **41**(16), 35 (1963).

356. D. Ragusa, *Plasma Jet for Chemistry,* Plasmadyne Circular, December 8, 1959.

357. P. S. Skell and L. D. Westcott, *J. Am. Chem. Soc.,* **85,** 1023 (1963).

358. J. J. Havel, M. J. McGlinchey, and P. S. Skell, *Acc. Chem. Res.,* **6,** 97 (1973).

359. M. J. McGlinchey and P. S. Skell, "Organometallic and Organic Syntheses Using Main Group Elemental Vapors," in *Cryochemistry,* M. Moskovits and G. A. Ozin, Eds., Wiley–Interscience, New York, 1976, p. 137.

360. P. L. Timms, "Techniques of Preparative Cryochemistry," in *Cryochemistry,* M. Moskovits and G. A. Ozin, Eds., Wiley–Interscience, New York, 1976, p. 61.

361. D. L. Perry and J. L. Margrave, *J. Chem. Educ.,* **53,** 696 (1976).

362. P. L. Timms, *Acc. Chem. Res.,* **6,** 118 (1973).

363. J. L. Margrave, K. G. Sharp, and P. W. Wilson, "Dihalides of the Group IVB Elements," *Topics Curr. Chem.,* **26,** 1 (1972).

364. J. L. Margrave and P. W. Wilson, *Acc. Chem. Res.,* **4,** 145 (1971).

365. D. A. White and J. P. Coleman, *J. Electrochem. Soc.,* **125,** 1401 (1978).

366. D. A. White, *J. Electrochem. Soc.,* **124,** 1177 (1977).

367. R. D. Pehlke, *Unit Processes of Extractive Metallurgy,* Amer. Elsevier, New York, 1973, p. 201.

368. G. J. Janz, "Molten Salts Handbook" p. 435, Academic Press, N.Y. (1967).

369. I. M. Frantsevich, I. N. Podorvan, and Y. V. Rudchenko, *Nauk Zap. Kiivs'k. Univ.,* **13,** 107 (1956).

370. P. D'Aragon, Can. Pat. 592,670 (Feb. 16, 1960).

371. J. L. Andrieux and J. Dauphin, *Compt. Rend.,* **245,** 1359 (1957).

372. J. L. Andrieux, *Compt. Rend.*, **184**, (1927).
373. J. L. Andrieux, *Compt. Rend.*, **189**, 1279 (1929).
374. M. Dodero, *Compt. Rend.*, **198**, 1593, **199**, 566 (1934).
375. J. L. Andrieux and M. Chêne, *Compt. Rend.*, **206**, 661 (1938).
376. J. L. Andrieux and G. Weiss, *Compt. Rend.*, **219**, 550 (1944).
377. J. L. Andrieux and A. Canaud, *Compt. Rend.*, **218**, 710 (1944).
378. J. L. Andrieux and J. Dauphin, *Compt. Rend.*, **245**, 1359 (1957).
379. J. Robinson and R. A. Osteryoung, *J. Electrochem. Soc.*, **125**, 1454 (1978).
380. C. S. Stokes and W. W. Knipe, *Ind. Eng. Chem.*, **52**, 287 (1960).
381. C. S. Stokes and W. W. Knipe, *Ind. Eng. Chem.*, **52**, 287 (1960).
382. Third International Symposium on Plasma Chemistry, ed., P. Fauchais, July 13–19, Universitie de Limoges, Limoges, France.
383. H. W. Leutner and C. S. Stokes, *Ind. Eng. Chem.*, **53**, 341 (1961).
384. M. P. Freeman and J. F. Skrivan, *Hydrocarbon Process Petrol. Refiner*, **41**,(8), 124 (1962).
385. R. F. Baddour and J. M. Iwasyk, *Ind. Eng. Chem., Process Design Develop.*, **1**, 169 (1962).
386. R. J. Lagow and J. A. Morrison, *Inorg. Chem.*, **16**, 1823 (1977).
387. C. W. Marynowski, J. R. Phillips, R. C. Phillips, and N. K. Kiester, *Ind. Eng. Chem., Fundam.*, **1**, 52 (1962).
388. A. V. Grosse, H. W. Leutner, and A. C. Stokes, "Plasma Jet Chemistry," The Research Inst. of Temple Univ. First Annual Report, Office of Naval Research Contract NONR 3085 (02), December 31, 1961.
389. R. J. Lagow, L. A. Shimp, D. K. Lam, and R. F. Baddour, *Inorg. Chem.*, **11**, 2568 (1972).
390. R. J. Lagow, R. F. Baddour, D. K. Lam, and L. A. Shimp, U.S. Pat. 3,904,501 (1975).
391. V. Harris, J. Holmgren, S. Korman, and C. Sheer, *J. Electrochem. Soc.*, **106**, 874 (1959).
392. E. M. Santsky and G. S. Burkhanov, *J. Cryst. Growth*, **43**, 457 (1978).
393. R. F. Baddour and B. R. Bronfin, *Chem. Eng. News*, **41**(16), 35 (1963).
394. W. L. Jolly, *The Use of Electric Discharges in Chemical Synthesis*, Contract W-7405-Eng-48, Lawrence Radiation Laboratory UCRL 9501, December 12, 1960.
395. W. L. Jolly, in *Technique of Inorganic Chemistry*, Vol. 1 H. B. Jonassen and A. Weissberger, Eds., Interscience, New York, 1963.
396. A. S. Kana'an and J. L. Margrave, *Adv. Inorg. Radiochem.*, **6**, pp. 143–206 (1964).
397. D. B. Hibbert, A. J. B. Robertson, and M. J. Perkins, *J. Chem. Soc. Faraday Trans, 1*, **77**, 1499 (1977).
398. A. Streitweiser, Jr., and H. R. Ward, *J. Am. Chem. Soc.*, **85**, 539 (1963).
399. S. L. Miller, *J. Am. Chem. Soc.*, **77**, 2351 (1955).
400. S. L. Miller, *Ann. N.Y. Acad. Sci.*, **69**, 260 (1957).
401. S. L. Miller, *Biochim. Biophys. Acta*, **23**, 480 (1957).
402. Chem. Eng. News, 16, Oct. 31 (1977).
403. C. S. Cleaver, U.S. Pat. 2,782,219 (February 19, 1957).

404. C. S. Cleaver, U.S. Pat. 2,833,822 (May 6, 1958).
405. C. S. Cleaver, L. G. Blosser, and D. D. Goffman, *J. Am. Chem. Soc.*, **81**, 1120 (1959).
406. T. M. Sugden, *Trans. Faraday Soc.*, **52**, 1465 (1956).
407. T. M. Sugden and R. C. Wheeler, *Disc. Faraday Soc.*, **19**, 76 (1955).
408. D. J. Meschi, W. A. Chupka, and J. Berkowitz, *J. Chem. Phys.*, **33**, 530 (1960).
409. R. A. Laudise, *The Growth of Single Crystals*, Prentice-Hall, Englewood Cliffs, N.J., 1970.
410. R. A. Laudise, *J. Cryst. Growth*, **13/14**, 27 (1972).
411. R. A. Laudise, *J. Cryst. Growth*, **24/25**, 32 (1974).
412. K. Nassau and J. Nassau, *Lapidary J.*, **32**, 490 (1978).
413. C. W. F. T. Pistorias, *Progr. Solid State Chem.*, Vol. 11, Pergamon Press, New York, 1976, pp. 1–120.
414. P. Cotterill and P. R. Mould "Recrystallization and Grain Growth in Metals", Wiley Interscience, New York (1976).
415. F. P. Bundy, *Science*, **146**, 1673 (1964).
416. R. H. Wentorf, Jr., *J. Phys. Chem.*, **69**, 3063 (1965).
417. H. T. Hall and S. S. Kistler, *Ann. Rev. Phys. Chem.*, **9**, 395 (1958).
418. H. T. Hall, *Science*, **169**, 868 (1970).
419. L. F. Vereschagin, E. N. Yakovler, T. D. Varfolomeeva, A. Y. Preobrazhensky, V. N. Slesarev, V. A. Stepanov, and L. E. Shterenberg, U.S. Pat. 4,089,933 (May 16, 1978).
420. H. T. Hall, *J. Chem. Educ.*, **38**, 484 (1961).
421. R. H. Wentorf, Jr., *J. Chem. Phys.*, **34**, 809 (1961).
422. F. P. Bundy, *J. Chem. Phys.*, **38**, 618 (1963).
423. C. D. Brandle, D. C. Miller, and J. W. Nielsen, *J. Cryst. Growth*, **12**, 195 (1972).
424. P. W. Bridgman, *Proc. Am. Acad. Arts Sci.*, **60**, 305 (1925).
425. J. Czochralski, *Z. Phys. Chem.* (Leipzig), **92**, 219 (1917).
426. W. G. Pfann, *Trans. AIME*, **194**, 747 (1952).
427. E. J. Scott, *J. Chem. Phys.*, **23**, 2459 (1955).
428. M. A. Verneuil, *Ann. Chim. Phys.*, **3**, 20 (1904).
429. R. G. Rudness and R. W. Kebler, *J. Am. Ceram. Soc.*, **43**, 17 (1960).
430. R. E. De La Rue and F. A. Halden, *Rev. Sci. Instrum.*, **31**, 35 (1960).
431. E. P. Kamivov and J. R. Carruthers, *Appl. Phys. Lett.*, **22**, 326 (1973).
432. A. A. Ballman, H. Brown, P. K. Tien, and R. J. Martin, *J. Cryst. Growth*, **20**, 251 (1973).
433. W. G. Pfann "Zone Melting" 2nd ed., J. Wiley, New York (1966).
434. R. D. Pehlke, *Unit Processes of Extractive Metallurgy*, Amer. Elsevier, New York, 1973, p. 124.
435. G. Chandron, Ed., *Monographics sur les Métaux de Haute pureté*, Vol. 1, Masson, Paris, 1972.
436. A. Berghazen and E. B. Simonsen, *Trans. AIME*, **221**, 1029 (1961).
437. P. H. Keck and M. J. E. Golay, *Phys. Rev.*, **89**, 1297 (1953).

438. T. Akashi, K. Matumi, T. Okada, and T. Mizutani, *IEEE Trans. Magnetics*, **5**, 285 (1969).
439. G. A. Irons, C. W. Chang, R. Guthrie, and J. Szckely, *Met. Trans.*, **9B**, 151 (1977).
440. R. W. Stracham, *Lepel Rev.*, **1**, No. 16, p. 10.
441. S. Hayshi, S. Ono, and H. Komatsu, *Krist. Technik*, **13**, 263 (1978).
442. R. Falckenberg, *J. Cryst. Growth*, **29**, 195, (1975); **13/14** 723 (1972).
443. B. R. Pamplin, *Crystal Growth*, Pergamon Press, New York, 1975, p. 343.
444. R. W. Bartlett, F. A. Halden, and J. W. Fowler, *Rev. Sci. Instrum.*, **38**, 1313 (1967).
445. E. D. Koll, K. Nassau, R. A. Laudise, E. E. Simpson, and K. M. Kroupa, *J. Cryst. Growth*, **36**, 93 (1976).
446. N. Lias, E. E. Grudarski, E. D. Koll, and R. A. Laudise, *J. Cryst. Growth*, **18**, 1 (1973).
447. R. Roy and E. F. Osborn, *Econ. Geol.*, **47**, 717 (1952).
448. R. G. Yalman and J. F. Corwin, *J. Phys. Chem.*, **61**, 1432 (1957).
449. R. Roy, *Z. anorg. allgem. Chem.*, **276**, 285 (1954).
450. R. Roy and S. Theokritoff, J. Cryst. Growth, 12, 69 (1972).
451. E. Kaldis, "Principles of the Vapour Growth of Single Crystals," pp. 49–193, in *Crystal Growth: Theory and Techniques,* Vol. 1, C. H. L. Goodman, Ed., Plenum Press, New York, 1974.
452. T. Hirai, *Bull. Jap. Inst. Metals*, **17**(4), 313 (1978).
453. F. S. Galasso and R. D. Veltri, *Bull. Am. Ceram. Soc.*, **57**, 453 (1978).
454. T. Takahashi and H. Kamiya, *High Temp.–High Press.*, **9**, 437 (1977).
455. J. R. Hollahan and R. S. Rosler, "Plasma Deposition of Inorganic Thin Films," in *Thin Film Processes*, J. L. Vossen and W. Kern, Eds., Academic Press, New York, 1978.
456. R. S. Rosler, W. C. Benzing, and J. Blado, *Solid State Technol.*, June, p. 45 (1976).
457. A. P. Levitt, Ed., *Whisker Technology,* Wiley–Interscience, New York, 1970.
458. L. R. McCreight, H. W. Ranch, W. H. Sutton, *Ceramic and Graphite Fibers and Whiskers*, Academic Press, New York, 1965.
459. D. R. Boyd and Y. T. Sihvonen, *J. Appl. Phys.*, **30**, 176 (1959).
460. G. R. Antell and D. Effer, *J. Electrochem. Soc.*, **106**, 509 (1959).
461. A. E. Van Arkel, and J. H. de Boer, U.S. Pat. 1,671,213 (1928).
462. C. M. Cheerier and J. Sucket, Fr. Pat. 1,158,930 (June 20, 1958).
463. B. V. Derjazuin and D. B. Fedoseen, *Sci. Am.*, **233**, 102 (1975).
464. W. G. Eversole, U.S. Pat. 3,030,187 (April 17, 1962).
465. R. Ginsburgh, D. L. Heald, and R. C. Neville, *J. Electrochem. Soc.*, **125**, 1557 (1978).
466. W. A. Bryant, *J. Electrochem. Soc.*, **125**, 1534 (1978).
467. K. C. Lisk, *Nuclear Power Plant Systems and Equipment,* Industrial Press, New York, 1972.
468. A. R. Foster and R. L. Wright, Jr., *Basic Nuclear Engineering*, Allyn and Bacon, Boston, MA 1968.

469. J. G. Wills, *Nuclear Power Plant Technology*, Wiley, New York, 1967.
470. S. Glasstone and A. Sesonske, *Nuclear Reactor Engineering*, Van Nostrand, New York, 1967.
471. A. S. Bishop, *Project Sherwood*, Addison–Wesley, Reading, Mass., 1956.
472. D. J. Rose and M. Clark, Jr., *Plasmas and Controlled Fusion*, M.I.T. Press, Cambridge, Mass., 1961.
473. S. Glasstone and R. H. Lovberg, *Controlled Thermonuclear Reactions*, Van Nostrand, Princeton, N.J., 1960.
474. J. R. Lamarsh, *Introduction to Nuclear Engineering*, Addison–Wesley, Reading, Mass., 1975.
475. R. Hurst and S. McLain, *Progress in Nuclear Energy: Technology and Engineering*, McGraw-Hill, New York, 1956.
476. E. J. Moniz and T. L. Neff, *Phys. Today*, **31**(4), 42 (April 1978).
477. APS Study Group on Nuclear Fuel Cycles and Waste Management, *Rev. Mod. Phys.*, **50**(Part 2), 51 (1978).
478. Office of Technology Assessment, *Nuclear Proliferation and Safeguards*, Praeger, New York, 1977.
479. D. M. Gruen, Ed., *The Chemistry of Fusion Technology*, Plenum Press, New York, 1972.
480. G. Yonas, *Sci. Am.*, **239**(5), 50 (Nov. 1978).
481. P. A. Miller, R. I. Butler, M. Cowan, J. R. Freeman, J. W. Poukey, T. P. Wright, and G. Yonas, *Phys. Rev. Lett.*, **39**, 92 (1977).
482. S. A. Goldstein and L. Roswell, *Phys. Rev. Lett.*, **35**, 1079 (1975).
483. R. F. Post, *Mirror Systems: Fuel Cycles, Loss Reduction and Energy Recovery*, UCRL-71753, Lawrence Livermore Laboratory, Livermore, California, 1969.
484. R. W. Moir, W. L. Freis, and R. F. Post, *Experimental and Computational Investigations of the Direct Conversion of Plasma Energy to Electricity*, UCRL-72879, Lawrence Livermore Laboratory, Livermore, California, 1971.
485. M. J. Lubin and A. P. Fraas, *Sci. Am.*, **224**, 21 (1971).
486. M. J. Lubin, "Experiments Leading to Laser-induced Fusion," in *The Chemistry of Fusion Technology*, D. M. Gruen, Ed., Plenum Press, New York, 1972, pp. 359–384.
487. T. E. Boothe and H. J. Ache, *J. Phys. Chem.*, **82**, 1362 (1978).
488. "Laser Fusion Devices at Livermore and Rochester Use Nd-glass," *Phys. Today*, **31**(4), pp. 17 (April 1978).
489. Y. S. Touloukian, Ed., Various publications of the Thermophysical Properties Research Center, Purdue University, Lafayette, Indiana 47906.
490. R. T. Frost, G. Wouch, E. L. Gray and A. E. Lord, Jr., High Temp. Sci. **10**, 241 (1978).
491. J. Thermophysics ed by A. Cezair Layan, Vol 1, 1980.

INDEX